This volume is one of many in the *Wheels of Learning* craft training program. This program, covering more than 20 standardized craft areas, including all major construction skills, was developed over a period of years by industry and education specialists. Sixteen of the largest construction and maintenance firms in the U.S. committed financial and human resources to the teams that wrote the curricula and planned the national accredited training process. These materials are industry-proven and consist of competency-based textbooks and instructor guides.

The *Wheels of Learning* was developed by the National Center for Construction Education and Research in response to the training needs of the construction and maintenance industries. The NCCER is a nonprofit educational entity affiliated with the University of Florida and supported by the following industry and craft associations:

Partnering Associations
- ABC Texas Gulf Coast Chapter
- American Fire Sprinkler Association
- American Society for Training and Development
- American Welding Society
- Associated Builders and Contractors, Inc.
- Associated General Contractors of America
- Association for Career and Technical Education
- The Business Roundtable
- Carolinas AGC, Inc.
- Carolinas Electrical Contractors Association
- Construction Industry Institute
- Design Build Institute of America
- Merit Contractors Association of Canada
- Metal Building Manufacturers Association
- National Association of Minority Contractors
- National Association of Women in Construction
- National Insulation Association
- National Ready Mixed Concrete Association
- National Utility Contractors Association
- National Vocational Technical Honor Society
- North American Crane Bureau
- Painting and Decorating Contractors of America
- Portland Cement Association
- Steel Erectors Association of America
- U.S. Army Corps of Engineers
- University of Florida
- Vocational Industrial Clubs of America
- Women Construction Owners and Executives, USA

Some of the features of the *Wheels of Learning* program include:
- A proven record of success over many years of use by industry companies.
- National standardization providing "portability" of learned job skills and educational credits that will be of tremendous value to trainees.
- Recognition: upon successful completion of training with an accredited sponsor, trainees receive an industry-recognized certificate and transcript from the NCCER.
- Each level meets or exceeds Bureau of Apprenticeship and Training (BAT) requirements for related classroom training (CFR 29:29).
- Well illustrated, up-to-date, and practical information. All standardized manuals are reviewed annually in a continuous improvement process.

Acknowledgments

This manual would not exist were it not for the dedication and unselfish energy of those volunteers who served on the Technical Review Committees. A sincere thanks is extended to:

Orrin Charm	Mike Friedman	Robert Ruyle
Steve Cloninger	Joe Jones	Marilyn Sanford
Art Couch	John Knox	Brad Shipp
Gary Dawkins	Ron Mengel	Ken Smith
Charles Erichsen	Don Norris	Randy Vaughan
	Richard Robinson	

Special Acknowledgments

We would especially like to thank these companies for their financial contributions that made the development of this curriculum possible:

Associated Locksmiths of America

Building Industry Consulting Services International

Canadian Alarm & Security Association

Consumer Electronics Manufacturers Association

Continental Automated Buildings Association

Custom Electronic Design & Installation Association

Home Automation Association

National Burglar & Fire Alarm Association

National Systems Contractors Association

Security Industry Association

Consortium for Electronic Systems Technician Training

Contents

Introduction to the Trade 33101
Construction Materials and Methods 33102
Pathways and Spaces 33103
Fasteners and Anchors 33104
Hand Bending of Conduit 33105
Electrical Theory One..33106
Electrical Safety 33107
Low-Voltage Cabling 33108

Introduction to the Trade
Module 33101

Electronic Systems Technician Trainee Task Module 33101

INTRODUCTION TO THE TRADE

NATIONAL
CENTER FOR
CONSTRUCTION
EDUCATION AND
RESEARCH

OBJECTIVES

Upon completion of this module, the trainee will be able to:

1. Understand the purpose of the electronic systems industry and describe the role of an electronic systems technician in the industry.
2. Understand the role played by industry associations and be able to identify key associations.
3. State the rules for professional and ethical conduct.
4. Describe the importance of codes and standards and explain how they affect the work of the electronic systems technician.
5. Recognize some of the tools used in the industry.
6. Fill out a time sheet.
7. Fill out a job sheet.

Prerequisites

Successful completion of the following Task Modules is recommended before beginning study of this Task Module: None.

Required Trainee Materials

1. Trainee Task Module
2. Appropriate Personal Protective Equipment

Note: The designations "National Electrical Code," "NE Code," and "NEC," where used in this document, refer to the *National Electrical Code®*, which is a registered trademark of the National Fire Protection Association, Quincy, MA. *All National Electrical Code (NEC) references in this module refer to the 1999 edition of the NEC.*

COURSE MAP

This course map shows all of the modules in the first level of the Electronic Systems Technician curricula. The suggested training order begins at the bottom and proceeds up. Skill levels increase as a trainee advances on the course map. The training order may be adjusted by the local Training Program Sponsor.

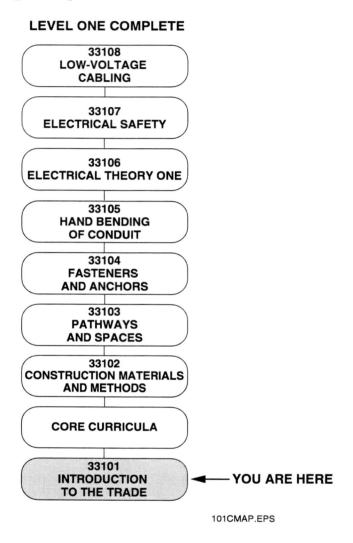

TABLE OF CONTENTS

Section	Topic	Page
	Course Map	2
1.0.0	Introduction	4
1.1.0	Electronic Systems	5
1.2.0	Opportunities In The Industry	5
1.3.0	Integrated Building Management Systems	6
1.4.0	Rules, Regulations, and Standards	7
2.0.0	Certification And Licensing	8
3.0.0	Your Responsibilities As An Employee	8
3.1.0	Professional Obligations	9
3.2.0	Company Standards	9
3.3.0	Obligations To Customers	9
3.4.0	Courtesy And Respect	10
3.5.0	Communicating As A Professional	11
3.5.1	Interpreting Instructions	12
3.5.2	Restate Written Instructions Orally	13
3.5.3	Clearly Relate A Customer Requirement	13
3.6.0	Teamwork	14
3.7.0	Conflict Resolution	15
3.7.1	Use Principled Negotiation In Solving A Conflict	15
4.0.0	Industry Standards And Building Codes	17
4.1.0	The National Electrical Code®	17
4.2.0	The Canadian Electrical Code, Part 1	18
4.3.0	National Fire Protection Association (NFPA)	18
4.4.0	National Building Codes	18
4.5.0	International Standards	18
4.6.0	Why Standards?	19
4.6.1	Standards Coordinate Team Activity	19
4.6.2	Standards Reduce Liability	19
5.0.0	Documentation And Paperwork	20
6.0.0	Types Of Training Programs	25
6.1.0	Standardized Training By The NCCER	25
6.1.1	Apprenticeship Training	25
6.1.2	Youth Training And Apprenticeship Programs	26
7.0.0	Tools Of The Trade	26
	Summary	30
	Review/Practice Questions	31
	Answers To Review/Practice Questions	34
	Appendix A	35
	Appendix B	40
	Appendix C	42
	Appendix D	43
	Appendix E	44

Trade Terms Introduced In This Module

ANSI: American National Standards Institute.

EIA: Electronic Industries Association.

High voltage: As defined by the National Electrical Code (NEC), the classification of high voltage covers voltages above 600V.

HVAC: Abbreviation for heating, ventilating, and air conditioning equipment.

Low voltage: As defined by the National Electrical Code (NEC), the classification of low voltage covers voltages below 600V.

Microprocessor: A computer chip that can be programmed to perform arithmetic and logic functions and process data.

Modem: Acronym for modulator/demodulator. An electronic device used to connect computers using communication lines. It converts digital signals to analog signals and vice versa.

Pathways: Facilities for the placement of cable.

Retrofit: Convert to new equipment or place new equipment in existing structures.

Takeoff: The process of surveying, measuring, and itemizing all materials and equipment needed for a project.

TIA: Telecommunications Industry Association.

1.0.0 INTRODUCTION

There has been an explosion in the development of communications, life safety (fire), security, entertainment, and building control technologies. Consumer demand for electronic systems (also known as **low-voltage** systems) that provide security and life safety, improve communications networking and control, and automate lighting and energy appliances has increased dramatically. Tens of thousands of specialty companies are fueling the need for competent electronic systems technicians (ESTs). In 1998, there were an estimated 165,000 to 230,000 ESTs employed by businesses in the United States and Canada. The electronic systems industry is among the fastest growing industries in the world.

The demand for qualified technicians is growing and is expected to outpace the supply of skilled workers. Employment opportunities in this field are affected by trends in construction, renovation, and new product development. As the economy grows, more ESTs will be needed to install and maintain systems in homes and businesses. New buildings will be wired during construction to accommodate networks of computers, telecommunications equipment, and control devices. Existing buildings will be modified to incorporate these systems as well. Overall, employment prospects for qualified ESTs should be excellent for decades to come.

The term *low-voltage system* is often used to describe the electronic systems used in the control of building functions such as lighting, security, and telecommunications. For some time, electronic systems were regarded as part of the electrical contracting industry and its technicians were viewed as electricians. While some licensed electricians install low-voltage systems, the trend has been for low-voltage specialists to handle the newer and more complicated systems, while electricians handle **high-voltage** work. Companies may specialize in security and life safety, entertainment, communications, temperature control, and other functions. They may specialize further in handling residential, commercial, or industrial facilities.

Electronic systems are now more common than ever in both residential and commercial establishments. New construction usually specifies the installation of such systems and there are many opportunities to **retrofit** existing structures to accommodate these systems. One of the most important changes to come along is the integration of control devices with computer systems and telecommunications equipment for remote control and maintenance of building systems.

1.1.0 ELECTRONIC SYSTEMS

Electronic systems are the components that distribute, carry, capture, and display voice, video, audio, and data signals. These products use electronic signals from **microprocessors** to control mechanical and electrical devices in and around residential and commercial premises. While electronic systems often use 120V power sources, the control devices in these systems typically operate at lower voltages.

1.2.0 OPPORTUNITIES IN THE INDUSTRY

Typical job roles for ESTs include technician, installer, and system designer. ESTs work with low-voltage products or systems designed to be installed and made permanent, rather than those that are portable. The work of an EST involves specialized systems and products used for:

- Entertainment (home and business video and audio media systems)
- Communications (telephone, fax, **modem**, computer networks, paging, and public address systems)
- Life safety (access control, burglar and fire alarms, and video surveillance)
- Environmental controls (heating, ventilation, air conditioning, and energy management)
- Automation controls for residential and commercial buildings

ESTs may be self-employed, or they can work for either large or small companies. Electronic systems companies differ from electrical contractors who install systems and service products that distribute and utilize electrical power. Installations occur both during and after building construction, indoors and outdoors. Technicians are often on call for troubleshooting and repair work.

The work of an EST is sometimes strenuous. It may involve standing for a long time, working in awkward or cramped positions, or working on ladders or scaffolds. There is a risk of injury from electrical shock, falls, and other construction hazards. To avoid injury, ESTs must learn and follow strict safety procedures.

ESTs install, connect, calibrate, and service products that carry voice, video, audio, and data within a premises (building, home, apartment, etc.). Some ESTs install and service fire and security alarm systems; others work with telecommunications equipment such as business telephone systems and computer networks.

An EST needs to have a basic understanding of electricity and electronics, microprocessors and computers, and signal/data communications. A knowledge of the very specialized terminology used in the industry is also essential. Strong familiarity with safety procedures and codes, as well as industry standards and building codes, is critical to success.

In your work, you will review and follow drawings, blueprints, and construction specifications.

At the job site, you will perform a variety of tasks, each requiring knowledge of specialized tools, standards, and procedures. These tasks include:
- Installing cable support structures
- Drilling wire **pathways**
- Pulling, securing, and terminating wire and cables
- Installing outlets and connection panels
- Installing, testing, and troubleshooting system electronic and mechanical components
- Programming digital components to perform in the manner specified by the client or by applicable codes

The business focus for each type of company may be different, but all ESTs must have an understanding of electricity and certain mechanical and structural knowledge. Core technical knowledge relates to electronic system interconnection, integration, networking, software programming, calibration of visual displays and sound systems, and user interface design.

1.3.0 INTEGRATED BUILDING MANAGEMENT SYSTEMS

One of the most exciting trends in the electronics industry is the integration of all the electronic systems for a single building into a common network controlled from a central computer. This approach allows one individual to monitor and control all building functions from a computer within the building or even in another location many miles away. A technician at a remote site can diagnose problems and help the customer restore the system after an alarm. Specialized computer software is used to accomplish the integration. Special cabling links the components of each subsystem and its controller, which is in turn linked to a central controller that manages the entire network. Selecting, installing, and terminating the network cabling and the components that make the network function are important parts of the job.

While these newer systems provide many benefits, they may also make installation and servicing more complicated. Each type of system has its own set of standards. Some of the standards may overlap, while others may seem to be in conflict. When faced with conflicting standards, it is best to check with your supervisor about how to proceed.

Figure 1 is a simplified schematic diagram of a building control system in which the **HVAC**, lighting, security, and fire alarm systems are integrated. Although each system has its own controller, central control of all systems is provided by a network controller and monitored/managed at a central facility containing computers, video monitors, and alarm monitors. This building automation system (BAS) allows building managers and life safety service organizations to monitor and manage building systems from distant locations using computers linked by telecommunication lines or satellite communication systems.

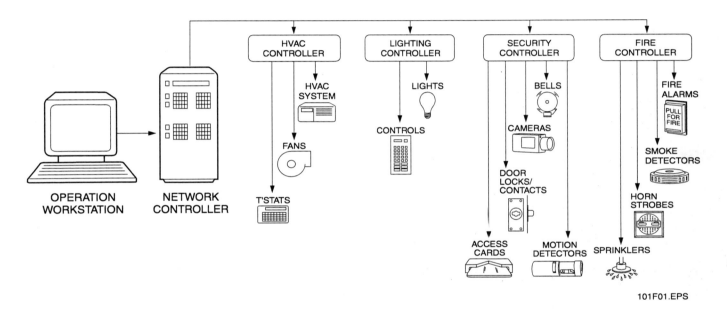

Figure 1. Building Automation System

1.4.0 RULES, REGULATIONS, AND STANDARDS

There are many standards, codes, regulations, and rules that govern the electronic systems industry. Each area of specialty will have its own set. For every installation and follow-up, there is probably a code, industry standard, manufacturer's instruction, laboratory requirement, or another written document which serves as the basis for the job. Many regions and localities have adopted their own set of regulations for specific systems. Compliance with codes and standards is required. Non-compliance can put you and your company at risk if something goes wrong.

In civil trials, the courts will frequently focus on the issue of compliance with recognized standards. In the case of a conflict between recognized standards and common industry practice, recognized standards will prevail.

INTRODUCTION TO THE TRADE — TRAINEE TASK MODULE 33101

2.0.0 CERTIFICATION AND LICENSING

Several trade and professional organizations issue certificates to ESTs. Some of the organizations provide training; others provide only testing. The certificate that is issued will certify that the EST has completed a course of instruction, has demonstrated the ability to perform specific job skills, or has passed a test that is recognized and accepted by the industry. A list of industry associations and organizations, including those that provide certification for ESTs, is included in *Appendix A*.

In addition to certification, some local or state authorities may require that you be licensed in order to install and service security and/or telecommunications equipment.

3.0.0 YOUR RESPONSIBILITIES AS AN EMPLOYEE

Companies that employ ESTs include manufacturers, distributors, contractors, dealers, and service organizations. Each employer is in business to make a profit. In basic terms, profit is the money left over after all expenses such as payroll, rent, insurance, taxes, and other costs of doing business have been covered.

The profit a company makes is used to help the company grow by providing the funds required to buy new equipment, tools, vehicles, and to hire and train new employees. It also helps the company create a reserve that can keep it going in a slow economy. In the case of large companies, the ability of a company to make a profit determines whether investors will buy its stock, supplying the capital necessary to expand and strengthen the company. The stronger a company is financially, the more secure its future.

An important part of your responsibility as an employee, therefore, is to do your part to help your employer be profitable. This means working efficiently and professionally.

The way you conduct yourself on the job has a major impact on your employer and on the electronic systems industry as a whole. Professional and ethical conduct is required of all ESTs, no matter how long you have been on the job. From the very first day, your ability to communicate, your technical competence, and your attention to detail will predict your success in this industry. The five basic rules for professional and ethical conduct are:

- Be professional at all times.
- Show respect and offend no one.
- Master the technology.
- Work safely.
- Respect and respond to customer requests.

3.1.0 PROFESSIONAL OBLIGATIONS

The company you work for has the right to expect you to work in the most efficient and practical way possible to keep labor costs down. You should be able to anticipate and solve problems in a way that maintains customer satisfaction and employer confidence. The best employees build the company image by working as though they own the company. Customers have a right to expect professionalism and respect from the company's employees.

As an employee, your first professional obligation is to your employer. Remember, you represent your company. While on the job, all of your actions reflect back on the company.

3.2.0 COMPANY STANDARDS

Many larger companies will provide you with a document that describes company standards for employees. Smaller companies may also have a set of guidelines or expectations. If you are not provided with a document related to company standards, it is a good idea to ask if one exists. Sometimes you will be expected to learn the standards during a period of on-the-job training. If this is the case, make a written note of the things your trainer points out to you as expectations or "the way we do things here."

Company standards differ from industry standards and codes that govern the industry. Company standards exist to provide a measure of consistency in how employees are treated and expected to act on the job. The standards serve to protect both the company and the employee. Company standards provide a set of guidelines or procedures for each employee to use when starting and following through on a job.

Company standards may be gathered from a number of different sources, including the employee handbook, statements of company policy, and instruction or procedure manuals.

3.3.0 OBLIGATIONS TO CUSTOMERS

No matter what their position, each employee who interacts with customers and others represents their company. Customers form an impression of the company based upon their initial interaction. If the employee does not seem to be attentive to their needs, that all-important first impression will be negative. Poor customer relations may encourage the customer to seek out a competitor's service. One of the keys to good customer relations lies in your ability to be a courteous and effective communicator.

In every business, good customer relations is the key to a company's success. The ability of a company to be profitable is directly related to the number of loyal customers it has. Customer service is the most important aspect of any business. Dissatisfied customers will seek out a competitor and are likely to speak disparagingly of your company. It does not take long for the negative effects of poor customer service to impact your company's profitability.

A company has an obligation to fill the customer's order in a timely manner by providing the service requested. The company must employ competent, licensed (where required) employees to do the work. The company must ensure that all employees are well trained and follow company and industry standards.

The employee has the same obligations to the customer that the company does. The customer has the right to expect that all interactions with company employees will be respectful and courteous. They depend upon you to work safely and to understand and work to the standards and codes related to the task. Sometimes customers will not understand the procedures or technology behind the system you are installing. Customers usually appreciate it when you take the time to explain what you are about to do. Customers need an opportunity to ask questions and become familiar with the new system.

3.4.0 COURTESY AND RESPECT

Showing respect and using common courtesy is the mark of a professional. Respectful, courteous interactions always make a favorable impression. When you are courteous, you demonstrate that you can put yourself into the other person's shoes. There are some common expectations for good customer relations. It does not matter what your position is in relationship to a work project. All forms of communication should be:

- Honest
- Timely
- Clear and specific
- Free from coarse language or profanity

An EST is expected to communicate certain things at certain times. You must do this to the best of your ability. For example, customers may expect you to communicate the following:

- Project schedules, progress reports, and potential problems or delays
- Safety and security issues
- Requirements or decisions that involve the customer

Supervisors, co-workers, and vendors expect you to clearly communicate:

- Schedules and progress reports
- Work responsibilities and specific techniques used to complete a task
- Location information and arrangements for deliveries
- Safety issues

Providing the right information to the right person when needed is a work practice that should not be ignored. You can never go wrong when being courteous.

Respect also extends to personal appearance and work habits. A professional appearance will go a long way towards establishing a positive impression. Pay attention to personal hygiene. Wear clothing that is appropriate for the job. Take the time to appear neat, organized, and safety conscious to create the best possible image. You should always be ready to start a job on time, with the right tools, and wearing clothing that is safe and free from offensive slogans.

Learn to plan ahead. Make sure the tools, equipment, and materials needed for the next day's work will be available on time.

Put on the appropriate safety gear for the job. Demonstrate your concern for safety and efficient work practices. At the end of the day or the job, leave the customer's site neat, knowing that all hidden work is safe and correctly installed. It always helps to use a respectful attitude toward co-workers, workers from other trades, vendors, and customers.

3.5.0 COMMUNICATING AS A PROFESSIONAL

It is rare to have a job where people skills do not count. The way you interact at the job site can make or break your success in this field. Most skilled ESTs recognize that good people skills are valuable tools of the trade. After all, first impressions count. It is not enough just to be technically competent. You must show the customer that you will handle the job and solve any problems efficiently and pleasantly. Every interaction is an opportunity to convey a professional image for you and your company.

Professionals pay attention to all the products of their work, including communication. They always act in a professional and ethical manner. This means demonstrating the traits of honesty, productivity, safety, cooperation, and civility. Just as it is dangerous to ignore safety concerns, it is also risky to ignore how your words and actions may be interpreted by others.

As an EST, you will be working with many different people. In one day, you could interact with a supervisor, co-workers, other tradespeople, customers, and an inspector. Every work site is different. You may be the lone technician or part of a larger team. When communications are positive and effective, the stage is set for a productive working environment. How you treat other people is your personal responsibility. The way you get treated is often a result of how well you interact with others.

Experienced ESTs realize that success in the workplace is determined by a combination of skill, knowledge, and behavior appropriate to the situation. It is important for every EST to convey a professional image. The level of professionalism demonstrated by each technician reflects back on the industry as a whole. Clear, effective communication makes work more productive and pleasant for everyone. If your natural people skills are not what they should be for this kind of work, take the time to learn how to improve them.

3.5.1 Interpreting Instructions

The interpretation and recall of instructions is a very common task in every workplace. Yet the misinterpretation of oral, written, or graphic instructions can lead to costly mistakes. Mistakes can negatively affect safety, quality of work, and the confidence others have in you to do the job correctly.

Communication experts (Moravian study) show that the words or content of a face-to-face interaction influence only about 7% of how the total communication is perceived. Body language (55%) and tone of voice (38%) provide over 90% of the speaker's message. These are the clues we look for when interpreting a message (*Figure 2*). Usually, we pick up on many things besides the actual words spoken. There are even more opportunities for misinterpretation when people communicate by phone or in writing. Just think how much facial expressions or hand gestures contribute to your understanding of what is being said.

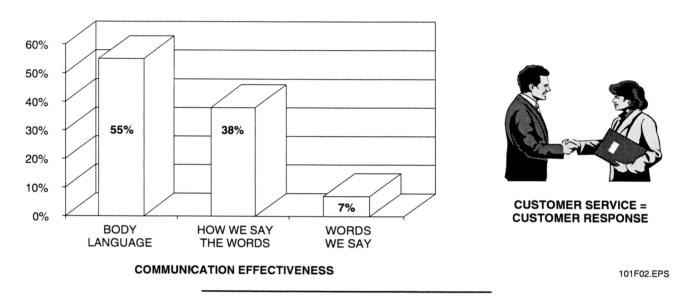

Figure 2. How Communication Is Interpreted

Good communication skills include active listening and the precise presentation of information. They are both essential for efficient work practices and avoiding costly mistakes. These people skills are not easy to achieve without a good deal of effort and practice.

Everyone has stories about how they were treated as a customer by a representative of a particular company. The situations that were most frustrating are usually the ones where you, as the customer, were not treated with respect. The situations that left you with a positive feeling were very different. If you recall those interactions, you may remember that the employee acted confidently, helpfully, and professionally. Most likely, the service person appeared to really concentrate and listen carefully to what you were asking. When he or she gave you information, it was probably presented clearly and concisely. Even if the situation itself was frustrating, the way you were treated as a customer made a difference. If a problem was handled well, you are more likely to continue to do business there.

The first rule of good communication is to concentrate on the speaker and actively listen to what is being said. This means not interrupting and avoiding the tendency to think about what you are going to say next. There is plenty of time to think of what to say after you are sure you have understood the message. Since what is being said could be open to interpretation, it helps to restate what you think you have been told. Restating the important parts of the message will help you remember it better. When you hear the information and repeat it in your own words, it gets placed into your memory twice. Writing the information down is a third way to keep it in focus.

The key to preventing a misunderstanding is to check out the most important information. Restate what you think are the most important points and watch how the speaker responds. Do not worry about sounding stupid. It is better to get the correct message before you start a job. If there is any doubt about what the speaker meant to say, asking questions is the best way to clear things up. The answer could mean the difference between doing it right the first time and having to waste time on rework.

3.5.2 Restate Written Instructions Orally

Written instructions can often be very short. You may have to read between the lines to understand the full message. The notes you read may be incomplete or difficult to understand. It is always better to get the correct information up front. This is especially true when instructions are critical to a successful installation or follow-up. Again, the trick is to restate the written instructions orally in your own words to check for accuracy.

3.5.3 Clearly Relate A Customer Requirement

Customers may ask you questions or request that you tell a co-worker or supervisor a requirement they have for the job. To convey the information accurately, be sure you understand what the requirement is first. Briefly restate the basics of the requirement back to the customer. He or she will be able to clear up any misunderstandings before you relay the message. If you cannot answer a customer's question or solve a problem, direct the customer to someone who can.

When relaying a message, be sure you have the co-worker or supervisor's full attention. Telling a person something important when they are distracted guarantees that only part of the message will be heard. Even less will be remembered. Gain their full attention by stating why the requirement is important and how it involves them. State the requirement clearly, using a logical sequence. Describe the desired outcome first, then detail the steps required.

State the most important information up front, then repeat it. If the message is complicated, be sure the listener understands each part before going on to the next. You can check for understanding by asking the listener to tell you what they understand about the requirement. Use this as an opportunity to correct any misinformation before mistakes are made.

Finally, follow up to make sure the customer's needs are fulfilled or questions have been answered. Later, ask your co-worker what action was taken. Find out if the customer knows the requirement was acted upon or if delays have been explained. The extra effort you make to get back to the customer will increase their confidence in you as a professional. The customer usually does not care who solves the problem, just that it was solved.

3.6.0 TEAMWORK

Teamwork is the ability of a group of workers to accomplish a common goal. The workers at any job site will be individuals with different levels of skill and different personalities. The team's ability to get the job done often depends on how well everyone communicates and works together. Each person at a job site will have a particular task to accomplish within a set period of time. With all the different tasks needed to complete a job, it is likely that you will run into a situation where teamwork is important.

The ideal situation is one in which a team of workers agree on the best way to achieve a common goal. They work together, helping to manage the work so that everyone is able to do their own job with the least amount of interference. Sometimes the team is directed by a strong leader who can plan and schedule the work so things run smoothly. However, it is always up to the team members to develop good working relationships so that communication flows easily among the team members. To develop good teamwork, work toward the goals shown in *Figure 3*.

GOOD TEAM PLAYERS:
- ARE RESPECTFUL AND MAINTAIN THE SELF ESTEEM OF OTHERS
- DEVELOP TRUSTFUL WORKING RELATIONSHIPS
- FOCUS ON THE SITUATION OR ISSUE, NOT ON THE PERSON
- TAKE THE INITIATIVE TO BE COOPERATIVE
- ARE PROFESSIONAL AT ALL TIMES

101F03.EPS

Figure 3. Elements Of Teamwork

Work situations are not always ideal. It is likely that the individuals will have competing priorities. You are required to figure out the best way to work with people you do not know very well. You may also have to deal with a frustrating situation without hurting the feelings of others. This is where your people skills and teamwork skills come into play.

People who work together usually have the same needs, but their methods of achieving results depends upon their personality. Treat supervisors, customers, and co-workers with respect as a first step toward gaining their cooperation. Some common needs are:

- The need to contribute, to know what your role is in relation to others
- The need to feel competent, to be confident you are doing what is right

- The desire to achieve results, to know how your work measures up to standards
- The desire to have work efforts recognized and rewarded in a timely fashion

3.7.0 CONFLICT RESOLUTION

It has been said that conflict is a growth industry. Every day, people from different backgrounds and with differing needs must work together to solve common problems. Whenever people work together, there is a potential for conflict. Conflict starts when two people want different things. They may each have a strong opinion about how a task should be completed.

Like it or not, part of your job will involve negotiating. If you are out in the world interacting with others, you are negotiating. You may not think of yourself as a negotiator, and many people are not good at it. Again, like the other people skills, becoming good at negotiating takes time and practice.

Everyone wants to have a say in the decisions that affect them. Few people are willing to accept decisions dictated by someone else. Negotiation is the basic way you get what you want from others. It does not necessarily result in a win or lose situation. Those skilled at conflict resolution recommend the proven techniques of principled negotiation as a way to manage the back and forth communication required on the job. The best possible outcome of a conflict is when everyone comes out satisfied.

Even though negotiation takes place every day, it is not easy to do it well. If a conflict is handled poorly, each side leaves unhappy and worn out. Unresolved conflicts worsen over time and can lead to negative work relationships. People in conflict start to resent each other. The whole situation can lead to more problems. That is why it is important to resolve conflicts before they grow bigger. Physical violence and intimidation are never acceptable methods of communication.

3.7.1 Use Principled Negotiation In Solving A Conflict

Principled negotiation recognizes that each person brings their own point of view to the table. Trying to understand exactly what these are is the first step in resolving a conflict. Sometimes a conflict is resolved easily when misunderstandings are corrected.

The first step in dealing with the other person is to listen carefully to his or her concerns. Then, try to restate their concerns in your own words. When you do this, the other person will have more confidence that they are being heard. Sometimes this is all it takes. Take the time to find out where the other person is coming from. Most conflicts can be resolved by mentally stepping into the other person's shoes. Take a few moments to examine the other person's needs from his or her point of view. If both your needs are reasonable, you can try to find ways to accommodate each other. Suggest things that work for both sides.

Sometimes, each person takes a stand and will not budge. The alternative to a standoff is to change the nature of the game. Good negotiators realize that it is important not to be overwhelmed by an apparent impasse. Instead, you should discuss the conflict calmly and rationally. Let the other person know that their point of view is important to you.

There are four basic steps (*Figure 4*) to negotiate in a way that is more likely to help people agree with one another.

NEGOTIATING STEPS:

1. SEPARATE THE PEOPLE FROM THE PROBLEM.
2. FOCUS ON INTERESTS, NOT POSITIONS.
3. IDENTIFY SEVERAL OPTIONS BEFORE DECIDING WHAT TO DO.
4. BASE EXPECTED RESULTS ON OBJECTIVE STANDARDS.

Figure 4. Four Basic Negotiation Steps

- *Separate the people from the problem* – We all have strong emotions about certain things that affect us. Some feelings come from past experiences, so two people often have very different ways of seeing something. In a conflict, the position we take may be tied up with our own ego. We may not communicate everything that is important because specific issues may not be important from our standpoint. When you separate the person from the problem, you focus on the issues involved. You are more likely to see yourself as working side by side on a problem, rather than working against the other person.

- *Focus on interests, not positions* – It is not easy to reach agreement when people are concerned about their position, rather than working to understand the real problem. A negotiating position may not be what you or the other person really wants. For example, someone may take a certain position because they don't want to look bad in front of others. Try to uncover and focus on the actual need or interest before taking a stand.

- *Identify several options before deciding what to do* – When we have a stake in the outcome of a decision, it is difficult to make a good decision under pressure. Before trying to reach an agreement, take some time to think of other options that may be agreeable to both parties.

- *Base expected results on objective standards* – When discussing possible outcomes, neither person needs to give in. You can decide that a fair solution will be based on a selected standard. The standard could be an expert opinion, completion by a certain time, or quality based on industry standards. This is more likely to lead to a favorable result.

4.0.0 INDUSTRY STANDARDS AND BUILDING CODES

All telecommunications, life safety, and security systems, as well as the work done by ESTs, are governed by industry standards and building codes. It is important that you become familiar with these standards and codes. The following is a list of standards and codes that govern the work you will do as an EST. During your training, you will become very familiar with these documents. It will become second nature to you to refer to some of these documents in the course of your work. A good rule of thumb is when in doubt, look it up.

In the event of a conflict between two or more standards, the more stringent requirement should be applied. In one sense, the only true conflict would be a situation in which there is no way to comply with all the provisions at the same time.

As an example, if a building code called for manual stations to be mounted at 36" to 56" above the floor, and another standard specified 42" to 60" above the floor, this would generally be considered as an apparent conflict. Because it is possible to comply with both requirements by installing the manual stations at 42" to 56" above the floor, it is not a true conflict.

4.1.0 THE NATIONAL ELECTRICAL CODE®

The National Fire Protection Association sponsors, controls, and publishes the *National Electrical Code (NEC)* within the United States' jurisdictional area. The NEC specifies the minimum provisions necessary to safeguard persons and property from electrical hazards.

Most federal, state, and local municipalities have adopted the NEC, in whole or in part, as their legal electrical code. Some states or localities adopt the NEC and add more stringent requirements. The NEC is used by:

- Lawyers and insurance companies to determine liability
- Fire marshals and electrical inspectors in loss prevention and safety enforcement
- Designers to ensure a compliant installation

The code sets the minimum standards that must be met to protect people and property from electrical hazards. It is revised every three years. The NEC is arranged by chapter, article, and section (e.g., ***NEC Chapter 8, Article 800, Section 800-52***). Portions of the code having significant importance to the electronic systems technician are listed in *Appendix B*.

It is important to remember that the NEC covers only the requirements necessary to ensure a safe environment, not an environment in which the telecommunications systems are guaranteed to operate free of any interference or errors.

4.2.0 THE CANADIAN ELECTRICAL CODE, PART 1

The Canadian Standards Association sponsors, controls, and publishes the *Canadian Electrical Code, Part 1 (CEC, Part 1)*. The intent of this code is to establish safety standards for the installation and maintenance of electrical equipment. As with the NEC, the CEC may be adopted and enforced by the provincial and territorial regulatory authorities.

Telecommunications installers must be familiar with the CEC. *Appendix C* lists sections of the CEC that are of particular interest to the EST.

4.3.0 NATIONAL FIRE PROTECTION ASSOCIATION (NFPA)

In addition to the NEC, the National Fire Protection Association (NFPA) develops and produces other codes that apply to telecommunications, life safety, and security systems. These are listed in *Appendix D*.

4.4.0 NATIONAL BUILDING CODES

The organizations listed below publish and maintain nationally accepted building codes. State and local governments generally adopt one of these codes. However, the state and local governments may also have additional requirements that must be met. It is essential that everyone involved in the installation of electronic systems be familiar with the local and national codes.

- International Code Council (ICC)
- Building Officials and Code Administrators International (BOCA), which publishes the *National Building Code*
- International Conference of Building Officials (ICBO), which publishes the *Uniform Building Code*
- Southern Building Code Congress International (SBCCI), which publishes the *Standard Building Code*

Note: Canadian codes must also be followed where applicable.

4.5.0 INTERNATIONAL STANDARDS

ANSI/TIA/EIA publishes standards for the manufacturing, installation, and performance of electronic and telecommunications equipment and systems. Five of these ANSI/TIA/EIA standards govern telecommunications cabling in buildings. Each standard covers a specific part of building cabling. They address the required cable, hardware, equipment, design, and installation practices. In addition, each ANSI/TIA/EIA standard lists related standards and other reference materials that deal with the same topics.

Most of the standards include sections that define important terms, abbreviations, and symbols.

The ANSI/TIA/EIA standards which govern telecommunications cabling in buildings are listed in *Appendix E*.

The Institute of Electrical and Electronic Engineers also publishes a variety of standards that apply to low-voltage systems. One such standard is IEEE Standard 1394, *High Performance Serial Bus*.

4.6.0 WHY STANDARDS?

Since the earliest recorded time, people have relied upon standards. Without a uniformly recognized system of weights and measures, for example, commerce would not exist.

It would be fairly simple and straightforward if every customer, every style of door, each type of building, or every type of signal could be handled in exactly the same way. Unfortunately, that is impossible. Customers each have a specific need that caused them to contract for our services. In many cases, these needs will be similar, but in others, they will differ. The trick is to develop standards and procedures that allow us to respond to different needs in the same way.

4.6.1 Standards Coordinate Team Activity

Everyone has experienced the frustration of not having the correct tool for a particular job or trying to fix a job completed by someone else. Standardization helps to solve these problems. Following the same procedures on each and every job makes it easier for others to follow up after you. Most companies agree on the specific procedures or devices that will be used. This enables us to have the tools that we need to do the job. Without a standard choice of equipment and a consistent way of doing things, we reinvent each and every job, wasting much time and effort.

4.6.2 Standards Reduce Liability

As mentioned earlier, in civil trials, the courts will frequently focus on the issue of compliance with recognized standards. In the case of a conflict between recognized standards and common industry practice, recognized standards will prevail. For example, an alarm system found to be in violation of the recognized standards after a fatal fire may create a liability for the installer.

Where no particular standards have been legally adopted or are in force in that jurisdiction, national standards may apply. In addition to the moral and social responsibility associated with conducting your work correctly, there are the added civil liability implications. To comply with or exceed the recognized codes and standards is your best protection from liability. When faced with justifying an action in a court proceeding, a company is better off if it can cite recognized standards instead of a common company or industry practice.

5.0.0 DOCUMENTATION AND PAPERWORK

Every job requires paperwork. At the beginning of the project, you will have drawings and other information describing the project and how it is to be accomplished. This documentation is normally in the form of drawings, including:

- *Work statement* – A detailed description of the work to be performed and the end result expected. It is part of the contract.

- *Floor plan* – A drawing showing the layout of each floor of the building. Separate plans based on the floor plan will show wiring and cable runs. You will learn more about construction drawings in the Core Curricula.

- *Bill of Materials* – A detailed list, including quantities, of all system components, cabling, wire, terminations, boxes, raceways, conduits, and other materials needed for an installation. This list is developed by the designer or estimator as part of the material **takeoff**.

- *Riser diagrams* – Diagrams that identify zones within the building, the sequence of device connection, end-of-line devices, and the conductor counts for each cabling run (*Figure 5*).

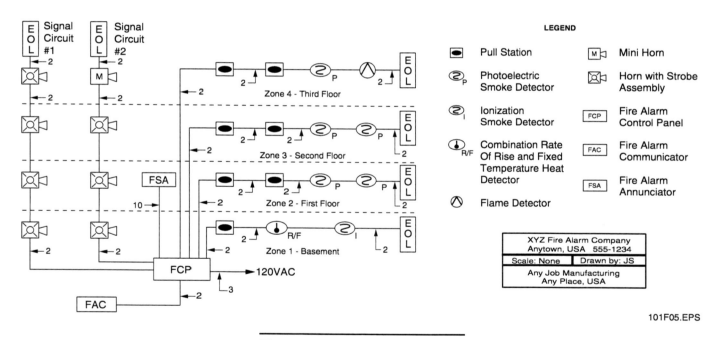

Figure 5. Typical Riser Diagram

- *Point-to-point drawings* – Drawings that indicate actual terminal connections and include the connections from the control through all devices to the terminal device (*Figure 6*). Some point-to-point drawings will show all devices; others will show only typical devices.

- *As-built diagrams* – Diagrams that reflect the finished job. It is not unusual for changes to be made during the course of a project. These changes are marked on a set of drawings, which then become known as the *as-builts*.

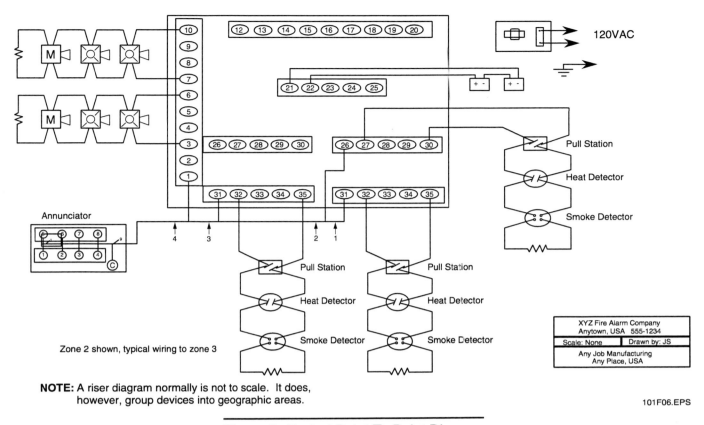

Figure 6. Typical Point-To-Point Diagram

Other paperwork that is an important part of any job includes:

- *Time sheet* – Each person who works on a project must keep an accurate record of the hours they work. This record provides the means for the employer to bill for the work and to verify the accuracy of their price quote. Some employers require employees to punch a time clock at the job site.
- *Job sheet* – The job sheet is used to keep track of all aspects of the job. Job sheets will vary from one company to another. An example is shown in *Figure 7*. In this example, the project leader will summarize the hours spent by each employee, along with the equipment, parts, and material used on the job.
- *Certificate of completion* – This document, which is completed at the end of the project, helps the installer prepare for inspection of the finished work by the client or by an inspection official. A sample from a fire alarm system project is shown in *Figure 8*.
- *Change order* – During the course of a project, the client may request changes, or it may turn out that the installation could not be done exactly as planned. (For example, air conditioning ductwork may have been placed where you planned to run your cable trays.) Each such change must be documented. Before the job is signed off, the changes must also be marked on the as-builts.

INTRODUCTION TO THE TRADE — TRAINEE TASK MODULE 33101

JOB SHEET

Work Order No. 1005001

Start Date 06 OCT. 1999

Customer:

Any Company
123 Fourth St.
Anytown, USA   12345
Phone: (123) 456-7899
Fax: (123) 456-7890

Shipped By:

Another Company
456 Fifth St.
Anothertown, USA   56789
Phone: (111) 555-5555
Fax: (111) 555-4444

Description of Work:

Employee	Hours	Overtime	Date	Quantity	Description	Part No.	Cost

101F07.EPS

Figure 7. Example Of A Job Sheet

Record of Completion

Name of Protected Property: _____
Address: _____
Rep. of Protected Prop. (name/phone): _____
Authority Having Jurisdiction: _____
Address/Phone Number: _____

1. Type(s) of System or Service
   _____ NFPA 72, Chapter 3 — Local
   If alarm is transmitted to location(s) off premises, list where received:
   _____

   _____ NFPA 72, Chapter 3 — Emergency Voice/Alarm Service
   Quantity of voice/alarm channels: _____ Single: _____ Multiple: _____
   Quantity of speakers installed: _____ Quantity of speaker zones: _____
   Quantity of telephones or telephone jacks included in system: _____

   _____ NFPA 72, Chapter 4 — Auxiliary
   Indicate type of connection:
   Local energy: _____ Shunt: _____ Parallel telephone: _____
   Location and telephone number for receipt of signals:
   _____

   _____ NFPA 72, Chapter 4 — Remote Station
   Alarm: _____
   Supervisory: _____

   _____ NFPA 72, Chapter 4 — Proprietary
   If alarms are retransmitted to public fire service communications center or others, indicate location and telephone number of the organization receiving alarm:
   _____

   Indicate how alarm is retransmitted:
   _____

   _____ NFPA 72, Chapter 4 — Central Station
   The Prime Contractor:
   _____

Central Station Location:
_____

Means of transmission of signals from the protected premises to the central station:
_____ McCulloh   _____ Multiplex   _____ One-Way Radio
_____ Digital Alarm Communicator   _____ Two-Way Radio   _____ Others

Means of transmission of alarms to the public fire service communications center:
(a) _____
(b) _____
System Location: _____

	Organization Name/Phone	Representative Name/Phone
Installer	_____	_____
Supplier	_____	_____
Service Organization	_____	_____

Location of Record (As-Built) Drawings:
_____

Location of Owners Manuals:
_____

Location of Test Reports:
_____

A contract, dated _____ , for test and inspection in accordance with NFPA standard(s) No(s). _____ ,
dated _____ , is in effect.

Figure 8. Example Of A Certificate Of Completion (1 Of 2)

2. Record of System Installation
   (Fill out after installation is complete and wiring checked for opens, shorts, ground faults, and improper branching, but prior to conducting operational acceptance tests.)
   This system has been installed in accordance with the NFPA standards as shown below, was inspected by _____
   on _____ , includes the devices shown below, and has been in service since _____
   _____ NFPA 72, Chapters 1 3 4 5 6 7 (circle all that apply)
   _____ NFPA 70, *National Electrical Code*, Article 760
   _____ Manufacturer's Instructions
   _____ Other (specify): _____
   Signed: _____ Date: _____
   Organization: _____

3. Record of System Operation
   All operational features and functions of this system were tested by _____ on _____ , and found to be operating properly in accordance with the requirements of:
   _____ NFPA 72, Chapters 1 3 4 5 6 7 (circle all that apply)
   _____ NFPA 70, *National Electrical Code*, Article 760
   _____ Manufacturer's Instructions
   _____ Other (specify): _____
   Signed: _____ Date: _____
   Organization: _____

4. Alarm-Initiating Devices and Circuits (use blanks to indicate quantity of devices)
   MANUAL
   (a) _____ Manual Stations _____ Noncoded, Activating _____ Transmitters _____ Coded
   (b) _____ Combination Manual Fire Alarm and Guard's Tour Coded Stations
   AUTOMATIC
   Coverage: Complete: _____ Partial: _____
   (a) _____ Smoke Detectors _____ Ion _____ Photo
   (b) _____ Duct Detectors _____ Ion _____ Photo
   (c) _____ Heat Detectors _____ FT _____ RR _____ FT/RR _____ RC
   (d) _____ Sprinkler Waterflow Switches: _____ Transmitters _____ Noncoded, Activating _____ Coded
   (e) _____ Other (list): _____

5. Supervisory Signal-Initiating Devices and Circuits (use blanks to indicate quantity of devices)
   GUARD'S TOUR
   (a) _____ Coded Stations
   (b) _____ Noncoded Stations, Activating _____ Transmitters
   (c) _____ Compulsory Guard Tour System Comprised of _____ Transmitter Stations and _____ Intermediate Stations
   NOTE: Combination devices recorded under 4(b) and 5(a).
   SPRINKLER SYSTEM
   (a) _____ Coded Valve Supervisory Signaling Attachments
       Valve Supervisory Switches, Activating _____ Transmitters
   (b) _____ Building Temperature Points
   (c) _____ Site Water Temperature Points
   (d) _____ Site Water Supply Level Points
   Electric Fire Pump:
   (e) _____ Fire Pump Power
   (f) _____ Fire Pump Running
   (g) _____ Phase Reversal
   Engine-Driven Fire Pump:
   (h) _____ Selector in Auto Position
   (i) _____ Engine or Control Panel Trouble
   (j) _____ Fire Pump Running
   Engine-Driven Generator:
   (k) _____ Selector in Auto Position
   (l) _____ Control Panel Trouble
   (m) _____ Transfer Switches
   (n) _____ Engine Running

Figure 8. Example Of A Certificate Of Completion (2 Of 2)

6.0.0 TYPES OF TRAINING PROGRAMS

There are two basic forms of training programs that most employers consider. The primary one is on-the-job training (OJT) to improve the competence of their employees in order to provide better customer service and for the continuity and growth of the company. The second is formal apprenticeship training, which provides the same type of training but also conforms to federal and state requirements under the Code of Federal Regulations (CFR), Titles 29:29 and 29:30.

6.1.0 STANDARDIZED TRAINING BY THE NCCER

The National Center for Construction Education and Research (NCCER) is an independent, private educational foundation founded and funded by the construction industry to solve the training problem plaguing the industry today. The basic idea of the NCCER is to replace governmental control and credentialing of the construction workforce with industry-driven training and education programs. The NCCER departs from traditional classroom or distance learning by offering a competency-based training regimen. Competency-based training means that instead of simply requiring specific hours of classroom training and set hours of OJT, you have to prove that you know what is required and demonstrate that you can perform the specific skill. All completion information on every trainee is sent to the NCCER and kept within the National Registry. The NCCER can confirm training and skills for workers as they move from company to company, state to state, or to different offices in the same company. These are portable credentials and are recognized nationally.

6.1.1 Apprenticeship Training

As stated earlier, formal apprenticeship programs conform to federal and state requirements under CFR Titles 29:29 and 29:30. All approved apprenticeship programs provide OJT as well as classroom instruction. The related training requirement is fulfilled by all NCCER craft training programs. The main difference between NCCER training and registered apprenticeship programs is that apprenticeship has specific time limits in which the training must be completed. Apprenticeship standards set guidelines for recruiting, outreach, and a specific time limit for each of a variety of OJT tasks. Additionally, there are reporting requirements and audits to ensure adherence to the apprenticeship standards. Companies and employer associations register their individual apprenticeship programs with the Bureau of Apprenticeship and Training (BAT) within the U.S. Department of Labor, and in some instances, with state apprenticeship councils (SAC):On-the-job Training (OJT) of 2,000 hours per year and a minimum of 144 hours of classroom-related training are required. Apprenticeship programs vary in length from 2,000 hours to 10,000 hours.

6.1.2 Youth Training And Apprenticeship Programs

Youth apprenticeship programs are available that allow students to begin their apprenticeship or craft training while still in high school. A student entering the program in the 11th grade may complete as much as one year of the NCCER *Wheels of Learning* program by high school graduation. In addition, programs (in cooperation with local construction industry employers) allow students to work in the craft and earn money while still in school. Upon graduation, students can enter the industry at a higher level and with more pay than someone just starting in a training program.

Students participating in the NCCER or youth apprenticeship training are recognized through official transcripts and can enter the second level or year of the program wherever it is offered. They may also have the option of applying credits at two-year or four-year colleges that offer degree or certificate programs in their selected field of study.

7.0.0 TOOLS OF THE TRADE

An EST uses a variety of hand and power tools, as well as common and special test equipment, for installation and troubleshooting of low-voltage equipment. The specific tools and equipment you use will vary to some extent depending on your system specialty (telecommunications, security, entertainment systems, etc.).

In the Core training modules, you will receive instruction on basic hand and power tools. In the *Construction Materials and Methods* module in this level, you will learn more about drilling and cutting tools and how they are used in creating cable pathways. In the *Electrical Theory* module, you will learn about basic electrical test instruments. You will also learn about conduit bending tools in this level.

The applications of tools and test equipment will be covered as you progress through your training. *Figures 9, 10,* and *11* show some representative examples of special tools and test devices you will use as an EST.

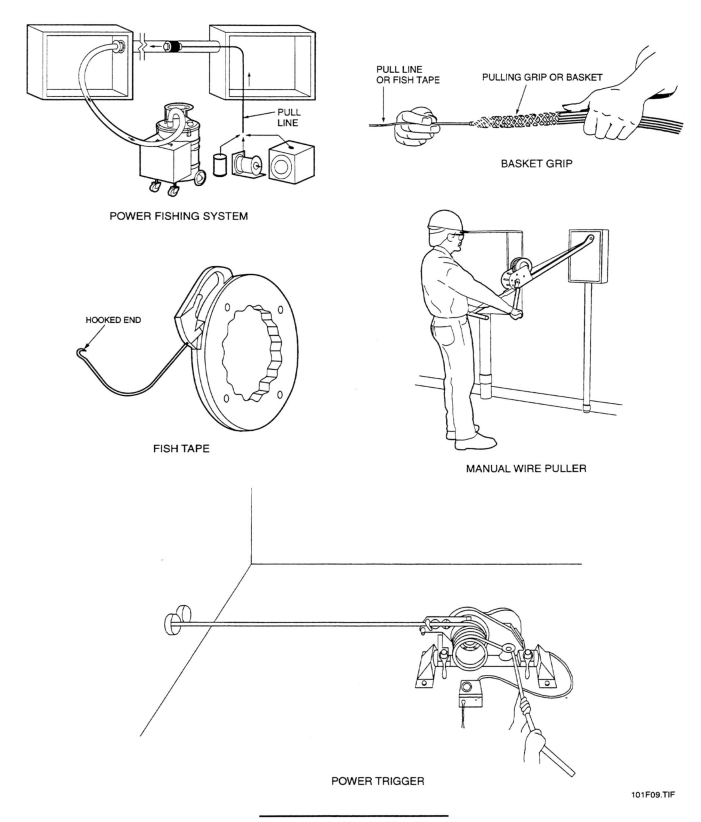

Figure 9. Cable-Pulling Tools

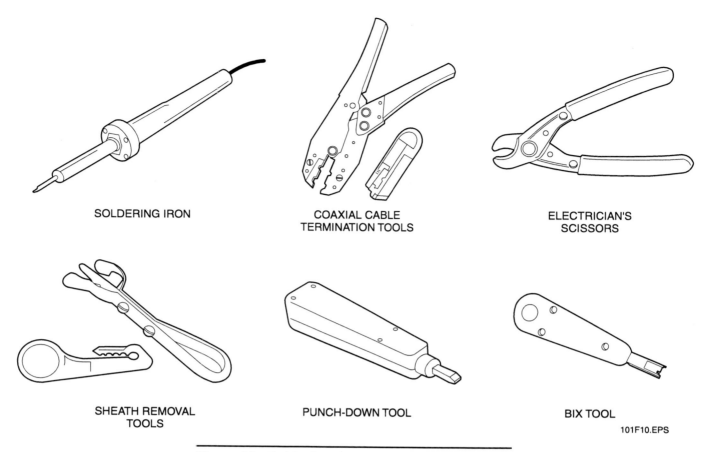

Figure 10. Cable-Stripping And Termination Tools

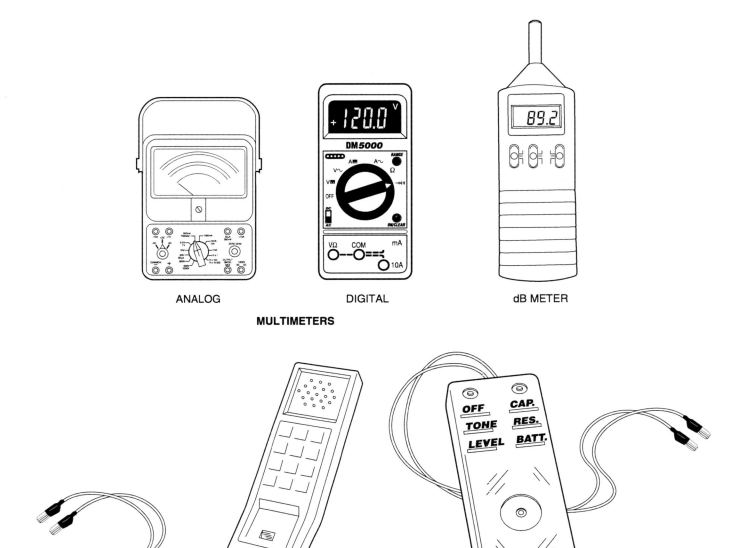

Figure 11. Test Instruments Used In Low-Voltage Work

SUMMARY

Electronic systems technicians install and service equipment and systems in a variety of technology areas, including: alarm and security, communications, entertainment, and integrated building management systems. ESTs work with computer networks, phone systems, audio and video systems, as well as home and business automation systems.

In 1998, there were approximately 36,000 companies employing at least 165,000 ESTs. The opportunities in this field can only grow as the population becomes more and more dependent on computers and the demand for instant global communication and ready-access entertainment increases.

In addition to learning the technical aspects of their profession, including the safe and efficient use of a variety of tools and test devices, ESTs must become familiar with the numerous installation standards and building codes that govern the industry.

References

For advanced study of topics covered in this task module, the following books are suggested:

National Electrical Code Handbook, Latest Edition, National Fire Protection Association, Quincy, MA.

TIA/EIA Telecommunications Building Wiring Standards, Latest Edition, Global Engineering, Englewood, CO.

REVIEW/PRACTICE QUESTIONS

1. Which of the following types of systems is *not* installed by ESTs?
   a. Voice and data
   b. Entertainment
   c. Communications
   d. Power distribution

2. A modem is a device used to _____.
   a. connect computers using communication lines
   b. amplify voice signals
   c. select the operating mode of a security system
   d. turn on emergency lighting during a power failure

3. Which of the following is likely to have a positive effect on your company's profit?
   a. Rework caused by improper installation
   b. Employee tardiness
   c. An increase in insurance costs
   d. Completing a job on time

4. If you are working with others toward a common objective, the best way to resolve a conflict is to _____.
   a. take control of the situation and dictate to the others
   b. threaten anyone who disagrees with you
   c. keep talking until everyone agrees with you
   d. negotiate a solution that everyone can agree with

5. If there is a conflict between two codes or standards that apply to a job, the best approach is to _____.
   a. apply the most demanding standard
   b. apply the least demanding standard
   c. call the organizations that publish the codes
   d. ask the customer for advice

6. The *National Electrical Code* is published by the _____.
   a. National Fire Protection Association
   b. National Electrical Code Association
   c. International Code Council
   d. American National Standards Institute

7. Which of the following is true regarding building codes?
    a. National codes always take precedence over local codes.
    b. Local codes are the same as national codes, so there is no conflict.
    c. Local codes may contain requirements that are not covered in national codes.
    d. Codes are for information only; they do not have to be followed.

8. The ANSI/TIA/EIA standard that governs residential and light commercial pathways and spaces is _____.
    a. ANSI/TIA/EIA-568-A
    b. ANSI/TIA/EIA-569-A
    c. ANSI/TIA/EIA-570-A
    d. ANSI/TIA/EIA-606

9. In a BAT-certified training program, it is necessary to complete a minimum of _____ in each year of the training program.
    a. 150 hours of classroom instruction
    b. 1,000 hours of OJT
    c. 144 hours of classroom training and 2,000 hours of OJT
    d. 200 hours of classroom training and 1,000 hours of OJT

10. A student entering an NCCER craft training program in the 11th grade can complete as much as _____ of the program by high school graduation.
    a. one year
    b. two years
    c. three years
    d. four years

notes

ANSWERS TO REVIEW/PRACTICE QUESTIONS

Answer	Section Reference
1. d	1.2.0
2. a	Terms, 1.2.0
3. d	3.1.0
4. d	3.7.0
5. a	4.0.0
6. a	4.1.0
7. c	4.4.0
8. b	4.5.0, Appendix E
9. c	6.1.1
10. a	6.1.2

APPENDIX A

TELECOMMUNICATIONS AND SECURITY ORGANIZATIONS

ALOA
Associated Locksmiths of America
303 Liveoak
Dallas, TX 75204
Phone: (214) 827-1701
Fax: (214) 827-1810
www.aloa.org

The Associated Locksmiths of America, Inc. is a trade association established in 1956 and dedicated to enhancing professionalism, education, and ethics among locksmiths and those in related sectors of the security industry.

BICSI
BICSI Headquarters
8610 Hidden River Parkway
Tampa, FL 33637
Phone: (813) 979-1991 or (800) 242-7405
Fax: (813) 971-4311
www.bicsi.org

BICSI, an international not-for-profit telecommunications association, was founded in 1974 to serve and support telephone company building industry consultants (BICs) responsible for the design and distribution of telecommunications wiring in commercial and multi-family buildings.

BICSI has grown dramatically since those early days and by the end of 1999 will serve nearly 17,000 members from every state in the U.S. and from over 75 countries around the world. Their programs and interests cover the broad spectrum of voice, data, and video technologies.

CABA
Continental Automated Buildings Association
1500 Montreal Rd., Bldg. M-20
Ottawa, ON, Canada
K1A 0R6
Phone: (613) 990-7407 or (888) 798-CABA (2222)
Fax: (613) 991-9990
www.caba.org

CABA (the Continental Automated Buildings Association) is North America's key source for information, education, and networking relating to home and building automation. Its mission is to encourage the development, promotion, and adoption of business opportunities in the home and building automation industry. Members include manufacturers, dealers, installers, telecommunications companies, energy utilities, builders, consultants, research organizations, publishers, educational institutions, governments, and associations. CABA's numerous publications are recognized by many in the industry for providing more information about the home and building automation market than any other single source in the United States and Canada.

CANASA
Canadian Alarm and Security Association
610 Alden Rd.
Suite 201
Markham, ON, Canada
L3R 9Z1
Phone: (905) 513-0622
Fax: (905) 513-0624
www.canasa.org

A non-profit, national association established in 1977, the Canadian Alarm and Security Association (CANASA) represents the interests of the electronic security alarm industry and helps dealers, distributors, manufacturers, and monitoring companies across Canada succeed in business. Its Professional Development Courses, Information Services, and Membership Advantage Program (MAP) are all part of a comprehensive network of services designed to help members save money and run their businesses faster, safer, and in a professional and ethical manner. CANASA's mission is to promote and protect the interests of member companies.

CANASA is very active on a wide range of issues affecting its members, consumers, and the industry at large. The Association maintains an active dialogue with government officials and regulatory bodies on false alarm policies and prevention methods, industry competition, and licensing. CANASA offers a wide range of benefits and programs for its members, including monthly news updates, consumer publications, benefit plans, opportunities for professional development and education, member meetings, social events, and conferences. CANASA also produces and manages the Security Canada trade shows—the largest security industry shows in Canada. In addition to its national efforts, 13 chapters and subchapters across Canada work to effect positive change on a local level.

CEDIA
Custom Electronic Design and Installation Association
9202 N. Meridan Street
Suite 200
Indianapolis, IN 46260
Phone: (317) 571-5602
Fax: (317) 571-5603
www.cedia.org

The Custom Electronic Design and Installation Association, or CEDIA, is an international, U.S.-based trade association. It is comprised of companies which specialize in planning or installing electronic systems in the home—such as home theaters, single- or multi-room entertainment systems, communications systems, alarm and surveillance systems, lighting control products, integrated systems, and other residential electronics. Founded in 1989, CEDIA's mission is to build recognition and acceptance for the specialized profession and to speak up for its interests by addressing industry, government, and the marketplace.

CEMA
Consumer Electronics Manufacturers Association
2500 Wilson Blvd.
Arlington VA 22201
Phone: (703) 907-7600
Fax: (703) 907-7601
www.cemacity.org

The Consumer Electronics Manufacturers Association (CEMA) is a trade association whose primary members are U.S. consumer electronic manufacturers. It is a sector of the Electronic Industries Alliance, which was founded in 1924. CEMA has shaped legislation and developed thousands of engineering standards that enable new technologies to be brought to market. CEMA sponsors the International Consumer Electronics Show and other trade shows which bring together manufacturers, distributors, retailers, and the press.

EIA
Electronic Industries Association
2500 Wilson Blvd.
Arlington, VA 22201

The Electronic Industries Association (EIA) is a trade association of the electronics industry that formulates technical standards, disseminates marketing data, and maintains contact with government agencies.

HAA
Home Automation Association
1444 I Street, NW, Suite 700
Washington, DC 20005
Phone: (202) 712-9050
Fax: (202) 216-9646
www.homeautomation.org

The Home Automation Association (HAA)—the trade association of the home control industry—was founded in 1988 and marked the recognition of the home control industry. HAA currently has over 300 company members, including manufacturers, distributors, dealers, installers, and service providers of home automation products. Home automation is defined as a process or system (using different methods or equipment) which provides the ability to enhance one's lifestyle and make a home more comfortable, safe, and efficient. Home automation can link lighting, entertainment, security, telecommunications, heating, and air conditioning into one centrally controlled system.

NBFAA
National Burglar and Fire Alarm Association
7101 Wisconsin Avenue, Suite 901
Bethesda, MD 20814-4805
Phone: (301) 907-3203
Fax: (301) 907-7897
www.alarm.org

Founded in 1948, the National Burglar and Fire Alarm Association (NBFAA) is organized to pursue the common interests of the industry. The NBFAA is a federation of state associations and is governed by a board of directors selected from each state. Information on the latest technology, techniques, and practices are shared throughout the industry through NBFAA publications, meetings, seminars, and trade shows. Several benefits are available, including reduced rate group health, auto, and liability insurance. The NBFAA is active in false alarm prevention and legislative issues at the federal, state, and local level. The organization has provided its resources and support and played an active role in fashioning alarm industry licensing and fire safety standards in many states. The NBFAA is a leading resource to the nation's news media, the fire service, law enforcement, government, and other organizations on the subject of alarm systems and related fields.

NCCER
The National Center for Construction Education and Research
P.O. Box 141104
Gainesville, FL 32614-1104
Phone: (352) 334-0911
Fax: (352) 334-0932
www.nccer.org

The National Center for Construction Education and Research provides a wide variety of construction training programs, maintains a national registry of the work completed by trainees, and issues certificates to the trainees as they complete each level of an NCCER-sponsored training program.

NFPA
National Fire Protection Association
One Batterymarch Park
Quincy, MA 02269

The National Fire Protection Association (NFPA) publishes the *National Electrical Code (NEC)* and its related materials, along with numerous other codes relating to fire protection and safety.

NICET
National Institute for Certification in Engineering Technologies
1420 King Street
Alexandria, Virginia 22314-2715
Phone: (703) 684-2835 or (800) 787-0034

The National Institute for Certification in Engineering Technologies (NICET) is an organization that administers testing and certification programs in the fire alarm and sound industry. NICET does not provide training.

NSCA
National Systems Contractors Association
419 First Street Southeast
Cedar Rapids, IA 52401
Voice: (800) 446-6722 or (319) 366-6722
Fax: (319) 366-4164
www.ncsa.org

The National Systems Contractors Association (NSCA) is a globally recognized trade association working continuously on improving the electronic systems contracting industry. It began as the National Sound and Communications Association but broadened its scope as members broadened their activities. Contractors are predominately commercial. NSCA's mission is to provide members of the NSCA and the electronic systems contracting industry with formal educational opportunities, professional development, information exchange, and member services in response to member-identified needs and priorities and to represent the industry in the external environment.

SIA
Security Industry Association
635 Slaters Lane, Suite 110
Alexandria, VA 22314
Phone: (703) 683-2075
Fax: (703) 683-2469
www.siaonline.org

The Security Industry Association (SIA) is an organization of manufacturers and distributors dedicated to the development of professionalism in the security industry. SIA develops equipment standards, has adopted warranty return and repair policies, sponsors market research, participates in false alarm prevention activities, and coordinates trade show exhibits.

TIA
Telecommunications Industry Association
www.tiaonline.org

TIA is a national trade organization that represents suppliers of communications and information technology products on public policy, standards, and market-development issues affecting its membership. TIA members provide communications and information technology products, materials, systems, distribution services, and professional services in the United States and around the world. The association's member companies manufacture or supply virtually all of the products used in global communication networks. TIA represents the telecommunications industry with its subsidiary, the Multimedia Telecommunications Association (MMTA), in conjunction with the Electronic Industries Alliance (EIA).

APPENDIX B

PORTIONS OF THE NEC AFFECTING TELECOMMUNICATIONS AND LIFE SAFETY SYSTEMS

NEC Reference	Title	Description
NEC Section 90-2	Scope	The Scope provides information about what is covered in the NEC. This section offers reference to the National Electrical Safety Code for industrial or multi-building complexes.
NEC Section 90-3	Code Arrangement	This section explains how the NEC chapters are positioned. Specifically, NEC Chapter 8, Communications Systems, is an independent chapter except where reference is made to other chapters.
NEC Article 100, Part A	Definitions	Definitions are those not commonly defined in English dictionaries. Some terms of interest include accessible, bonding, explosion-proof apparatus, ground, premises wiring, and signaling circuit.
NEC Section 110-26	Working Space About Electric Equipment (600V, Nominal, or Less)	This section explains the space for working clearances around electrical equipment. This information is useful when placing a terminal in an electrical closet or electronic components on a communications rack.
NEC Article 250	Grounding	This article is referenced from *NEC Article 800*. It contains specific requirements for the communications grounding and bonding network.
NEC Article 500	Hazardous (Classified) Locations	All of *NEC Article 500* is referenced in *NEC Article 800 (NEC Section 800-7)*. This article covers hazardous locations such as gasoline stations and industrial complexes. Additionally, healthcare facilities *(NEC Section 517-80)* are of particular importance. Theaters and marinas are also included in this article.
NEC Article 725	Class 1, Class 2, and Class 3 Remote-Control, Signaling, and Power-Limited Systems	*NEC Article 725* specifies circuits other than those used specifically for electrical light and power.
NEC Article 760	Fire Alarm Systems	*NEC Article 760* contains requirements for the wiring and equipment used in fire alarm systems.

NEC Reference	Title	Description
NEC Article 770	Optical Fiber Cables and Raceways	*NEC Article 770* pertains to optical fiber cables and raceways. Within this section are the requirements for listing of cable, marking, and installation.
NEC Article 800	Communications Systems	*NEC Article 800* contains the requirements for communications systems.
NEC Article 810	Radio and Television Equipment	*NEC Article 810* contains the requirements for radio and television.
NEC Article 820	Community Antenna Television and Radio Distribution Systems	*NEC Article 820* contains requirements for community antenna television and radio distribution systems.

APPENDIX C

PORTIONS OF THE CEC AFFECTING TELECOMMUNICATIONS AND LIFE SAFETY SYSTEMS

CEC Reference	Title	Description
2	General Rules	Provides information on: • Permits • Marking of cables • Flame spread requirements for electrical wiring and cables
10	Grounding and Bonding	Contains detailed grounding and bonding information and requirements for using and identifying grounding and bonding conductors.
12	Wiring Methods	Involves the requirements for installing wiring systems. It outlines: • Raceway systems • Boxes • Other system elements
56	Optical Fiber Cables	Contains the requirements for installing optical fiber cables.
60	Electrical Communication Systems	Contains the requirements for installing communications circuits.

APPENDIX D

ANSI/NFPA STANDARDS APPLICABLE TO LOW-VOLTAGE SYSTEMS

Code Number	Title
ANSI/NFPA-13, 13D, 13R	Installation of Sprinkler Systems
ANSI/NFPA-70	National Electrical Code
ANSI/NFPA-72	National Fire Alarm Code
ANSI/NFPA-75	Protection of Electronic Computer Data Processing Equipment
ANSI/NFPA-101	Life Safety Code
ANSI/NFPA-780	Standard for Installation of Lightning Protection Systems

APPENDIX E

STANDARDS GOVERNING TELECOMMUNICATIONS CABLING

ANSI/TIA/EIA-568-A, *Commercial Building Telecommunications Cabling Standard:*

- ANSI/TIA/EIA-568-A-1, *Propagation Delay and Delay Skew Specifications for 100Ω Four-Pair Cable*
- ANSI/TIA/EIA Telecommunications Systems Bulletin (TSB-67), *Transmission Performance Specifications for Field Testing of Unshielded Twisted-Pair Cabling Systems*
- ANSI/TIA/EIA Telecommunications Systems Bulletin (TSB-72), *Centralized Optical Fiber Cabling Guidelines*
- ANSI/TIA/EIA Telecommunications Systems Bulletin (TSB-75), *Additional Horizontal Cabling Practices for Open Offices*

ANSI/TIA/EIA-569-A, *Commercial Building Standard for Telecommunications Pathways and Spaces*

ANSI/TIA/EIA-570-A, *Residential and Light Commercial Telecommunications Wiring Standard*

ANSI/TIA/EIA-606, *Administration Standard for the Telecommunications Infrastructure of Commercial Buildings*

ANSI/TIA/EIA-607, *Commercial Building Grounding and Bonding Requirements for Telecommunications*

Note: The Institute of Electrical and Electronic Engineers also publishes a variety of standards that apply to low-voltage systems. One such standard is IEEE Standard 1394, *High Performance Serial Bus.*

NCCER CRAFT TRAINING USER UPDATES

The NCCER makes every effort to keep these manuals up-to-date and free of technical errors. We appreciate your help in this process. If you have an idea for improving this manual, or if you find an error, a typographical mistake, or an inaccuracy in the NCCER's Craft Training Manuals, please write us, using this form or a photocopy. Be sure to include the exact module number, page number, a description of the problem, and the correction, if possible. Your input will be brought to the attention of the Technical Review Committee. Thank you for your assistance.

Instructors – If you found that additional materials were necessary in order to teach this module effectively, please let us know so that we may include them in the Equipment/Materials list in the Instructor's Guide.

Write: Curriculum Revision and Development Department
National Center for Construction Education and Research
P.O. Box 141104
Gainesville, FL 32614-1104
Fax: 352-334-0932

Craft _____ Module Name _____

Copyright Date _____ Module Number _____ Page Number(s) _____

Description of Problem
_____
_____
_____
_____
_____

(Optional) Correction of Problem
_____
_____
_____

(Optional) Your Name and Address
_____
_____
_____

notes

Construction Materials and Methods

Module 33102

Electronic Systems Technician Trainee Task Module 33102

CONSTRUCTION MATERIALS AND METHODS

NATIONAL
CENTER FOR
CONSTRUCTION
EDUCATION AND
RESEARCH

OBJECTIVES

Upon completion of this module, the trainee will be able to:

1. Describe the composition and uses of the common types of residential building materials.
2. Identify the major structural components of a residential building.
3. Describe the composition and uses of the common types of commercial building materials.
4. Describe common methods of residential and commercial construction.
5. State the major steps in the construction of a frame residence.
6. Explain common terms used in construction.
7. Identify various types of suspended ceilings.
8. Describe how cable is run from building to building.
9. Select appropriate drills, bits, and cutting tools and make openings in various types of construction materials, including:
    - Lumber
    - Concrete
    - Concrete block
    - Steel
10. Install plywood on a gypsum board wall.

Prerequisites

Successful completion of the following Task Modules is recommended before beginning study of this Task Module: Core Curricula; Electronic Systems Technician Level One, Module 33101.

Required Trainee Materials

1. Trainee Task Module
2. Appropriate Personal Protective Equipment

Copyright © 1999 National Center for Construction Education and Research, Gainesville, FL 32614-1104. All rights reserved. No part of this work may be reproduced in any form or by any means, including photocopying, without written permission of the publisher.

COURSE MAP

This course map shows all of the modules in the first level of the Electronic Systems Technician curricula. The suggested training order begins at the bottom and proceeds up. Skill levels increase as a trainee advances on the course map. The training order may be adjusted by the local Training Program Sponsor.

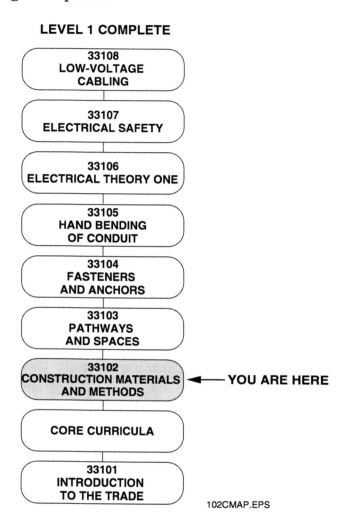

TABLE OF CONTENTS

Section	Topic	Page
	Course Map	2
1.0.0	Introduction	7
2.0.0	Building Materials	8
2.1.0	Lumber	8
2.1.1	Pressure-Treated Lumber	9
2.2.0	Plywood	9
2.3.0	Building Boards	10
2.3.1	Softboard	11
2.3.2	Hardboard	11
2.3.3	Particleboard	11
2.3.4	Oriented Strand Board (OSB)	12
2.3.5	Mineral Fiberboards	12
2.3.6	High-Density Overlay (HDO) And Medium-Density Overlay (MDO) Plywood	13
2.4.0	Engineered Lumber	13
2.5.0	Gypsum Board	16
2.5.1	Types Of Gypsum Products	16
2.6.0	Masonry Materials	19
2.6.1	Concrete	19
2.6.2	Concrete Masonry Units (CMU)	21
2.6.3	Brick	22
2.6.4	Stone	23
2.6.5	Metal	24
3.0.0	Residential Frame Construction	24
3.1.0	Floor Construction	28
3.1.1	Girders	28
3.1.2	Floor Joists	29
3.1.3	Wood I-Beams	31
3.1.4	Trusses	31
3.1.5	Notching And Drilling Of Wooden Joists	32
3.1.6	Bridging	33
3.1.7	Subflooring	34
3.2.0	Wall Construction	34
3.2.1	Corners	36
3.2.2	Partition Intersections	37
3.2.3	Window And Door Openings	37
3.2.4	Firestops	39
3.2.5	Bracing	40
3.2.6	Sheathing	41
3.3.0	Ceiling Construction	41
3.4.0	Roof Construction	43
3.4.1	Roof Components	44
3.4.2	Roof Sheathing	46
3.4.3	Truss Construction	47
3.4.4	Dormers	48
3.5.0	Plank-And-Beam Framing	50
3.6.0	Wall Framing In Masonry	51
3.7.0	Walls Separating Occupancies	53

TABLE OF CONTENTS (Continued)

Section	Topic	Page
4.0.0	Commercial Construction Methods	53
4.1.0	Floors	55
4.2.0	Exterior Walls	56
4.3.0	Interior Walls And Partitions	58
4.3.1	Metal Framing Materials	59
4.3.2	Bracing Walls	64
4.3.3	Metal Joists And Roof Trusses	65
4.4.0	Ceilings	68
4.4.1	Exposed Grid Systems	69
4.4.2	Metal Pan Systems	71
4.4.3	Direct-Hung Concealed Grid Systems	72
4.4.4	Integrated Ceiling Systems	75
4.4.5	Luminous Ceiling Systems	76
4.4.6	Suspended Drywall Furring Ceiling Systems	77
4.4.7	Special Ceiling Systems	79
4.4.8	General Guidelines For Accessing Suspended Ceilings	82
4.5.0	Firestopping	82
4.5.1	Non-Mechanical Firestopping Materials	83
5.0.0	Tools Used For Running Cable	85
5.1.0	Guidelines For Using All Power Tools	85
5.1.1	Safety Rules Pertaining To All Power Tools	85
5.1.2	Guidelines Pertaining To The Care Of All Power Tools	86
5.2.0	Drilling Tools	87
5.2.1	Portable Drills And Screwguns	87
5.2.2	Hammer Drills And Rotary Hammers	89
5.2.3	Special Drilling Equipment	91
5.3.0	Cutting Tools	95
5.3.1	Reciprocating Saws	95
5.3.2	Jig Saws	96
5.3.3	Power Cutout Tool	97
5.3.4	Light Box Cutter	97
5.3.5	Metal Stud Punches	97
5.4.0	Powder-Actuated Fastening Tools	100
5.5.0	Stud Finders	101
5.6.0	Fish Tapes	102
6.0.0	Project Schedules	102
	Summary	104
	Review/Practice Questions	105
	Answers To Review/Practice Questions	108

Trade Terms Introduced In This Module

APA-rated: Building material that has been rated by the American Plywood Association for a specific use.

Blocking: A wood block used as a filler piece and support member between framing members.

Bridging: Wood or metal pieces placed diagonally between joists to provide added support.

Cantilever: A beam, truss, or floor that extends beyond the last point of support.

Corrugated: Material formed with parallel ridges or grooves.

Cripple stud: In wall framing, a short framing stud that fills the space between the header and the top plate or between the sill and the soleplate.

Crosscutting: Cutting across the grain of lumber.

Dimension lumber: Any lumber within a range of 2" to 5" thick and up to 12" wide.

Dormer: A framed structure that projects out from a sloped roof.

Double top plate: A length of lumber laid horizontally over the top plate of a wall to add strength to the wall.

Fire rating: A classification indicating in time (hours) the ability of a structure or component to withstand fire conditions.

Firestop: A piece of lumber or fire-resistant material installed in an opening to prevent the passage of fire.

Firestopping: A material or mechanical device used to block openings in walls, ceilings, floors, etc., to prevent the passage of fire and smoke.

Footing: The foundation for a column or the enlargement placed at the bottom of a foundation wall to distribute the weight of the structure.

Furring strips: Strips of wood or metal applied to a wall or other surface to make it level, form an air space, and/or provide a fastening surface for finish covering.

Gable: The triangular wall enclosed by the sloping ends of a ridged roof.

Girder: The main steel or wood supporting beam for a structure.

Green concrete: Concrete that has hardened but has not yet gained its full structural strength.

Gypsum: A chalky type of rock that serves as the basic ingredient of plaster and gypsum wallboard.

Gypsum wallboard: A generic term for gypsum core panels covered with paper on both sides. It is commonly used to finish walls.

Header: A thick horizontal member that supports the load over an opening such as a door or window. Also known as a *lintel*.

Joists: Equally spaced framing members that support floors and ceilings.

Kerf: A groove or notch made by a saw.

Millwork: Various types of manufactured wood products such as doors, windows, and moldings.

Oriented strand board (OSB): Panels made from layers of wood strands bonded together.

Plastic concrete: Concrete in a liquid or semiliquid workable state.

Plenum: A sealed chamber for moving air under slight pressure at the inlet or outlet of an air conditioning system. In some commercial buildings, the space above a suspended ceiling often acts as a return air plenum.

Post-stressed concrete: Concrete placed around steel reinforcement such as rods or cables that are isolated from the concrete. After the concrete has cured, tension is applied to the rods or cables to provide greater structural strength.

Pre-stressed concrete: Concrete that is placed around pre-stressed reinforcing steel in a casting bed. This type of concrete cannot be cut without first consulting a structural engineer.

Rabbeted: A board or panel with a groove cut into one or more of its edges.

Rafter: A sloping structural member of a roof frame to which sheathing is attached.

Reinforced concrete: Concrete that has been placed around some type of reinforcing material, usually steel.

Ribband: A 1 × 4 nailed to ceiling joists to prevent twisting and bowing of the joists.

Ripping: Cutting with the grain of the lumber.

Shakes: A handsplit wood shingle.

Sheathing: The sheet material or boards used to close in walls and roofs.

Shiplap: Lumber with edges that are shaped to overlap adjoining pieces.

Sill plate: A horizontal timber that supports the framework of a building. It forms the transition between the foundation and the frame.

Soleplate: The lowest horizontal member of a wall or partition. It rests directly on the rough floor.

Striated: A surface design that has the appearance of fine parallel grooves.

Stringer: The support member at the sides of a staircase; also, a timber used to support formwork for a concrete floor.

Strongback: An L-shaped arrangement of lumber used to support ceiling joists and keep them in alignment. In concrete work, it represents the upright support for a form.

Stucco: A type of plaster used to coat exterior walls.

Subfloor: Panels or boards fastened to the tops of floor joists.

Substrate: The underlying material to which a finish is applied.

Top plate: The upper horizontal member of a wall or partition frame.

Trimmer joist: A full-length horizontal member that forms the sides of a rough opening in a floor. It provides stiffening for the frame.

Trimmer stud: The vertical framing member that forms the sides of a rough opening for a door or window. It provides stiffening for the frame and supports the weight of the header.

Truss: An engineered assembly made of wood or metal that is used in place of individual structural members such as the joists and rafters used to support floors and roofs.

Underlayment: A material such as plywood or particleboard that is installed on top of a subfloor to provide a smooth surface for the finish flooring.

Vaulted ceiling: A high, open ceiling that generally follows the roof pitch.

Veneer: The covering layer of material for a wall or the facing materials applied to a substrate.

1.0.0 INTRODUCTION

Why would someone who installs and services electronic systems need to know how buildings are constructed? The simple answer is that you have to know how the building is put together in order to run cable from place to place within the building, and from one building to another, without doing any damage. Part of that is understanding building materials and how to drill through them.

This module provides an overview of the various types of building materials used in residential and commercial construction and the tools used to penetrate them. We will also cover the construction sequence, so you will know how your work in running and terminating cables, installing distribution boxes, etc., fits into the larger scheme of the construction process.

Residential and commercial (including multifamily) construction methods are very different, so they will be covered separately. When we speak of residential, we are referring to single-family and two-family dwellings. Keep in mind, however, that some small commercial buildings (e.g., apartment buildings and townhouses) may use the same construction techniques and materials as those used in residential buildings.

2.0.0 BUILDING MATERIALS

Many different materials are used in the construction of a building. Wood frame construction is most common in residential work. Concrete block and brick are also used in residential and light commercial construction.

The construction of large commercial buildings such as office buildings, warehouses, apartment buildings, parking garages, etc., generally involves the use of a steel or concrete support structure and walls made of concrete or steel and glass.

2.1.0 LUMBER

The framework of a single-family or two-family dwelling is usually built from lumber, which is divided into five categories:

- *Boards* – Members up to 1½" thick and 2" wide or wider.
- *Light framing (L.F.)* – Members 2" to 4" thick and 2" to 4" wide.
- *Joists and planks (J&P)* – Members 2" to 4" thick and 6" wide or wider.
- *Beams and **stringers** (B&S)* – Members 5" and thicker by 8" and wider.
- *Posts and timbers (P&T)* – Members 5" × 5" and greater, approximately square.

The vast majority of lumber used in framing a house is softwood such as pine or fir. Hardwoods such as oak and maple are used primarily in furniture and decorative pieces.

Light framing lumber, studs, joists, and planks are all classified as **dimension lumber**.

You are probably familiar with the terms *2 by 4, 1 by 6*, etc. What you may not know is that these numbers represent the nominal (rough) size of the lumber in inches. Once the lumber is dressed (finished) at the lumber mill, it is somewhat smaller, typically ½" to ¾" less than the nominal size in each dimension. *Table 1* shows the final dimensions for some standard sizes of softwood dimension lumber. Note that these dressed dimensions apply only to softwoods; hardwoods have different conversion tables.

Nominal	Dressed
2 × 2	1½ × 1½
2 × 4	1½ × 3½
2 × 6	1½ × 5½
2 × 8	1½ × 7¼
2 × 10	1½ × 9¼
2 × 12	1½ × 11¼

Table 1. Nominal And Dressed Sizes Of Dimension Lumber (In Inches)

2.1.1 Pressure-Treated Lumber

Pressure-treated lumber is softwood lumber protected by chemical preservatives forced deep into the wood through a vacuum-pressure process. Pressure-treated lumber has been used for many years in on-ground and below-ground applications such as landscape timbers, **sill plates**, and foundations. In some parts of the country, it is also used extensively in the building of decks, porches, docks, and other outdoor structures. It is popular for these uses in areas where structures are exposed to snow for several months of the year. A major advantage of pressure-treated lumber is its relatively low price in comparison with redwood and cedar. When natural woods such as these are used, only the more expensive heartwood will resist decay and insects.

Because the chemicals used in pressure-treated lumber present some hazards to people and the environment, special precautions apply to its use:

- When cutting pressure-treated lumber, always wear eye protection and a dust mask.
- Wash any skin that is exposed while cutting or handling the lumber.
- Wash clothing that is exposed to sawdust separately from other clothing.
- Do not burn pressure-treated lumber, as the ash poses a health hazard. Bury it or put it in with the trash.
- Be sure to read and follow the manufacturer's safety instructions as defined in the Material Safety Data Sheet (MSDS).

One place to look for pressure-treated lumber is any location where wood comes into contact with the ground, or outdoors where the wood is exposed to moisture.

2.2.0 PLYWOOD

Plywood is made by gluing together thin layers of wood known as **veneers**. Plywood can have three or more plies (layers). These are bonded together at right angles with glue and heat under tremendous pressure. Putting the plies together at right angles increases the strength; also, the more plies there are, the greater the strength. The ply that is in the center is called the *core* and each of the exposed plies is called a *veneer* or *face* (*Figure 1*). All other plies between the core and veneer are called the *crossbands*. Constructing the plywood with the grain of adjacent plies running at right angles reduces the possibility of warping.

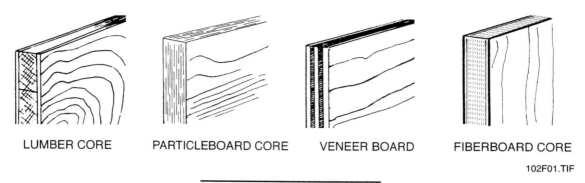

Figure 1. Types Of Plywood

The average or standard size of plywood is 4'-0" × 8'-0". A few companies produce plywood from 6' to 8' widths and up to 16' in length. **Sheathing**-grade plywood is nominally sized by the manufacturer to allow for expansion (e.g., 4'-0" × 8'-0" is really 47¾" × 95¾").

The thickness of plywood will vary from ³⁄₁₆" to 1¼". The common sizes are ¼", ½", and ⅝" for finish paneling and ⅜", ½", ⅝", and ¾" for structural purposes.

Plywood is rated by the American Plywood Association (APA) for interior or exterior use. Exterior-rated plywood is used for sheathing, siding, and other applications where there may be exposure to moisture or wet weather conditions. Exterior plywood panels are made of high-grade veneers bonded together with a waterproof glue that is as strong as the wood itself.

Interior plywood uses lower grades of veneer for the back and inner plies. Although the plies may be bonded with a water-resistant glue, waterproof glue is normally used. The lower-grade veneers reduce the bonding strength, however, which means that interior-rated panels are not suitable for exterior use.

2.3.0 BUILDING BOARDS

The ingenuity and technology that helped develop the plywood industry also assisted in the development of other materials in sheet form. The main ingredients for these products, known as *building boards*, are vegetable or mineral fibers. After mixing these ingredients with binder, the mixture becomes very soft.

At this point, the mixture passes through a press, which uses heat and pressure to produce the required thickness and density of the finished board.

Sawdust, wood chips, and wood scraps are the major waste materials at sawmills. These scrap materials are softened with heat and moisture, mixed with a binder and other ingredients, and then run through presses that produce the desired density and thickness.

The finished wood products that come off the presses are classified as *softboard*, *hardboard*, *particleboard*, or **oriented strand board (OSB)**.

2.3.1 Softboard

Softboard, also known as *fiberboard*, comes in various thicknesses; the most common is ½" thick. It is also manufactured in thicknesses of ⅝", ¾", and 1". There are some fiberboards made to a thickness of 3" for special purposes. The standard sheet sizes for fiberboard are 2'-0" × 8'-0" and 4'-0" × 8'-0". The major uses for fiberboard are insulating, sheathing, and sound control. Because air cells are trapped within the fibers, no structural value can be obtained from fiberboard. Some boards are coated with asphalt or impregnated during the manufacturing process in order to shed water. Other boards are finished on one side for use as decorative paneling on ceilings or walls.

2.3.2 Hardboard

Hardboard is a manufactured building material, sometimes called *tempered board* or *pegboard*. Hardboards are water-resistant and extremely dense. The common thicknesses for hardboards are ³⁄₁₆", ¼", and ⁵⁄₁₆". The standard sheet size for hardboards is 4'-0" × 8'-0". However, they can be made in widths up to 6' and lengths up to 16' for specialized uses.

These boards are susceptible to breaking at the edges if they are not properly supported. Holes must be predrilled for nailing; direct nailing into the material will cause it to fracture.

Three grades of hardboard are manufactured:

- *Standard* – Standard grade hardboard is suitable only for interior use such as cabinets.
- *Tempered* – Tempered grade hardboard (Masonite) is the same as standard grade except that it is denser, stronger, and more brittle. Tempered hardboard is suitable for either interior or exterior uses such as siding, wall paneling, and other decorative purposes.
- *Service* – Service grade hardboard is not as dense, strong, or heavy as standard grade. It can be used for basically everything for which standard or tempered hardboard is used. Service grade hardboard is manufactured for items such as cabinets, parts of furniture, and perforated hardboard.

2.3.3 Particleboard

The main composition of this type of material is small particles or flakes of wood.

Particleboard is pressed under heat into panels. The sheets range in size from ¼" to 1½" in thickness and from 3' to 8' in width. There are also thicknesses of 3" and lengths ranging up to 24' for special purposes. Particleboard has no grain, is smoother than plywood, is more resilient, and is less likely to warp.

Some types of particleboard can be used for **underlayment** if permitted by the local building codes. If particleboard is used as underlayment, it is laid with the long dimension across the joists and the edges staggered. Particleboard can be nailed, although some types will crumble or crack when nailed close to the edges.

2.3.4 Oriented Strand Board (OSB)

Oriented strand board (OSB) is a manufactured structural panel used for wall and roof sheathing and single-layer floor construction. *Figure 2* shows two kinds of OSB panels. OSB consists of compressed wood strands arranged in three perpendicular layers and bonded with phenolic resin. Some of the qualities of OSB are dimensional stability, stiffness, fastener holding capacity, and no core voids. Before cutting into OSB, be sure to check the applicable MSDS for safety hazards. The MSDS is the most reliable source of safety information.

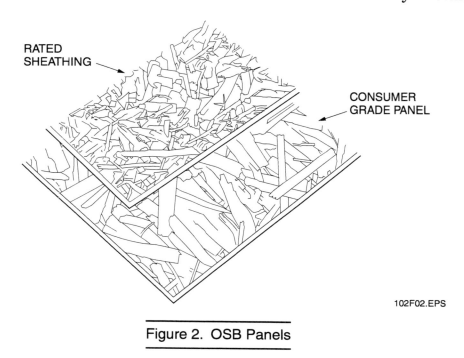

Figure 2. OSB Panels

2.3.5 Mineral Fiberboards

The building boards just covered are classified as *vegetable fiberboards*. Mineral fiberboards fall into the same category as vegetable fiberboards. The main difference is that they will not support combustion. Glass and **gypsum** rock are the most common minerals used in the manufacture of these fiberboards. Fibers of glass or gypsum powder are mixed with a binder and pressed or sandwiched between two layers of asphalt-impregnated paper, producing a rigid insulation board.

There are some types of chemical foam mixed with glass fibers that will also make a good, rigid insulation. However, this mineral insulation will crush and should not be used when it must support a heavy load.

CAUTION: Whenever working with older materials that may be made with asbestos, contact your supervisor for the company's policies on safe handling of the material. State and federal regulations require specific procedures to follow prior to removing, cutting, or disturbing any suspect materials. Also, some materials emit a harmful dust when cut. Check the MSDS before cutting.

2.3.6 High-Density Overlay (HDO) And Medium-Density Overlay (MDO) Plywood

High-density overlay (HDO) plywood panels have a hard, resin-impregnated fiber overlay heat-bonded to both surfaces. HDO panels are resistant to both abrasion and moisture and can be used for concrete forms, cabinets, countertops, and similar high-wear applications. HDO also resists damage from chemicals and solvents. HDO is available in four common thicknesses: $\frac{3}{8}$", $\frac{1}{2}$", $\frac{5}{8}$", and $\frac{3}{4}$".

Medium-density overlay (MDO) panels are coated on one or both surfaces with a smooth, opaque overlay. MDO accepts paint well and is suitable for use as structural siding, exterior decorative panels, and soffits. MDO panels are available in eight common thicknesses ranging from $\frac{11}{32}$" to $\frac{23}{32}$".

Both HDO and MDO panels are manufactured with waterproof adhesive and are suitable for exterior use. If MDO panels are to be used outdoors, however, the panels should be edge-sealed with one or two coats of a good-quality exterior housepaint primer. This is easier to do when the panels are stacked.

2.4.0 ENGINEERED LUMBER

In the past, the primary source of structural beams, timbers, joists, and other weight-bearing lumber was old-growth trees. These trees, which need more than 200 years to mature, are tall and thick and can produce a large amount of high-quality, tight-grained lumber. Extensive logging of these trees to meet demand resulted in higher prices and conflict with forest conservation interests.

The development of wood laminating techniques by lumber producers has permitted the use of younger-growth trees in the production of structural building materials. These materials are given the general classification of *engineered lumber products*.

Engineered lumber products fall into five categories: laminated veneer lumber (LVL), parallel strand lumber (PSL), laminated strand lumber (LSL), wood I-beams, and glue-laminated lumber or glulam (*Figure 3*).

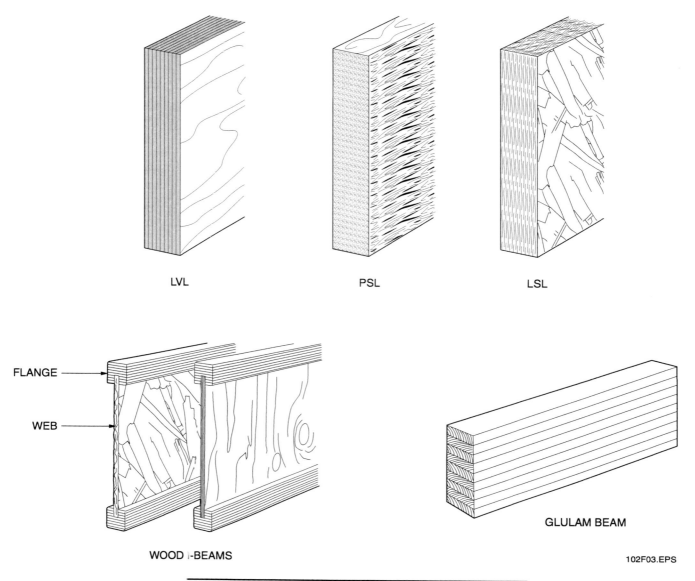

Figure 3. Examples Of Engineered Lumber Products

Engineered lumber products provide several benefits:

- They can be made from younger, more abundant trees.
- They can increase the yield of a tree by 30% to 50%.
- They are stronger than the same size of structural lumber. Therefore, the same size piece of engineered lumber can bear more weight than that of solid lumber. Or, looked at another way, a smaller piece of engineered lumber can bear equal weight.
- Greater strength allows the engineered lumber to span a greater distance.
- A length of engineered wood is lighter than the same length of solid lumber. It is therefore easier to handle.

LVL is used for floor and roof beams and for **headers** over windows and doors. It is also used in scaffolding and concrete forms. No special cutting tools or fasteners are required.

PSL is used for beams, posts, and columns. It is manufactured in thicknesses up to 7". Columns can be up to 7" wide, while beams range up to 18" in width.

LSL (not shown) is used for **millwork** such as doors and windows and any other product that requires high-grade lumber. However, LSL will not support as much of a load as a comparable size of PSL because PSL is made from stronger wood.

Wood I-beams consist of a web with flanges bonded to the top and bottom. This arrangement, which mimics the steel I-beam, provides exceptional strength. The web can be made of OSB or plywood. The flanges are grooved to fit over the web.

Wood I-beams are used as floor **joists**, **rafters**, and headers. Because of their strength, wood I-beams can be used in greater spans than a comparable length of dimension lumber. Lengths of up to 80' are available.

Glulam is made from several lengths of solid lumber that have been glued together. It is popular in architectural applications where exposed beams are used (*Figure 4*). Because of its exceptional strength and flexibility, glulam can be used in areas subject to high winds or earthquakes.

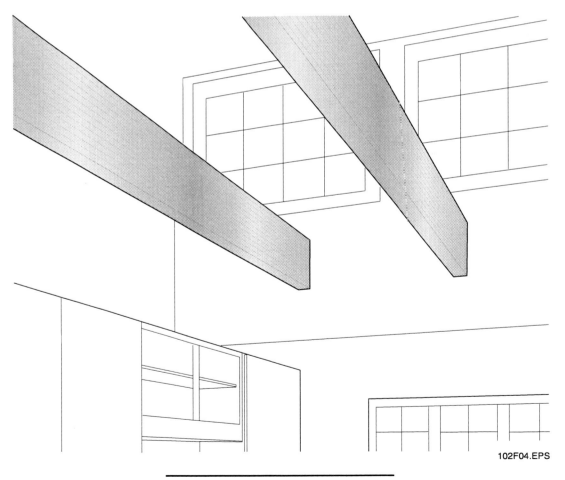

Figure 4. Glulam Beam Application

Glulam beams are available from 2½" to 8¾". Depths range from 5¹⁄₂₀" to 28½". They are available in lengths up to 40'. They are used for many purposes, including: ridge beams; basement beams; headers of all types; stair treads, supports, and stringers; and **cantilever** and **vaulted ceiling** applications.

2.5.0 GYPSUM BOARD

Gypsum wallboard, also known as *gypsum drywall*, is one of the most popular and economical methods of finishing the interior walls and ceilings of wood-framed and metal-framed buildings. Properly installed and finished, drywall can give a wall or ceiling made from many panels the appearance of being made from one continuous sheet.

Gypsum board is a generic name for products consisting of a noncombustible core. This product is made primarily of gypsum with a paper surfacing covering the face, back, and long edges. A typical board application is shown in *Figure 5*.

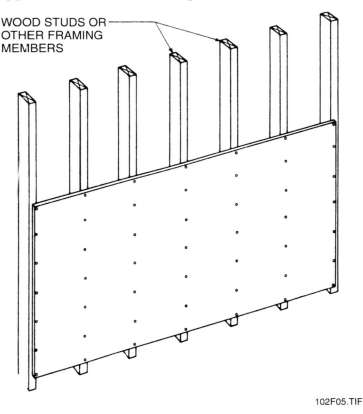

Figure 5. Typical Gypsum Wallboard Application

2.5.1 Types Of Gypsum Products

Many types of gypsum board are available for a variety of building needs (see *Table 2*). Gypsum board panels are mainly used as the surface layer for interior walls and ceilings; as a base for ceramic, plastic, and metal tile; for exterior soffits; for elevator and other shaft enclosures; and to provide fire protection for architectural elements.

Type	Thickness	Sizes	Use
Regular, paper faced	¼"	4' × 8' to 10'	Recovering old gypsum walls
	⅜"	4' × 8' to 14'	For double-layer installation
	½", ⅝"	4' × 8' to 16'	For standard single-ply installation
Regular with foil back	½", ⅝"	4' × 8' to 14'	Use as a vapor barrier or radiant heat retarder
Type X, fire-retardant	⅜", ½", ⅝"	4' × 8' to 16'	Use in garages, workshops, and kitchens, as well as around furnaces, fireplaces, and chimney walls; ⅝" is 3- or 4-hour fire rated
Moisture-resistant	½", ⅝"	4' × 6' to 16'	For tile backing and around kitchens, spas, baths, showers, and laundry rooms
Decorator panels	5/16"	4' × 8'	Any room in the house
Gypsum lath	⅜", ½", ⅝"	16" × 4' 2' × 8' to 12'	Use as a base for plaster Use ⅜" for 16" on center (OC) stud spacing; ½" or ⅝" for 24" OC stud spacing

Table 2. Types And Uses Of Gypsum Wallboard

Gypsum board products are available with reflective aluminum foil backing, which provides an effective vapor barrier for exterior walls. When applied with the foil surface against the framing, with a minimum of ¾" enclosed air space adjacent to the foil, additional insulating efficiency is achieved.

This combination effectively reduces radiant heat loss in the cold season and radiant heat gain in the warm season. However, foil-backed gypsum board is not used as a backing material for tile, as a second face ply on a two-ply system, in conjunction with heating cables, or when laminating directly to masonry, ceiling, and roof assemblies.

Various thicknesses of gypsum wallboard are available in regular, Type X, water-resistant, and predecorated boards.

- *¼" gypsum board* – A lightweight, low-cost board used as a base in a multilayer application for improving sound control, to cover existing walls and ceilings in remodeling, for curved walls, and for barrel ceilings.
- *5/16" gypsum board* – A lightweight board developed for use in manufactured construction, primarily mobile homes.
- *⅜" gypsum board* – A lightweight board principally applied in a double-layer system over wood framing and as a face layer in repair and remodeling.
- *½" gypsum board* – Generally used for single-layer wall and ceiling construction in residential work and in double-layer systems for greater sound and **fire ratings**.

- *5/8" gypsum board* – Used in quality single-layer and double-layer wall systems. The greater thickness provides additional fire resistance, higher rigidity, and better impact resistance. It is also used to separate occupied and unoccupied areas (e.g., a house from a garage or an office from a warehouse).
- *1" gypsum board (either a single 1" board or two 1/2" factory-laminated boards)* – Used as a liner or as a core board in shaft walls and in semi-solid or solid gypsum board partitions. It is also known as *coreboard*.

Standard gypsum boards are 4' wide and 8', 10', or 14' long. The width is compatible with the standard framing of studs or joists spaced 16" or 24" on center.

Regular gypsum board is used as a surface layer on walls and ceilings. Type X gypsum board is available in 1/2" and 5/8" thicknesses and has an improved fire resistance made possible by the use of special core additions. It is also available with a predecorated finish. Type X gypsum board is used in most fire-rated assemblies.

Predecorated gypsum board has a decorated surface which does not require further treatment. The surfaces may be coated, printed, or have a vinyl film. Textured patterns are also available. It requires additional trim, dividers, and corners.

Water-resistant gypsum board, also known as *green board*, has a water-resistant gypsum core and water-repellent paper. The facing typically has a light green color. It is available with a regular or Type X core and in 1/2" or 5/8" thicknesses. Water-resistant gypsum board is not recommended for use in tubs and shower enclosures and other areas exposed to water. The tile backer is now preferred for high-moisture applications.

A special type of wallboard is replacing water-resistant gypsum wallboard as a backing for tile in damp areas such as baths and shower stalls. One type is known as *cement board*. It is made from a slurry of portland cement mixed with glass fibers. It is colored light blue for easy recognition. These backer boards, in addition to their use as a tile backer, can be used as a floor underlayment, countertop base, heat shield for stoves, and as a base for exterior finishes such as **stucco** and brick veneer. They are available in 4' × 8' and 3' × 5' panels. Common thicknesses are 1/4", 7/16", and 1/2".

Gypsum backing board is designed to be used as a base layer or backing material in multilayer systems. It is available with aluminum foil backing and with regular or Type X cores.

Gypsum core board is available as a 1" thick solid core board or as a factory-laminated board composed of two 1/2" boards. It is used in shaft walls and laminated gypsum partitions with additional layers of gypsum board applied to the core board to complete the wall assembly. It is available in a width of 24" and with a variety of edges (square and tongue-and-groove are the most common).

Gypsum sheathing is used as a protective, fire-resistive membrane under exterior wall-surfacing materials such as wood siding, masonry veneer, stucco, and shingles. It also provides protection against the passage of water and wind and adds structural rigidity to the framing system. The noncombustible core is surfaced with firmly bonded, water-repellent paper. In addition, a water-repellent material may be incorporated in the core. It is available in 2' and 4' widths, and ½" to ⅝" thicknesses. The latter is also available with Type X core.

Gypsum board **substrate** for floor or roof assemblies has a ½" thick Type X core and is available in 24" or 48" widths. It is used under combustible roof coverings to protect the structure from fires originating on the roof. It can also serve as an underlayment when applied to the top surfaces of floor joists and under **subflooring**. It may also be used as a base for built-up roofing applied over steel decks.

Gypsum form board has a fungus-resistant paper and is used as a support and permanent form for poured-in-place reinforced gypsum concrete roof decks.

Gypsum base for veneer plaster is used as a base for thin coats of hard, high-strength gypsum veneer plaster.

Gypsum lath is a board product used as a base to receive hand- or machine-applied plaster. It is available in ⅜" or ½" thicknesses and in widths of 16" or 24". Gypsum lath is normally available in 48" lengths. Other lengths are available by special order.

2.6.0 MASONRY MATERIALS

For the purpose of this module, the term *masonry* includes construction using stone, brick, concrete block, and poured concrete.

2.6.1 Concrete

Concrete is a mixture of four basic materials: portland cement, fine aggregates, coarse aggregates, and water. When first mixed, concrete is in a semi-liquid state and is referred to as **plastic concrete**. When the concrete hardens but has not yet gained structural strength, it is called **green concrete**. After the concrete has hardened and gained its structural strength, it is called *cured concrete*. Various types of concrete can be obtained by varying the basic materials and/or by adding other materials to the mix. These added materials are called *admixtures*.

The desirable properties of concrete in the plastic state are:

- *Moldability* – Plastic concrete may be molded by forms into almost any shape. This is often used to obtain a decorative effect.
- *Portability* – Plastic concrete may be moved in mixing trucks, motorized buggies, wheelbarrows, or by belt conveyors or hydraulic pumps.

The desirable properties of cured concrete are:

- *High structural strength* – Unreinforced concrete has great compressive strength. **Reinforced concrete**, **pre-stressed concrete**, or **post-stressed concrete** has high structural strength under compression, tension, and lateral pressure.
- *Watertightness* – Although water is used to prepare concrete and concrete can harden under water, properly proportioned and mixed concrete is virtually watertight in most cases.
- *Durability* – Properly mixed and placed concrete usually continues to gain strength for several years and becomes almost as durable and abrasion-resistant as the hardest natural stone.

Portland cement is a finely ground powder consisting of varying amounts of lime, silica, alumina, iron, and other trace components. While dry, it may be moved in bulk or can be bagged in moisture-resistant sacks and stored for relatively long periods of time. Portland cement is a hydraulic cement because it will set and harden by reacting with water with or without the presence of air. This chemical reaction is called *hydration* and can occur even when the concrete is submerged in water. The reaction creates a calcium silicate hydrate gel and releases heat. This reaction begins the instant water is mixed with the cement and continues as the mixture hardens and cures. The reaction occurs rapidly at first, depending on how finely the cement is ground and what admixtures are present. Then, after its initial cure and strength are achieved, a cement mixture continues to slowly cure over a longer period of time until its ultimate strength is attained.

Because it is in a semi-liquid form when poured, concrete is placed in reinforced forms made of wood, metal, or other materials (*Figure 6*). Concrete floors, walls, and columns can be poured on site. Walls and other structural concrete components are sometimes prefabricated off site and moved to the site on a truck. They are then lifted into place with cranes.

In residential construction, concrete may be used in foundation walls and **footings**, basement floors, or as the foundation slab if the house has no basement.

In commercial construction, the entire structure, including floors, walls, and support columns, may be made of concrete. Walls can be anywhere from a few inches to several feet thick.

The ratio of basic ingredients in concrete is determined by a number of variables such as the application, weather conditions, etc. A common mix for do-it-yourself applications is 3:2:1 (i.e., one part portland cement, two parts sand, three parts aggregate, with enough water to make the mix workable). In the construction trades, the correct ratio for a given situation is determined much more scientifically, and is usually done by an engineer. Admixtures may be added to affect drying time, increase strength, and add color.

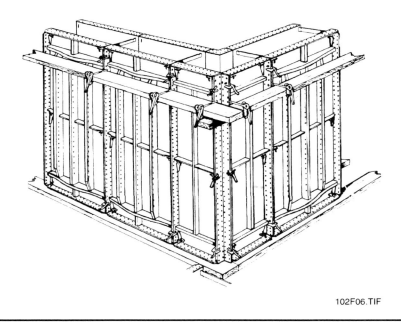

Figure 6. Wall Form Made From EFCO Hand-E-Form® Panel System Components

CAUTION: Those working with cement should be aware that it is harmful. Dry cement dust can enter open wounds and cause blood poisoning. When the cement dust comes in contact with body fluids, it can cause chemical burns to the membranes of the eyes, nose, mouth, throat, or lungs. Wet cement or concrete can also cause chemical burns to the eyes and skin. Make sure that appropriate personal protective equipment is worn when working with dry cement or wet concrete. If wet concrete enters waterproof boots from the top, remove the boots and rinse your legs, feet, boots, and clothing with clear water as soon as possible. Repeated contact with cement or wet concrete can cause an allergic reaction in certain individuals.

2.6.2 Concrete Masonry Units (CMU)

Commonly known as *concrete block*, concrete masonry units (CMU) are one of the most common building materials in both residential and commercial construction. *Figure 7* shows samples of these blocks. They are made from a mixture of portland cement, aggregates such as sand and gravel, and water.

Hollow concrete block is used in all kinds of residential and commercial applications. Residential basement walls are usually made of concrete block and it is often used as a base for finish materials such as brick and stucco.

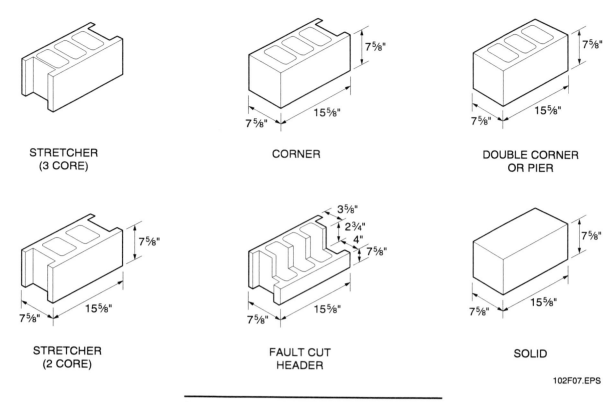

Figure 7. Examples Of Concrete Blocks

The typical size of a concrete block used in loadbearing construction is 7⅝" wide, 7⅝" high, and 15⅝" long. This is known as an 8" × 8" × 16" unit because it is designed for a ⅜" mortar joint. Mortar is a bonding agent made of cement, fine aggregate such as sand, and water. It is used to provide a watertight bond between blocks.

Sometimes, tiny foam beads are poured into concrete block walls to provide insulation. If you drill a hole though the block, the beads will begin to pour out and will not stop until the beads in that vertical channel have escaped. When installing low-voltage conduit, be sure you have the conduit ready to insert into the hole or have plugs available when you drill.

2.6.3 Brick

Brick is commonly used as a veneer for residential and commercial buildings. Brick is made from pulverized clay that is mixed with water and then molded into various shapes, primarily rectangular. Once the brick hardens and dries, it is fired in a furnace to provide the necessary hardness. Although there are many sizes available, a standard brick is 2¼" × 3¾" × 8".

Like cement block, bricks are bonded together with mortar. Brick is typically laid against a supporting structure such as a concrete block wall or a frame wall sheathed with plywood (*Figure 8*). An air space is maintained between the two walls to allow moisture to escape. A weephole is provided to drain condensation that develops in the air space. The main difference between conventional frame construction and brick facing is that the foundation wall is extended to provide support for the brick.

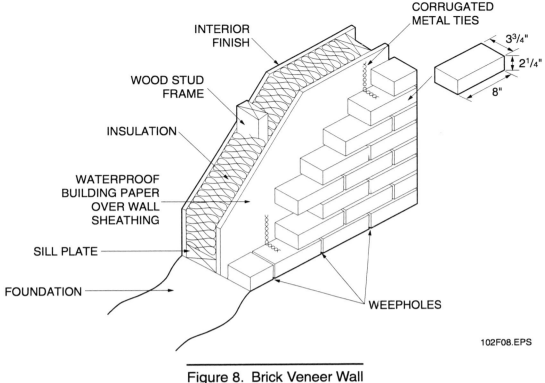

Figure 8. Brick Veneer Wall

Although concrete block and brick are hard, they can be readily penetrated with the correct drill and bit in order to install cable runs. These tools are covered later in this module.

2.6.4 Stone

Like brick, stone is used primarily as a facade over block or frame walls. Stone used for this purpose can be as much as 6" thick. However, in renovating very old homes, you may find stone foundations and walls a foot or more thick, and they are very difficult to drill through. *Figure 9* shows a diagram of a stone veneer wall with the stone laid in a random pattern.

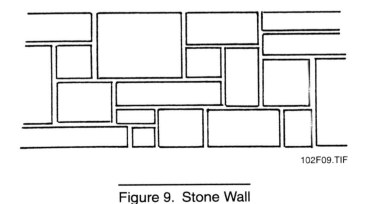

Figure 9. Stone Wall

2.6.5 Metal

Metals have a variety of applications, especially in commercial construction. Lightweight steel and aluminum studs are used in framing walls, floors, and roofs. Metal sheet material is common in walls and roofs of commercial buildings. **Corrugated** steel decking is used as a base for poured concrete floors in multistory commercial buildings.

Heavy-gauge structural steel **girders** and beams are used as the horizontal and vertical support members in many commercial buildings. Steel reinforcing bars and mesh are used to strengthen poured concrete in all applications.

3.0.0 RESIDENTIAL FRAME CONSTRUCTION

Wood frame construction (*Figure 10*) has been in common use since the 1800s. Frame construction begins by building a foundation, which usually consists of a poured concrete footing. In cold climates, the footing must be built below the frost line to prevent it from cracking. If there is no basement, a short foundation wall of poured concrete is set onto poured concrete footings.

Figure 10. Example Of Rough Carpentry (Western Platform Framing)

It is common to use wooden forms to shape the poured concrete footing (*Figure 11*), and then to use edge forms to make the slab floor. In some cases, the two pours are combined to make a monolithic pour (*Figure 12*). Reinforcing bars are embedded in the concrete both to reinforce it and to provide a connection to the adjoining concrete.

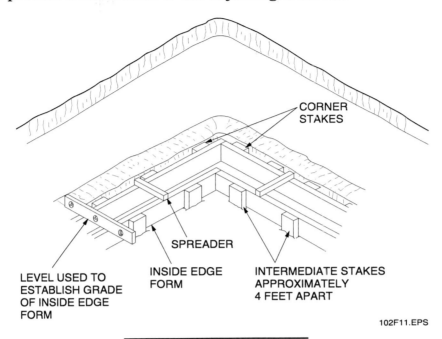

Figure 11. Foundation Form

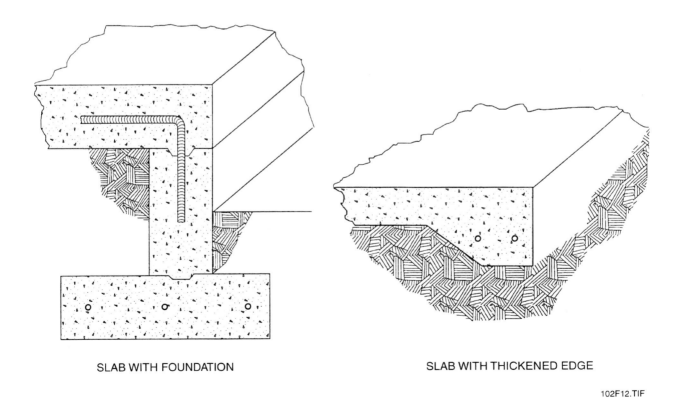

Figure 12. Types Of Slabs

If the house is to have a basement, the basement walls, usually made of concrete block, are set onto the footings (*Figure 13*), which are below ground level.

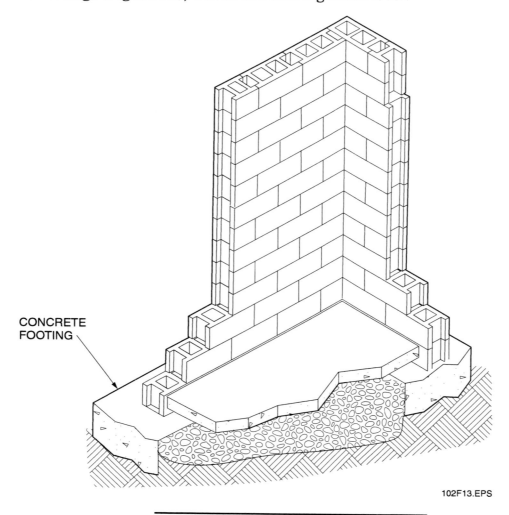

Figure 13. Concrete Block Basement Wall

Reinforced polystyrene foam wall forms (*Figure 14*) have become popular for basement walls because they are easy to build and can be left in place after the wall is poured. In addition, they provide substantial insulation. The concrete walls constructed with these forms can range from 3" or 4" to 10" thick.

A sill plate, which acts as the anchor for the wood framing, is installed onto the foundation wall (*Figure 15*). Anchor bolts or straps embedded in the concrete provide the means of attaching the sill plate to the foundation. The sills are often made from pressure-treated lumber. Otherwise, a vapor barrier must be placed between the sill and the foundation.

The sills provide a means of leveling the top of the foundation wall and also prevent the other wood framing lumber from making contact with the concrete or masonry, which can cause the lumber to rot.

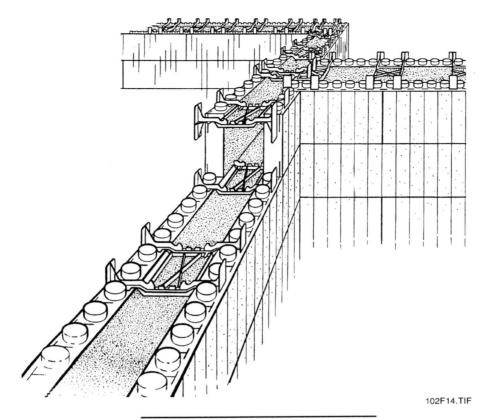

Figure 14. Polystyrene Form System

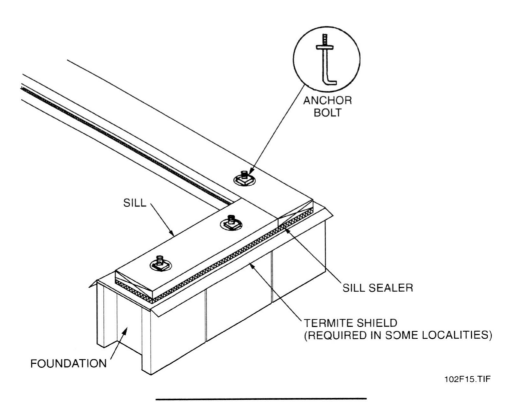

Figure 15. Typical Sill Installation

Today, sills are normally made using a single layer of 2 × 6 lumber. Local codes normally require that pressure-treated lumber and/or foundation-grade redwood lumber be used for the sill whenever it comes into direct contact with any type of concrete. However, where codes allow, untreated softwood can be used.

3.1.0 FLOOR CONSTRUCTION

Floor systems provide a base for the remainder of the structure to rest on. They transfer the weight of people, furniture, materials, etc., from the subfloor, to the floor framing, to the foundation wall, to the footing, then finally to the earth. Floor systems are constructed over basements or crawl spaces. Single-story structures built on slabs do not have floor systems; however, multilevel structures may have both a slab and a floor system. *Figure 16* shows a typical platform floor system and identifies the various parts.

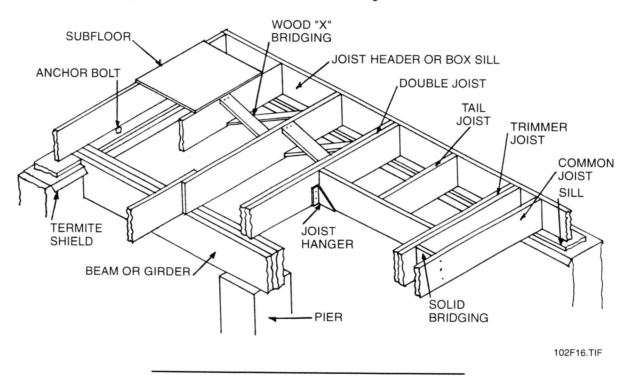

Figure 16. Typical Platform Frame Floor System

3.1.1 Girders

Floor joists rest on the sill and provide the support for the floor, as well as an attaching surface for the ceiling of the floor below, if applicable.

The distance between two outside walls is frequently too great to be spanned by a single joist. When two or more joists are needed to cover the span, support for the inboard joist ends must be provided by one or more beams, commonly called *girders*. Girders carry a very large portion of the weight of the building. They must be well-designed, rigid, and properly supported at the foundation walls and on the supporting posts or columns. They must also be installed so that they will properly support the floor joists. Girders may be made of solid timbers, built-up lumber, engineered lumber, or steel beams. In some instances, precast reinforced concrete girders may be used.

Girders and beams must be properly supported at the foundation walls, and at the proper intervals in between, either by supporting posts, columns, or piers (*Figure 17*). Solid or built-up wooden posts installed on pier blocks are commonly used to support floor girders, especially for floors built over a crawl space.

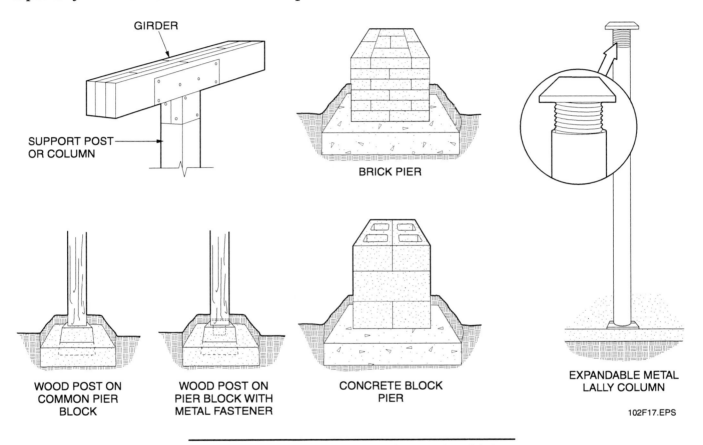

Figure 17. Typical Methods Of Supporting Girders

Four-inch round steel columns filled and reinforced with concrete, called *lally columns*, are commonly used as support columns in floors built over basements. Some types of lally columns must be cut to the required height, while others have a built-in jack screw that allows the column to be adjusted to the proper height. Metal plates are installed at the top and bottom of the column to distribute the load over a wider area. The plates normally have predrilled holes so that they may be fastened to the girder.

3.1.2 Floor Joists

Floor joists are a series of parallel, horizontal framing members that make up the body of the floor frame (*Figure 16*). They rest on and transfer the building load to the sills and girders.

Joists are normally placed 16" on center (OC). However, there are applications when joists can be set as close as 12" OC or as far apart as 24" OC. These distances are used because they accommodate 4' × 8' subfloor panels and provide a nailing surface where two panels meet. Joists can be supported by the top of the girder or may be framed to the side. *Figure 18* shows several methods for joist framing at the girder. If joists are lapped over the girder, the minimum amount of lap is 4" and the maximum amount of lap is 12".

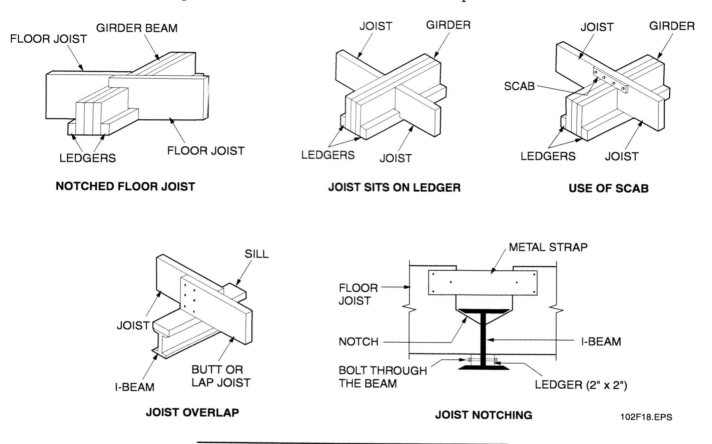

Figure 18. Methods Of Joist Framing At A Girder

There are many different types of joist hangers that can be used to fasten joists to girders and other support framing members. Joist hangers are used where the bottom of the girder must be flush with the bottoms of the joists. At the sill end of the joist, the joist should rest on at least 1½" of wood. In platform construction, the ends of all the joists are fastened to a header joist, also called a *band joist* or *rim joist*, to form the box sill.

Joists are doubled where extra loads need to be supported. When a partition runs parallel to the joists, a double joist is placed underneath. Joists must also be doubled around all openings in the floor frame for stairways, chimneys, etc., to reinforce the rough opening in the floor. These additional joists used at such openings are called **trimmer joists**.

In residential construction, floors traditionally have been built using wooden joists. However, the use of prefabricated engineered wood products, such as wood I-beams and various types of **trusses**, is also becoming common.

3.1.3 Wood I-Beams

Wood I-beam joists are typically manufactured with 1½" diameter, pre-stamped knockout holes in the web about 12" OC that can be used to accommodate wiring. Other holes or openings can be cut into the web, but these can only be of a certain size and at the locations specified by the I-beam manufacturer. Under no circumstances should the flanges of I-beam joists be cut or notched.

3.1.4 Trusses

Trusses are manufactured joist assemblies made of wood or a combination of steel and wood (*Figure 19*). Solid light-gauge steel and open-web steel trusses are also made, but these are used mainly in commercial construction. Like the wood I-beams, trusses are stronger than comparable lengths of dimension lumber, allowing them to be used over longer spans. Longer spans allow more freedom in building design because interior load bearing walls and extra footings can often be eliminated. Trusses are generally faster and easier to erect, with no need for trimming or cutting in the field. They also provide the additional advantage of permitting ducting, plumbing, and wiring to be run easily between the open webs. *Figure 20* shows a typical floor system constructed with trusses.

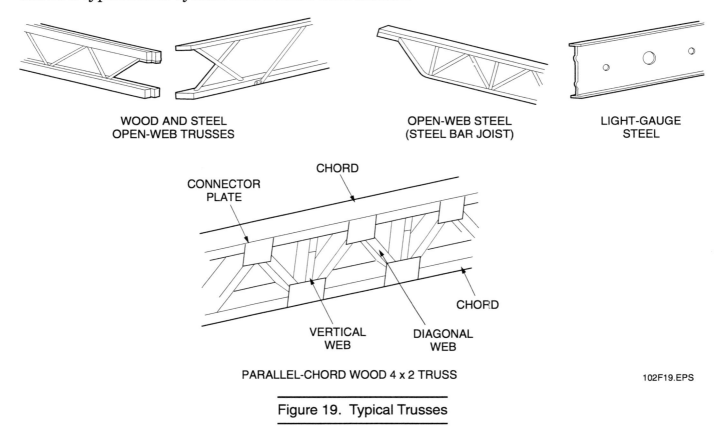

Figure 19. Typical Trusses

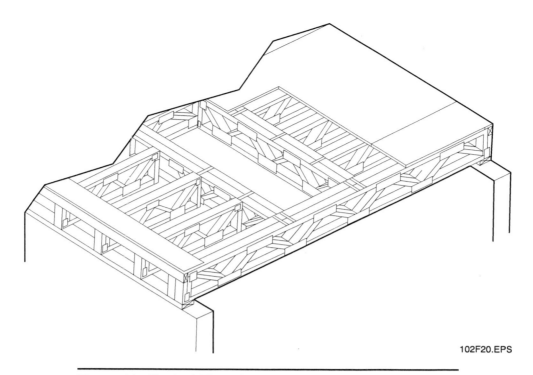

Figure 20. Typical Floor System Constructed With Trusses

3.1.5 Notching And Drilling Of Wooden Joists

When it is necessary to notch or drill through a floor joist, most building codes will stipulate how deep a notch can be made. For example, the Standard Building Code specifies that notches on the ends of joists shall not exceed one-fourth the depth. Therefore, in a 2×10 floor joist, the notch could not exceed 2½" (see *Figure 21*).

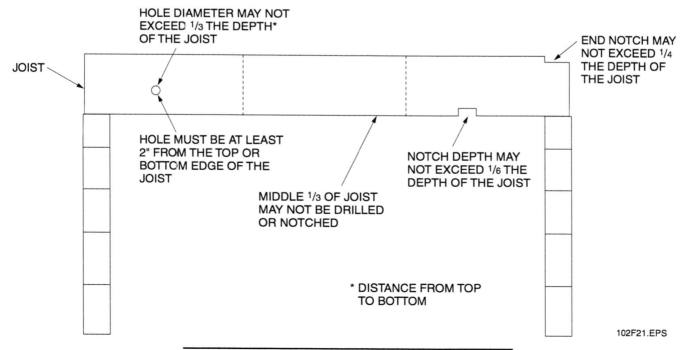

Figure 21. Notching And Drilling Of Wooden Joists

This code also states that notches for pipes in the top or bottom shall not exceed one-sixth the depth, and shall not be located in the middle third of the span. Therefore, when using a 2 × 10 floor joist, a notch cannot be deeper than 1⅝". This notch can be made either in the top or bottom of the joist, but it cannot be made in the middle third of the span. This means that if the span is 12', the middle span from 4' to 8' could not be notched.

This code further requires that holes bored for pipe or cable shall not be within 2" of the top or bottom of the joist, nor shall the diameter of any such hole exceed one-third the depth of the joist. This means that if a hole needs to be drilled, it may not exceed 3" in diameter if a 2 × 10 floor joist is used. Always check the local codes.

Some wood I-beams are manufactured with perforated knockouts in their web, approximately 12" apart. Never notch or drill through the beam flange or cut other openings in the web without checking the manufacturer's specification sheet.

Also, do not drill or notch other types of engineered lumber (e.g., LVL, PSL, and glulam) without first checking the specification sheets.

3.1.6 Bridging

Bridging is used to stiffen the floor frame and to enable an overloaded joist to receive some support from the joists on either side. Most building codes require that bridging be installed in rows between the floor joists, at intervals of not more than 8'. For example, floor joists with spans of 8' to 16' need one row of bridging in the center of the span.

Three types of bridging (*Figure 22*) are commonly used: wood cross-bridging, solid wood bridging, and metal cross-bridging. Wood and metal cross-bridging are composed of pieces of wood or metal set diagonally between the joists to form an X. Wood cross-bridging is typically 1 × 4 lumber placed in double rows that cross each other in the joist space.

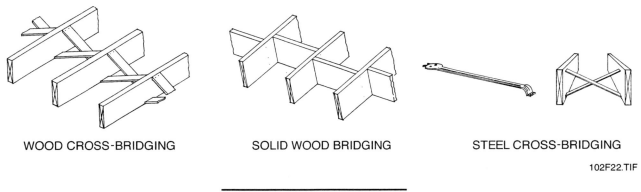

Figure 22. Types Of Bridging

Metal cross-bridging is installed in a similar manner. Metal cross-bridging comes in a variety of styles and different lengths for use with a particular joist size and spacing. It is usually made of 18-gauge steel and is ¾" wide. Solid bridging, also called **blocking**, consists of solid pieces of lumber (usually the same size as the floor joists) installed between the joists. The bridging pieces are offset from one another to enable end nailing.

3.1.7 Subflooring

Subflooring consists of panels or boards laid directly on and fastened to floor joists (*Figure 23*) in order to provide a base for underlayment and/or the finish floor material. Underlayment is a material, such as particleboard or plywood, laid on top of the subfloor to provide a smoother surface for finish flooring. The subfloor adds rigidity to the structure and provides a surface upon which walls and other framing can be laid out and constructed. Subfloors also act as a barrier to cold and dampness, thus keeping the building warmer and drier in winter. Subflooring can be constructed of plywood, OSB or other manufactured board panels, or common wooden boards.

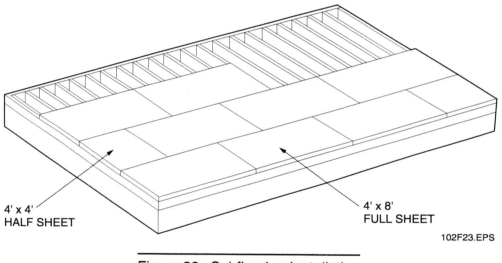

Figure 23. Subflooring Installation

3.2.0 WALL CONSTRUCTION

Wall framing is generally done with 2 × 4 studs spaced 16" on center. In a one-story building, 2 × 4 spacing can be 24" on center. If 24" spacing is used in a two-story building, the lower floor must be framed with 2 × 6 lumber.

Figure 24 identifies the structural members of a wood frame wall. Each of the members shown on the illustration is described below.

- *Blocking (spacer)* – A wood block used as a filler piece and support between framing members. Blocking also provides a surface for attaching equipment, etc.
- ***Cripple stud*** – In wall framing, a short framing stud that fills the space between a header and a **top plate** or between the sill and the **soleplate**.

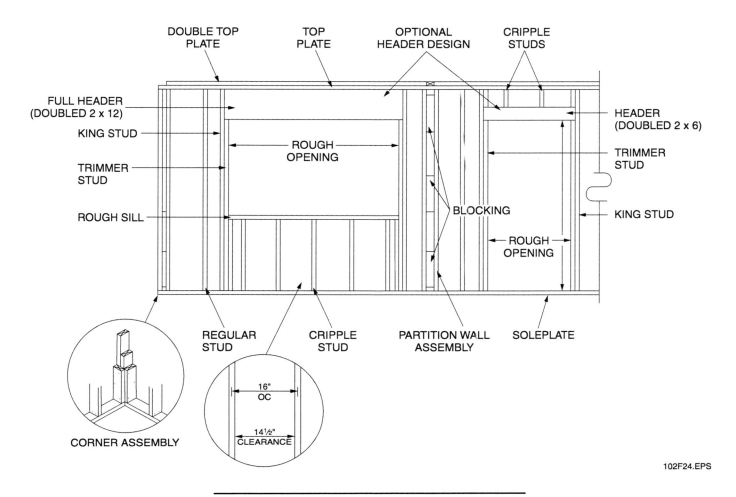

Figure 24. Wall And Partition Framing Members

- **Double top plate** – A plate made of two members to provide better stiffening of a wall. It is also used for connecting splices, corners, and partitions that are at right angles (perpendicular) to the wall.
- *Header (lintel)* – A horizontal structural member that supports the load over an opening such as a door or window.
- *King stud* – The full-length stud next to the **trimmer stud** in a wall opening.
- *Partition* – A wall that subdivides space within a building. A bearing partition or wall is one that supports the floors and roof directly above in addition to its own weight.
- *Rough opening* – An opening in the framing formed by framing members, usually for a window or a door.
- *Rough sill* – The lower framing member attached to the top of the lower cripple studs to form the base of a rough opening for a window.
- *Soleplate* – The lowest horizontal member of a wall or partition to which the studs are nailed. It rests on the rough floor.
- *Stud* – The main vertical framing member in a wall or partition.

- *Top plate* – The upper horizontal framing member of a wall used to carry the roof trusses or rafters.
- *Trimmer stud* – The vertical framing member that forms the sides of rough openings for doors and windows. It provides stiffening for the frame and supports the weight of the header.

3.2.1 Corners

A wall must have solid corners that can take the weight of the structure. In addition to contributing to the strength of the structure, corners must provide a good nailing surface for sheathing and interior finish materials. Building contractors generally select the straightest, least defective studs for corner framing.

There are many methods for constructing corners. Two of these methods are shown in *Figure 25*. Some builders will construct the corner in place, then plumb and brace it before raising the wall frames. This approach makes it easier to plumb and brace the frame, but it prevents installation of the sheathing before the frame is erected. If the corners are included in the frame, then a portion of the corner is included with each of the mating frame sections.

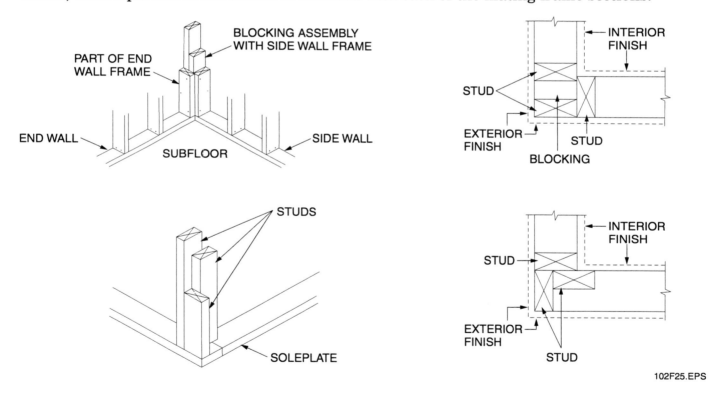

Figure 25. Corner Construction

3.2.2 Partition Intersections

Interior partitions must be securely fastened to outside walls. For that to happen, there must be a solid nailing surface where the partition intersects the exterior frame. There are several methods used to construct framing for partition Ts. Some of them are shown in *Figure 26*.

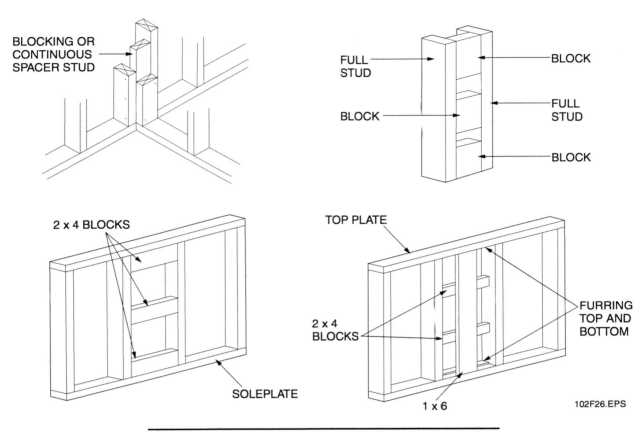

Figure 26. Constructing Nailing Surfaces For Partitions

3.2.3 Window And Door Openings

When wall framing is interrupted by an opening such as a window or door, a method is needed to distribute the weight of the structure around the opening. This is done by the use of a header. The header is placed so that it rests on the trimmer studs, which transfer the weight to the soleplate or subfloor and then to the foundation.

Headers are made of solid or built-up lumber. Laminated lumber and beams have become popular as header material, especially where the load is heavy.

Built-up headers are usually made from 2" lumber separated by ½" plywood spacers (*Figure 27*). A full header is used for large openings and fills the area from the rough opening to the bottom of the top plate. A small header with cripples is suitable for average-size windows and doors and is usually made from 2 × 4 or 2 × 6 lumber. Built-up headers are sometimes made by gluing and nailing ½" plywood the entire length of the header, instead of inserting plywood blocks. This method allows the framing crew to make a long section (16') of built-up header, then cut what they need for each opening from that section. The crew may use the same header for all openings. This saves time because it eliminates the need for cutting and installing cripple studs.

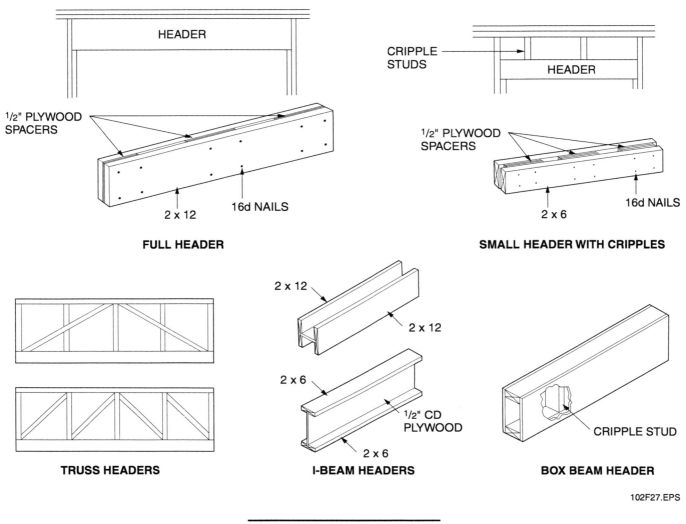

Figure 27. Types Of Headers

Truss headers are used when the load is especially heavy or the span is extra wide. The design of the trusses is generally included in the architect's plans.

Other types of headers used for heavy loads are wood or steel I-beams and box beams. The latter are made of plywood webs connected by lumber flanges in a box configuration.

The width of a header is equal to the rough opening plus the thickness of the trimmer studs. For example, if a rough opening for a 3' wide window is 38", the width of the header would be 41".

Figure 28 shows cross sections of typical wood-framed walls.

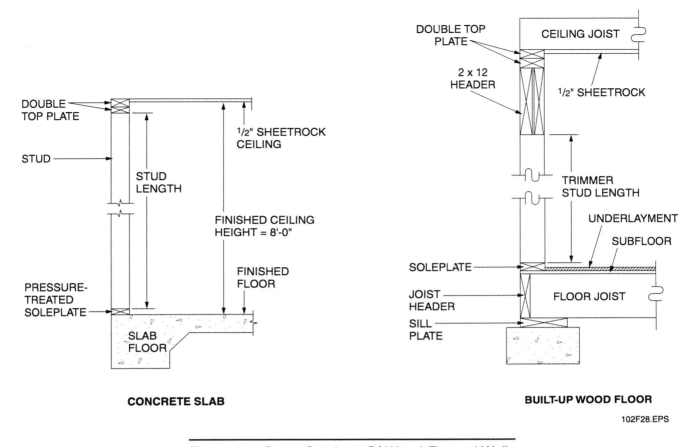

Figure 28. Cross Sections Of Wood-Framed Walls

3.2.4 Firestops

In some areas, local building codes will require **firestops**, which are used in concealed wall spaces to retard the spread of fire. In building a structure with a basement, a firestop or draftstop must be placed midpoint between the stud wall plates. Without firestops, the space between the studs will act like a flue in a chimney. There are two methods of installing firestops. The first method is to cut pieces of 2 × 4 material to fit horizontally between the studs, as shown in *Figure 29*.

The best method of installing firestops is at an angle. Without oxygen, there is no fire. By placing the firestop at an angle, the oxygen will be at the uppermost part of the firestop. As the fire burns the uppermost part of the firestop, it will continue to drop, thus retarding the fire spread. This may retard the spread of the fire by an extra 10 to 15 minutes. Firestops are used in a variety of areas such as walls, floors, and roofs. Later in the module, we will cover **firestopping** materials and devices used to seal openings that are made in walls, ceilings, and floors in order to pass cabling from one location to another.

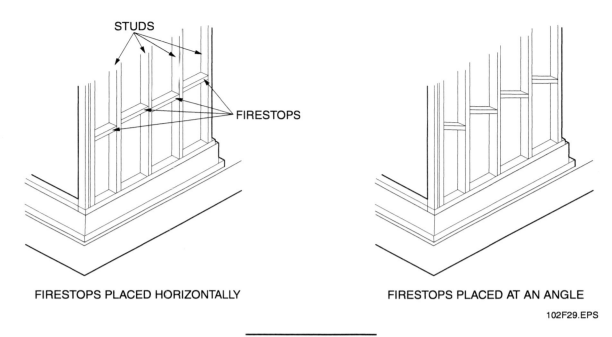

Figure 29. Firestops

3.2.5 Bracing

Bracing is important in the construction of exterior walls. Many local building codes require bracing if certain types of sheathing are used. In areas where high winds are a factor, lateral bracing may be required even when ½" plywood is used as the sheathing.

Several methods of bracing have been used since the early days of construction. One method is to cut a notch (let-in) for a 1 × 4 or 1 × 6 at a 45° angle on each corner of the exterior walls. Another method is to cut 2 × 4 braces at a 45° angle for each corner. Still another type of bracing (used where permitted by the local code) is metal strap bracing (*Figure 30*). This product is made of galvanized steel.

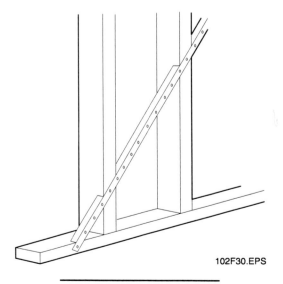

Figure 30. Metal Bracing

Metal strap bracing is easier to use than let-in wood bracing. Instead of notching out the studs for a 1 × 4 or 2 × 4, a circular saw is used to make a diagonal groove in the studs, top plate, and soleplate for the rib of the bracing strap. The strap is then nailed to the framing.

With the introduction of plywood, some areas of the country have done away with corner bracing. However, along with plywood came different types of sheathing that are by-products of the wood industry and do not have the strength to withstand wind pressures. When these are used, permanent bracing is needed. Building codes in some areas will allow a sheet of ½" plywood to be used on each corner of the structure in lieu of diagonal bracing when the balance of the sheathing is fiberboard. In other areas, the use of bracing is required regardless of the type of sheathing used. Always check local codes.

3.2.6 Sheathing

Sheathing is the material used to close in the walls. **APA-rated** material such as plywood and non-veneer panels such as OSB and other reconstituted wood products are generally used for sheathing.

When plywood is used, the panels will range from $\frac{5}{16}$" to $\frac{3}{4}$" thick. A minimum thickness of $\frac{3}{8}$" is recommended when siding is to be applied. The higher end of the range is recommended when the sheathing acts as the exterior finish surface. The panels may be placed with the grain running horizontally or vertically. If they are placed horizontally, local building codes may require that blocking be used along the top edges.

Typical nailing requirements call for 6d (6 penny) nails for panels ½" thick or less and 8d nails for thicker panels. Nails are spaced 6" apart at the panel edges and 12" apart at intermediate studs.

Other materials that are sometimes used as sheathing are fiberboard (insulation board), exterior-rated gypsum wallboard, and rigid foam sheathing. A major disadvantage of these materials is that siding cannot be nailed to them. It must either be nailed to the studs or special fasteners must be used.

When material other than rated panels is used as sheathing, rated plywood panels may be installed vertically at the corners to eliminate the need for corner bracing.

3.3.0 CEILING CONSTRUCTION

Ceiling joists are usually laid across the width of a building at the same positions as the wall studs.

The length of a joist is the distance from the outside edges of the double top plates. The ends of the joists are cut to match the rafter pitch so that the roof sheathing will lay flush on the framing (*Figure 31*). If the joist exceeds the allowable span, two pieces of joist material must be spliced over a bearing wall or partition. *Figure 32* shows two splicing methods. There should be a minimum overlap of 6". Another method of splicing is to place the two joists on either side of the rafter with a piece of blocking between the joists at the splice.

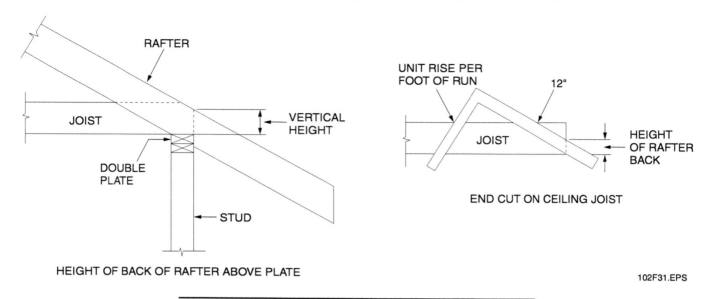

Figure 31. Cutting Joist Ends To Match The Roof Pitch

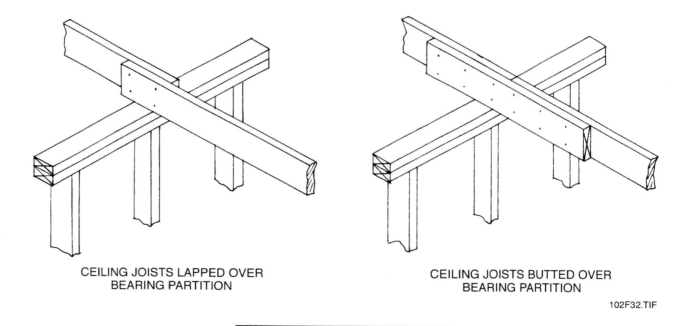

Figure 32. Spliced Ceiling Joists

If the spacing is the same as that of the wall studs, the joists are nailed directly above the studs. This makes it easier to run ductwork, piping, flues, and wiring above the ceiling. Metal joist hangers can be used in place of nailing.

After the joists are installed, a **ribband** or **strongback** is nailed across them to prevent twisting or bowing (*Figure 33*). The strongback is used for larger spans. In addition to holding the joists in line, it provides support for the joists at the center of the span.

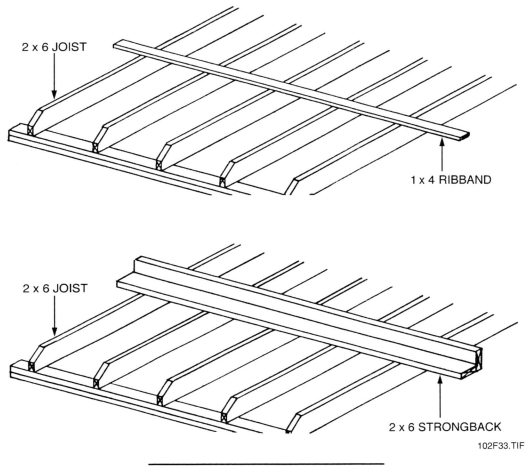

Figure 33. Reinforcing Ceiling Joists

3.4.0 ROOF CONSTRUCTION

The most common types of roofs used in residential construction are shown in *Figure 34* and described below.

- *Gable roof* – A gable roof has two slopes that meet at the center (ridge) to form a gable at each end of the building. It is the most common type of roof because it is simple, economical, and can be used on any type of structure.
- *Hip roof* – A hip roof has four sides or slopes running toward the center of the building. Rafters at the corners extend diagonally to meet at the ridge. Additional rafters are framed into these rafters.
- *Gable and valley roof* – This roof consists of two intersecting gable roofs. The part where the two roofs meet is called a *valley*.

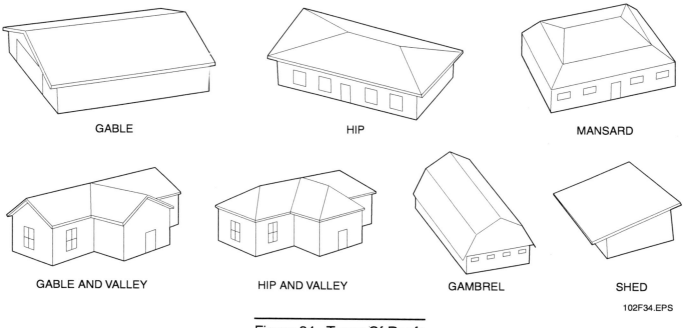

Figure 34. Types Of Roofs

- *Mansard roof* – The mansard roof has four sloping sides, each of which has a double slope. As compared with a gable roof, this design provides more available space in the upper level of the building.
- *Gambrel roof* – The gambrel roof is a variation on the gable roof in which each side has a break, usually near the ridge. The gambrel roof provides additional space in the upper level.
- *Shed roof* – Also known as a *lean-to roof*, the shed roof has a flat, sloped construction. It is commonly used in high-ceiling contemporary residences and for additions.

There are two basic roof framing systems. In stick-built framing, ceiling joists and rafters are laid out and cut by the builder on the site and the frame is constructed one stick at a time.

In truss-built construction, the roof framework is prefabricated off site. The truss contains both the rafters and the ceiling joist. Trusses and truss construction will be discussed later in this module.

3.4.1 Roof Components

Rafters and ceiling joists provide the framework for all roofs. The main components of a roof are shown in *Figure 35* and described below.

- *Ridge (ridgeboard)* – The highest horizontal roof member. It helps to align the rafters and tie them together at the upper end. The ridgeboard is one size larger than the rafters.
- *Common rafter* – A structural member that extends from the top plate to the ridge in a direction perpendicular to the wall plate and ridge. Rafters often extend beyond the roof plate to form the overhang (eaves) that protect the side of the building.

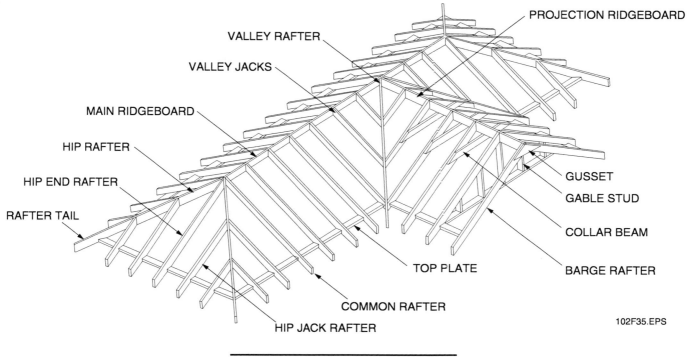

Figure 35. Roof Framing Members

- *Hip rafter* – A roof member that extends diagonally from the corner of the plate to the ridge.
- *Valley rafter* – A roof member that extends diagonally from the plate to the ridge along the lines where two roofs intersect.
- *Jack rafter* – A roof member that does not extend the entire distance from the ridge to the top plate of a wall. Hip jacks and valley jacks are shown in *Figure 35*. A rafter fitted between a hip rafter and a valley rafter is called a *cripple jack*. It touches neither the ridge nor the plate.
- *Plate* – The wall framing member that rests on top of the wall studs. It is sometimes called the *rafter plate* because the rafters rest on it. It is also referred to as the *top plate*.

On any pitched roof, rafters rise at an angle to the ridgeboard. Therefore, the length of the rafter is greater than the horizontal distance from the plate to the ridge. In order to calculate the correct rafter length, the builder must factor in the slope of the roof. Here are some additional terms that apply to roof layout (see *Figure 36*).

- *Span* – The horizontal distance from the outside of one exterior wall to the outside of the other exterior wall.
- *Run* – The horizontal distance from the outside of the top plate to the centerline of the ridgeboard (usually one-half of the span).
- *Rise* – The total height of the rafter from the top plate to the ridge. This is stated in inches per foot of run.

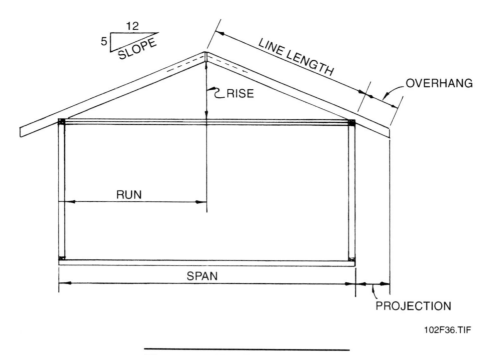

Figure 36. Roof Layout Factors

- *Pitch* – The angle or degree of slope of the roof in relation to the span. Pitch is expressed as a fraction. For example, if the total rise is 6' and the span is 24', the pitch would be ¼ (6 over 24).

- *Slope* – The inclination of the roof surface expressed as the relationship of rise to run. It is stated as a unit of rise to so many horizontal units. For example, a roof that has a rise of 5" for each foot of run is said to have a 5 in 12 slope (*Figure 36*). The roof slope is sometimes referred to as the *roof cut*.

3.4.2 Roof Sheathing

The sheathing is applied as soon as the roof framing is finished. The sheathing provides additional strength to the structure and a base for the roofing material. Some of the materials commonly used for sheathing are plywood, OSB, waferboard, **shiplap**, and common boards. When composition shingles are used, the sheathing must be solid. If wood **shakes** are used, the sheathing boards may be spaced. When solid sheathing is used, a ⅛" space is left between panels to allow for expansion.

Once the sheathing has been installed, an underlayment of asphalt-saturated felt or other specified material is installed to keep moisture out until the shingles are laid. For roofs with a slope of 4" or more, 15-pound roofer's felt is commonly used.

The underlayment is applied horizontally with a 2" top lap and a 4" side lap, as shown in *Figure 37*. A 6" lap should be used on each side of the centerline of hips and valleys. A metal drip edge is installed along the rakes and eaves to keep out wind-driven moisture.

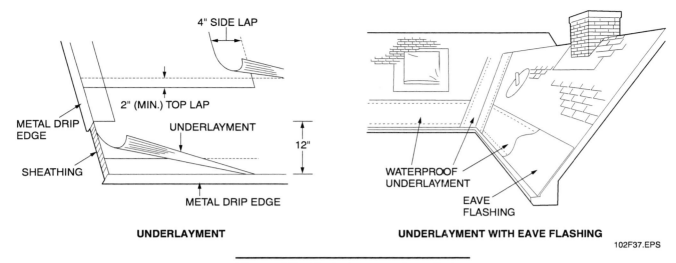

Figure 37. Underlayment Installation

In climates where snow accumulates, a waterproof underlayment should be used at roof edges and around chimneys, skylights, and vents. This underlayment has an adhesive backing that adheres to the sheathing. It protects against water damage that can result from melting ice and snow that backs up under the shingles. Sheet metal or other material is used at roof intersections and around chimneys, vents, skylights, etc., to prevent water from entering. In snowy climates, sheet metal eave flashing is often installed at the edge of a roof to prevent ice dams from forming.

3.4.3 Truss Construction

In most cases, it is much faster and more economical to use prefabricated trusses in place of rafters and joists. Even if a truss costs more to buy than the comparable framing lumber (and this is not always the case), it takes significantly less labor than stick framing. Another advantage is that a truss will span a greater distance without a bearing wall than stick framing. Just about any type of roof can be framed with trusses. Some of the special terms used to identify the members of a truss are shown in *Figure 38*.

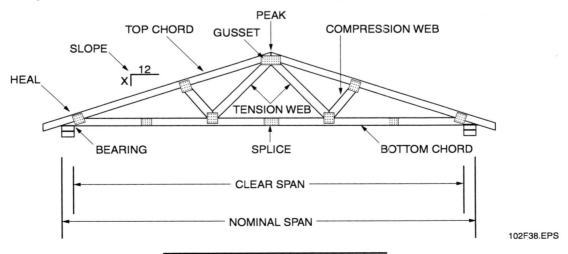

Figure 38. Components Of A Truss

A truss is a framed or jointed structure. It is designed so that when a load is applied at any intersection, the stress in any member is in the direction of its length. *Figure 39* shows some of the many kinds of trusses.

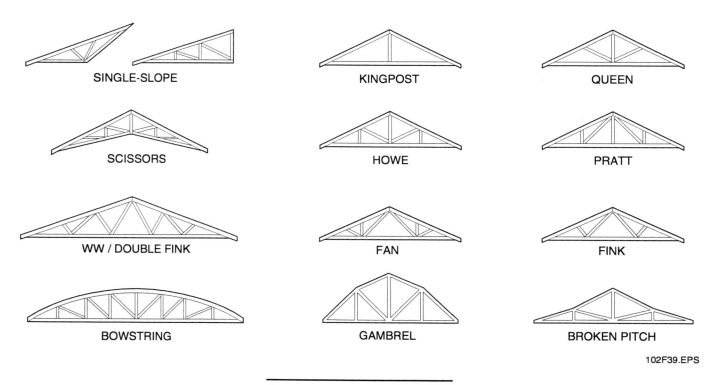

Figure 39. Types Of Trusses

Even though some trusses look nearly identical, there is some variation in the interior (web) pattern. Each web pattern distributes weight and stress a little differently, so different web patterns are used to deal with different loads and spans. The decision of which truss to use for a particular application will be made by the architect or engineer and will be shown on the blueprints.

3.4.4 Dormers

A **dormer** is a framed structure that projects out from a sloped roof. A dormer provides additional space and is often used in a Cape Cod style home, which is a single-story dwelling in which the attic is often used for sleeping rooms.

A shed dormer (*Figure 40*) is a good way to obtain a large amount of additional living space. If it is added to the rear of the house, it will not affect the appearance of the house from the front.

A gable dormer (*Figure 41*) serves as an attractive addition to a house. It provides a little extra space, as well as some light and ventilation. They are sometimes used over garages to provide a small living area or studio.

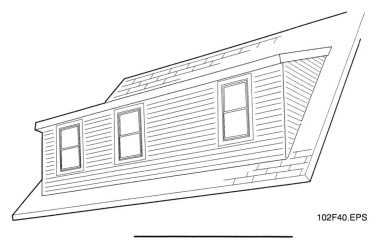

Figure 40. Shed Dormer

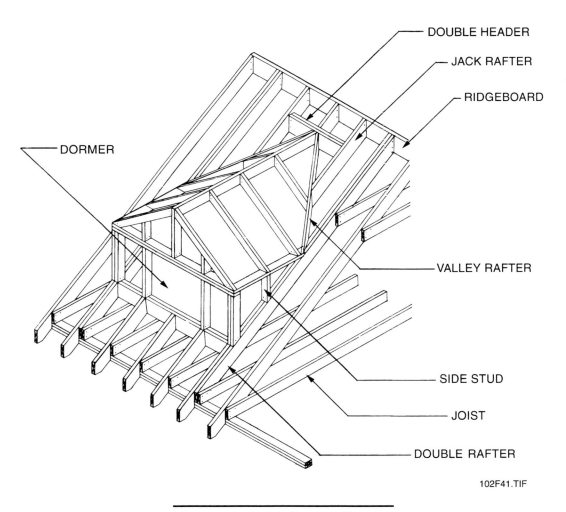

Figure 41. Gable Dormer Framing

CONSTRUCTION MATERIALS AND METHODS — TRAINEE TASK MODULE 33102

3.5.0 PLANK-AND-BEAM FRAMING

Plank-and-beam framing, also known as *post-and-beam framing* (*Figure 42*), employs much sturdier framing members than common framing. It is often used in framing roofs for luxury residences, churches, and lodges, as well as other public buildings where a striking architectural effect is desired.

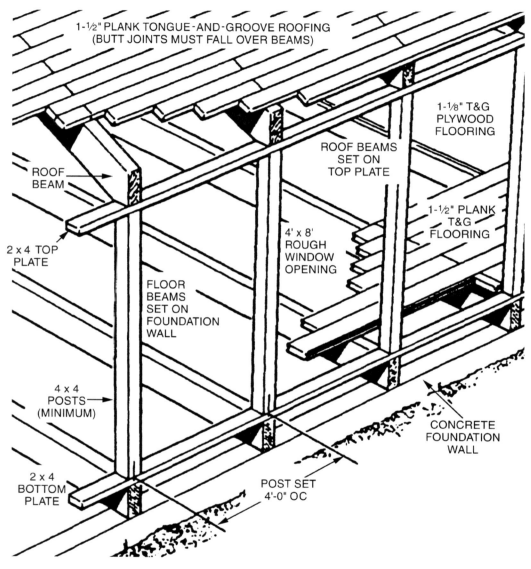

POST-AND-BEAM CONSTRUCTION FEATURES WIDELY SPACED (4'-0" OC) HEAVY POSTS AND BEAMS.

Figure 42. Post-And-Beam Framing

Because the beams used in this type of construction are very sturdy, wider spacing may be used. When plank-and-beam framing is used for a roof, the beams and planking can be finished and left exposed. The underside of the planking takes the place of an installed ceiling (*Figure 43*).

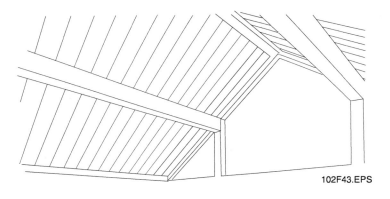

Figure 43. Example Of Post-And-Beam Framing

In lighter construction, solid posts or beams such as 4 × 4s are used. In heavier construction, laminated beams made of glulam, LVL, and PSL are used.

In post-and-beam framing, plank subfloors or roofs are usually of 2" nominal thickness, supported on beams spaced up to 8' apart. The ends of the beams are supported on posts or piers. Wall spaces between posts are provided with supplementary framing as required for attachment of exterior and interior finishes. This additional framing also provides lateral bracing for the building.

If local building codes allow end joints in the planks to fall between supports, planks of random lengths may be used and the beam spacing adjusted to fit the house dimensions. Windows and doors are normally located between posts in the exterior walls, eliminating the need for headers over the openings. The wide spacing between posts permits ample room for large glass areas.

A combination of conventional framing with post-and-beam framing is sometimes used where the two adjoin each other.

Where a post-and-beam floor or roof is supported on a stud wall, a post is usually placed under the end of the beam to carry a conventional load. A conventional roof can be used with post-and-beam construction by installing a header between posts to carry the load from the rafters to the posts.

3.6.0 WALL FRAMING IN MASONRY

In order to install cabling in buildings constructed of masonry, you must be aware of the methods used in furring masonry walls. As a general rule, furring of masonry walls is done on 16" centers. Some contractors will apply 1 × 2 **furring strips** 24" OC. This may save material, but it does not provide the same quality as a wall done on 16" centers.

In addition to the furring strips, 1 × 4 and 1 × 6 stock is used. All material that comes in contact with concrete or masonry must be pressure-treated.

Backing for partitions against a masonry block wall is done using one of the methods shown in *Figures 44* and *45*.

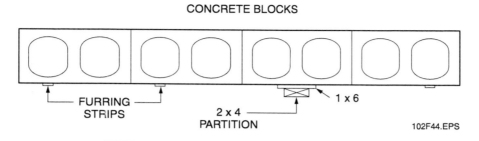

Figure 44. Partition Backing Using A 1 × 6

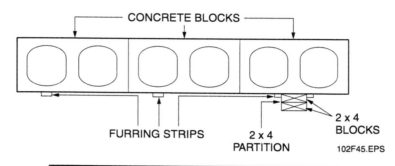

Figure 45. Partition Backing Using 2 × 4 Blocks

When preparing the corners of the block wall to receive the furring strips, enough space is allowed for the drywall to slip by the furring strips (*Figure 46*).

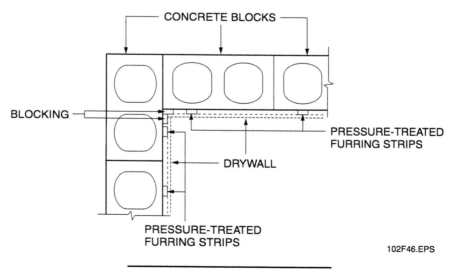

Figure 46. Corner Construction

A 1 × 4 is used at floor level to receive the baseboard. Either a narrow or wide baseboard can be used. Some builders will install a simple furring strip at floor level and depend on the vertical strips for baseboard nailing. Once the drywall has been installed, it is difficult to find the strips when nailing the baseboard.

3.7.0 WALLS SEPARATING OCCUPANCIES

Every wall, floor, and ceiling in a building is rated for its fire resistance, as established by building codes. The fire rating is stated in terms of hours (e.g., one-hour wall, two-hour wall, etc.). The rating denotes the length of time an assembly can withstand fire and give protection from it as determined under laboratory conditions. The greater the fire rating, the thicker the wall is likely to be. This subject is covered further in the section on commercial construction.

In multifamily residential construction (apartments, townhouses, etc.), the walls and ceilings dividing the occupancies must meet special fire and soundproofing requirements. The code requirements will vary from one location to another and may even vary within areas of a jurisdiction (i.e., dwellings in high-risk areas may have stiffer standards than those in other areas of the same city or county). In some cases, the code may require a masonry wall between occupancies. This masonry wall may even be required to penetrate the roof of the building so that if a fire occurs, it is contained within the unit in which it started, because it is unable to travel through the walls or across the attic space.

There are many different construction methods for multidwelling (party) walls. Each is designed to meet different fire and soundproofing standards. The wall is likely to be more than 3" thick and contain several layers of gypsum wallboard and insulation.

The important thing to note is that there are many variations, so you must know what you are drilling into before you start drilling. It is also possible that the codes will not permit you to run a cable from one unit to another in a multifamily occupancy. In addition, if drilling is permitted, you may have to use firestopping materials to seal off the opening.

4.0.0 COMMERCIAL CONSTRUCTION METHODS

The structural framework of large buildings such as office buildings, hospitals, apartment houses, hotels, etc., is usually made from concrete or structural steel. The exterior finish is often concrete panels that are either prefabricated and raised into place or poured into forms constructed at the site. Floors are usually made of concrete that is poured at the site using metal or fiberglass forms. Exterior walls (curtain walls) may also be made of glass in a metal or concrete framework. Before the concrete is poured, provision must be made for cabling pathways.

In some buildings, the framework is made of structural steel. Concrete panels fabricated off site are lifted into place and bolted or welded to the steel (*Figure 47*).

Figure 48 shows the structure of a building in which all the structural framework is made of concrete poured at the site. Each component of the structure requires a different type of form. In this case, the floor and beams were made in a single pour using integrated floor and beam forms, which were removed once the concrete hardened.

Figure 47. Concrete Wall Panels Over A Structural Steel Frame

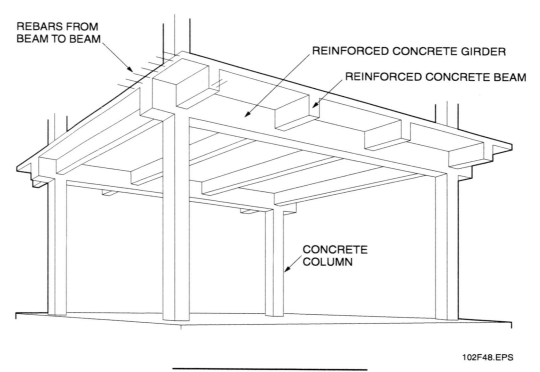

Figure 48. Concrete Structure

In some specialized commercial applications, tilt-up concrete construction is used. In tilt-up construction, the wall panels are usually poured on the concrete floor slab, then tilted into place on the footing using a crane. The panels are welded together.

The main difference between tilt-up and other types of large commercial construction is that there is no steel or concrete framework in tilt-up construction. The walls and floor slab bear the entire load. Tilt-up is most common in one- or two-story buildings with a slab at grade (no below-grade foundation). It is popular for warehouses, low-rise offices, churches, and a variety of other commercial and multifamily residential applications. Tilt-up panels of 50' in height are not uncommon. They typically range from 5" to 8" thick, but thicker walls can be obtained when using lighter-weight concrete.

4.1.0 FLOORS

Once the framework is in place, the concrete floors are poured using deck forms. Shoring is placed under the form to support it until the concrete hardens. *Figure 49* shows cellular floors poured over corrugated steel forms, which remain in place, providing channels through which cabling can be run.

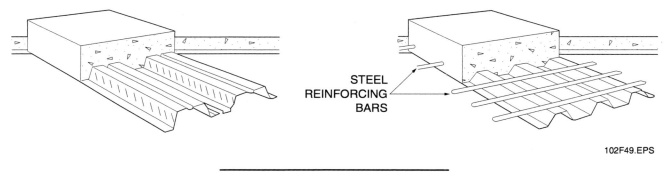

Figure 49. Corrugated Steel Forms

A section of metal, plastic, or fiber sleeve is often inserted vertically into the form before the concrete is poured to allow for electrical, communications, and other cabling to pass through the floor.

In some installations, underfloor duct systems are embedded in the concrete floor and are used to provide horizontal distribution of cables. Vertical access ports (handholes) are embedded in the form so that cable can be fished to various locations in the space. *Figure 50* shows a single-level feeder duct system. In two-level systems, one level carries electrical power cables and the other carries low-voltage cables.

Trench ducts are metal troughs that are embedded in the concrete floor and used as feeder ducts for electrical power and telecommunication lines. *Figure 51* shows a trench duct in a cellular floor.

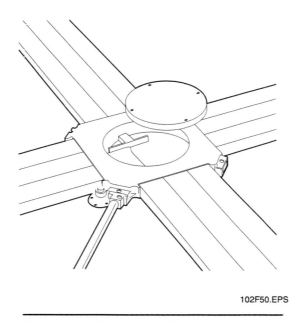

Figure 50. Flushduct Underfloor System

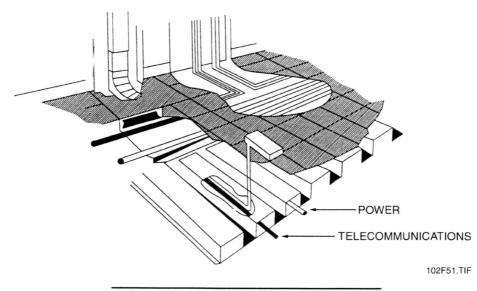

Figure 51. Trench Duct In A Cellular Floor

Access floors consist of modular floor panels supported by pedestals. They may or may not have horizontal bracing in addition to the pedestals. This type of structure is used in computer rooms, intensive care facilities, and other areas where a lot of cabling is required. In some applications, such as a factory, a trench may be formed in the concrete floor to accommodate cabling and other services.

4.2.0 EXTERIOR WALLS

Once the steel or concrete framework is built and the floors are poured, the walls can be installed. In some buildings, the curtain walls are made of lightweight steel and glass.

In others, concrete curtain walls are used. These are either prefabricated panels or poured-in-place concrete walls (*Figure 52*) that are built with forms.

Figure 52. Example Of A Poured-In-Place Concrete Wall

When walls are formed of concrete, openings for doors and windows are made by inserting wooden or metal bucks in the form, as shown in *Figure 53*. Openings for services such as piping and cabling are accommodated with fiber, plastic, or metal tubes inserted into the form.

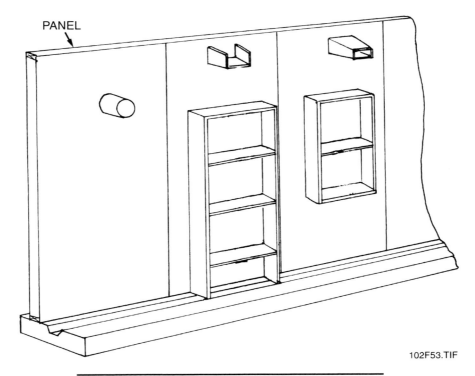

Figure 53. Framing Openings In Concrete Walls

4.3.0 INTERIOR WALLS AND PARTITIONS

The construction of walls and partitions in commercial applications is driven by the fire and soundproofing requirements specified in local building codes. In some cases, a frame wall with ½" gypsum drywall on either side is satisfactory. In extreme cases, such as the separation between offices and manufacturing space in a factory, it may be necessary to have a concrete block (CMU) wall combined with fire-resistant gypsum wallboard, along with rigid and/or fiberglass insulation, as shown in *Figure 54*. This is especially true if there is any explosion or fire hazard.

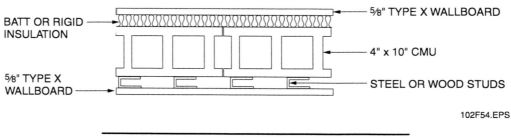

Figure 54. High Fire/Noise Resistance Partition

While they are sometimes used in residential construction, steel studs are the standard for framing walls and partitions for commercial construction. Once the studs are installed, one or more layers of gypsum wallboard and insulation are applied. The type and thickness of the wallboard and insulation depend on the fire rating and soundproofing requirements. Soundproofing needs vary from one use to another and are often based on the amount of privacy required for the intended use. For example, executive and physician offices may require more privacy than general offices.

The requirements for sound reduction and fire resistance can significantly affect the thickness of a wall. For example, a wall with a high sound transmission class (STC) and fire resistance might have a total thickness of nearly 6", while a low-rated wall might have a thickness of only 3" using steel studs and 2¼" using wooden studs. (See *Figure 55*.)

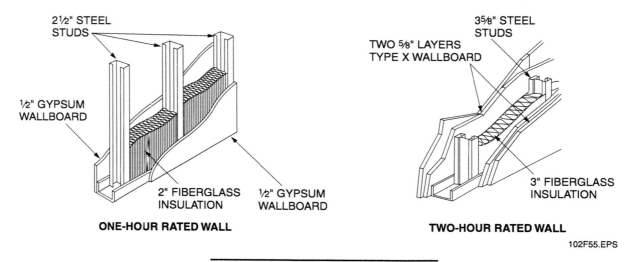

Figure 55. Partition Wall Examples

As discussed previously, the fire rating specified by the applicable building code determines the types and amount of material used in a wall or partition. As shown in *Figure 55*, a one-hour rated wall might be constructed of single sheets of ⅝" gypsum wallboard on wooden or 25-gauge metal studs. A two-hour rated wall requires heavy-gauge metal studs and two layers of fire-resistant gypsum wallboard.

4.3.1 Metal Framing Materials

Metal framing components include metal studs and, in some cases, metal joists and metal roof trusses. The vertical and horizontal framing members serve as structural load-carrying components for a variety of low- and high-rise structures. Metal stud framing is compatible with all types of surfacing materials.

The advantages of metal framing include noncombustibility, uniformity of dimension, lightness of weight, freedom from rot and moisture, and ease of construction. The components of metal frame systems are manufactured to fit together easily. There are a variety of metal framing systems, both loadbearing and nonloadbearing types. Some nonloadbearing partitions are designed to be demountable or moveable and still meet the requirements of sound insulation and fire resistance when covered with the proper gypsum system.

When metal studs are used for drywall framing systems, the channel stock features knurled sides for positive screw settings and comes in two grades. The first grade is the standard drywall stud (*Figure 56*). Standard studs come in widths of 1⅝" to 6". The flanges are 1⅜" × 1¼". Lengths of 6' to 16' are commercially available. The standard drywall stud is 25-gauge steel (the higher the gauge, the lighter the metal). Depending on the product number, lengths range from 8' to 12'; other lengths are available by special order.

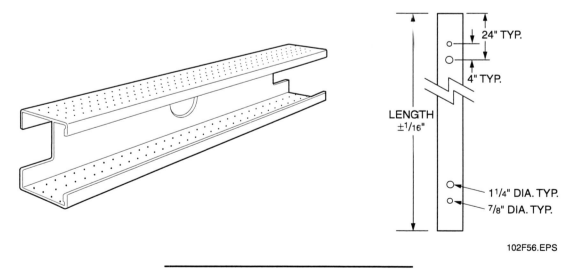

Figure 56. Standard Metal Stud Stock

The second grade is the extra heavy drywall stud (*Figure 57*). These studs have knurled sides for positive screw settings. They also have cutouts and utility knockout holes 12" from each end and at the midpoint of the stud.

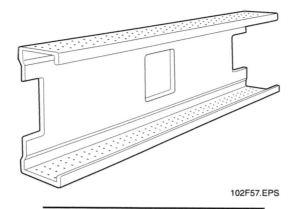

Figure 57. Heavy-Duty Stud Stock

The width of the extra heavy studs will also vary from 1⅝" to 6". The flanges are 1⅜" × 1¼". The extra heavy drywall stud is 20 gauge. This type of stud can be ordered in any length that is needed. Be sure the grade of stud being used meets prevailing code requirements.

Metal studs are also available in greater strengths of 18 gauge, 16 gauge, 12 gauge, and 10 gauge. These strengths are classified as *structural steel studs*. They are available in widths of 2½" to 8". The flanges are 1⅝" × 2" or 2½". They can be ordered in whatever length is needed.

Although different materials are used, the general approach to framing with metal studs is the same as that used for wooden studs. In fact, metal studs can be used with either wood plates or metal runners.

Like wood framing, metal framing is installed 12", 16", or 24" on center, openings are framed with headers and cripples, and special framing is needed for corners and partition Ts.

Depending on the load, reinforcement may be needed when framing openings. Bracing of walls to keep them square and plumb is also required. The illustrations in this section show examples of common framing techniques. *Table 3* shows the framing spacing for various gypsum drywall applications.

The erection of metal studs typically starts by laying metal tracks in position on the floor and ceiling and securing them (*Figure 58*). If the tracks are being applied to concrete (*Figure 59*), a low-velocity powder-actuated fastener is generally used. If the tracks are being applied to wood joists, such as in a residence, screws can be driven with a screw gun.

WARNING!   The use of a powder-actuated fastener requires special training and certification.

	Single-Ply Gypsum Board (Thickness)	Application To Framing	Maximum OC Spacing Of Framing
Ceilings:	⅜"	Perpendicular	16"
	½"	Perpendicular	16"
		Parallel	16"
	*½"	Perpendicular	24"
	⅝"	Perpendicular	24"

* Only ⅝" thick gypsum board should be used, applied perpendicularly (horizontally) on ceilings to receive a spray-applied water-based texture finish.

Sidewalls:	⅜"	Perpendicular or Parallel	16"
	½" or ⅝"	Perpendicular or Parallel	24"

Fasteners Only – No Adhesive Between Plies

	Multi-Ply Gypsum Board (Thickness)		Application To Framing		Maximum OC Spacing Of Framing
	Base	Face	Base	Face	
Ceilings:	⅜"	⅜"	Perpendicular	Perpendicular	16"
	½"	*⅜"	Parallel	Perpendicular	16"
		*½"	Parallel	Perpendicular	16"
	½"	*½"	Perpendicular	Perpendicular	24"
	⅝"	*½"	Perpendicular	Perpendicular	24"
		⅝"	Perpendicular	Perpendicular	24"

* Only ⅝" thick gypsum board should be used on the face layer, applied perpendicularly (horizontally) on ceilings to receive a spray-applied water-based texture finish.

Sidewalls:

For two-layer applications with no adhesive between plies, ⅜", ½", or ⅝" thick gypsum board may be applied perpendicularly (horizontally) or parallel (vertically) on framing spaced a maximum of 24" OC. Maximum spacing should be 16" OC when ⅜" thick board is used as the face layer.

Table 3. Maximum Framing Spacing

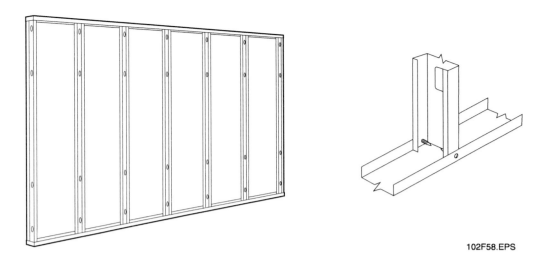

Figure 58. Metal Framing

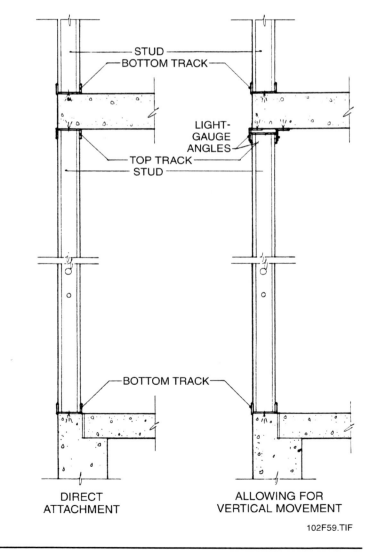

Figure 59. Metal Studs With Concrete Floors And Ceiling

Once the tracks are in place, the studs and openings are laid out in the same way as a wood frame wall. The studs may be secured to the tracks with screws or they may be welded. In some cases, the entire wall will be laid out on the floor, then raised and secured. When heavy-gauge walls are used, they may be constructed and welded in a shop and brought to the site.

There are some differences between installing metal nonloadbearing partitions and wooden nonloadbearing partitions. When constructing with wood, all partitions must be nailed together. With metal studs, this is not required.

As shown in *Figure 60*, the partitions are held back from the other partitions so that the drywall will slide past. Note the conduit fed through the openings in the studs.

Figure 60. Partition Held Back To Allow Drywall To Slide By

When metal studs are used to frame around steel beams, the metal studs are secured to the metal beam with powder-actuated fasteners (if allowed) and the support members are screwed to the metal studs (*Figure 61*). Note the hanger for the suspended ceiling at the right of the picture.

When the metal studs are installed against metal channels or flanges, the studs are secured to the channel with scrap pieces of metal studs, as shown in *Figure 62*.

Figure 61. Plate Attached To A Beam

Figure 62. Studs Secured To A Channel

4.3.2 Bracing Walls

Different forms of bracing are used to support metal stud walls. Lateral bracing using continuous metal strapping is always recommended as the minimum support for metal stud walls (*Figure 63*). Diagonal bracing using metal strapping is sometimes required (*Figure 64*). This is done with 20-gauge, 2" wide metal straps placed close to the end of the wall. Lateral and diagonal braces can be screwed and/or welded to the studs.

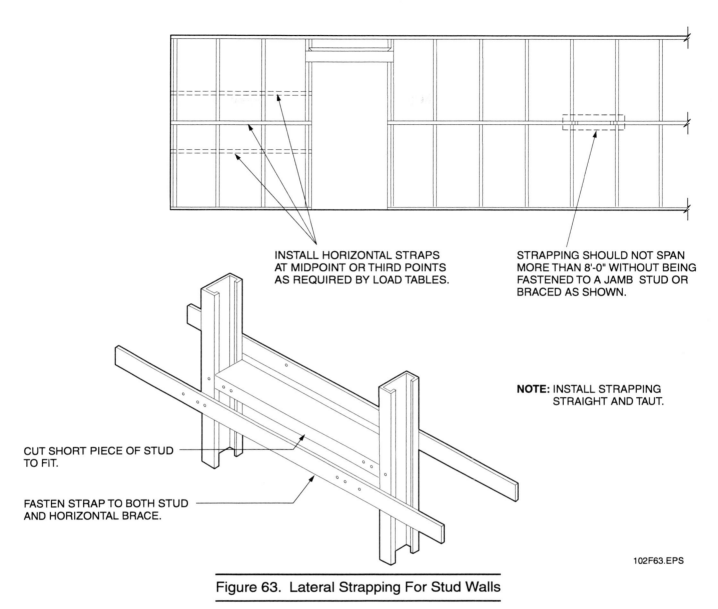

Figure 63. Lateral Strapping For Stud Walls

For heavier studs (6" and wider), steel channel threaded through the openings in the studs and welded to angle clips is sometimes required (*Figure 65*). Fine-gauge lateral bracing, fed through the openings in the studs and welded to each stud, is also used in some applications, as shown in *Figure 66*.

4.3.3 Metal Joists And Roof Trusses

In commercial work, metal studs are commonly used to frame interior nonloadbearing walls and partitions. In residential work, an entire house can be framed with steel studs, joists, and roof trusses.

Steel joists are available in the same sizes as wood joists. Joists can rest directly on concrete or masonry or they can be attached to a wood sill plate or top plate. See *Figure 67*.

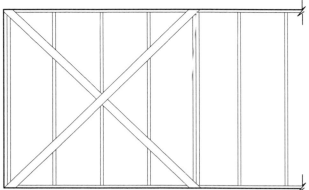

NOTES:

- INSTALL STRAPPING AS CLOSE TO 45° AS POSSIBLE.
- PLACE STRAPS ON BOTH SIDES OF STUD WALL IN ORDER TO PREVENT ECCENTRIC LOADING.
- CHECK FOR INCREASED AXIAL LOAD APPLIED TO WALL STUDS DUE TO TENSION IN STRAPS. DOUBLE STUDS WILL TYPICALLY BE REQUIRED AT STRAP ENDS.
- INSTALL WALL STRAPS STRAIGHT AND TAUT.
- FASTEN STRAPS TO EVERY STUD THEY PASS OVER.

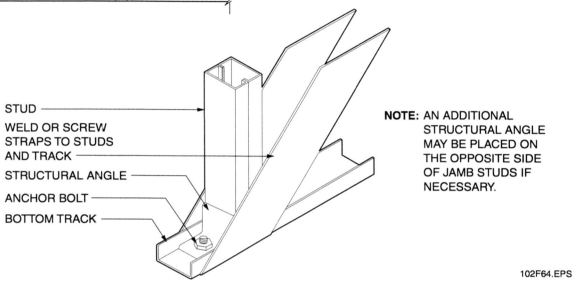

Figure 64. Diagonal Strapping For Shear Walls

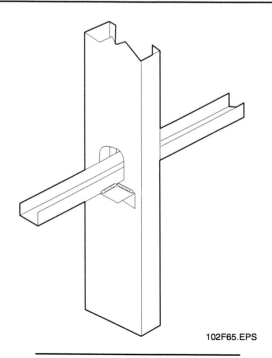

Figure 65. Heavy-Duty Bridging

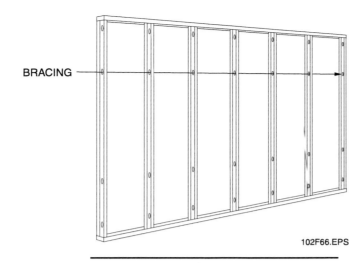

Figure 66. Fine-Gauge Lateral Bracing

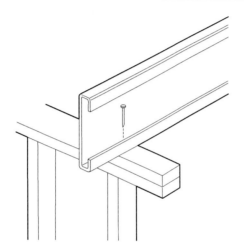

JOIST APPLIED DIRECTLY TO
DOUBLE TOP PLATE

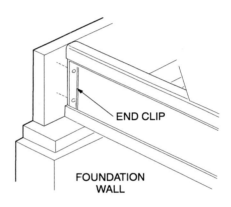

JOIST ATTACHED TO WOOD SILL
PLATE AND HEADER JOIST

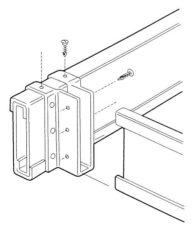

METAL JOIST ATTACHED TO
METAL HEADER JOISTS

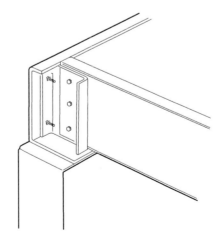

METAL JOIST ATTACHED TO
CONCRETE FOUNDATION

Figure 67. Examples Of Metal Joist Installations

One method of installing floor joists in a poured concrete foundation wall is to form slots in the wall to accept the joists (*Figure 68*).

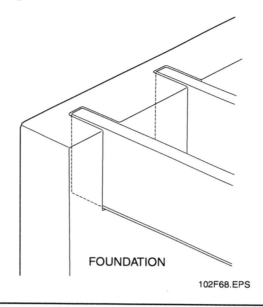

Figure 68. Installing Metal Floor Joists In A Slotted Foundation Wall

Metal roofs are framed with prefabricated trusses in which the framing members are welded together.

4.4.0 CEILINGS

Although suspended ceilings are sometimes found in residential applications, they are most commonly used in commercial construction. Suspended ceilings have a number of advantages in commercial work:

- They provide excellent noise suppression.
- They provide an area in which horizontal runs of cabling, piping, heating and cooling ducts, and other services can be readily accessed.
- In many commercial buildings, the area between the suspended ceiling and the floor above acts as the return air **plenum** for air conditioning and heating, eliminating the need for some of the sheet metal ductwork.
- The use of suspended ceilings eliminates the need for ceiling framing, as well as the need to box in horizontal runs of ductwork, piping, etc.

There are a wide variety of suspended acoustical ceiling systems. They use the same basic materials, but their appearances are completely different. The focus in this module is on the following systems:

- Exposed grid systems
- Metal pan systems

- Direct-hung concealed grid systems
- Integrated ceiling systems
- Luminous ceiling systems
- Suspended drywall furring ceiling systems
- Special ceiling systems

4.4.1 Exposed Grid Systems

For an exposed grid suspended ceiling, also called a *direct-hung system*, a light metal grid is hung by wire from the original ceiling or the deck above. Ceiling panels, which are usually 2' × 2' or 2' × 4', are then placed in the frames of the metal grid. Exposed grid systems are constructed using the components and materials described below and shown in *Figure 69*.

- *Main runners* – Primary support members of the grid system for all types of suspended ceiling systems. They are 12' in length and are usually constructed in the form of an inverted T. When it is necessary to lengthen the main runners, they are usually spliced together using extension inserts. However, the method of splicing the main runners will vary with the type of system being used.

- *Cross runners (cross ties or cross tees)* – Supports that are inserted into the main runners at right angles and spaced an equal distance from each other, forming a complete grid system. They are held in place by either clips or automatic locking devices. Typically, they are either 4' or 2' in length and are usually constructed in the form of an inverted T. Note that 2' cross runners are only required for use with 2' × 2' ceiling panels.

- *Wall angle* – Supports that are installed on the walls to support the exposed grid system at the outer edges.

- *Ceiling panels* – Panels that are laid in place between the main runners and cross ties to provide an acoustical treatment. Acoustical panels used in suspended ceilings stop sound reflection and reverberation by absorbing sound waves. These panels are typically designed with numerous tiny sound traps consisting of drilled or punched holes or fissures, or a combination of both. A wide variety of ceiling panel designs, patterns, colors, facings, and sizes are available, allowing most environmental and appearance demands to be met. Panels are typically made of glass or mineral fiber. Generally, glass panels have a higher sound absorbency than mineral fiber panels. Panel facings are typically made of embossed vinyl and are available in a variety of patterns such as fissured, pebbled, **striated**, etc. The specific ceiling panels used must be compatible with the ceiling suspension system, however, because there are variations among manufacturers' standards and not all panels fit all systems.

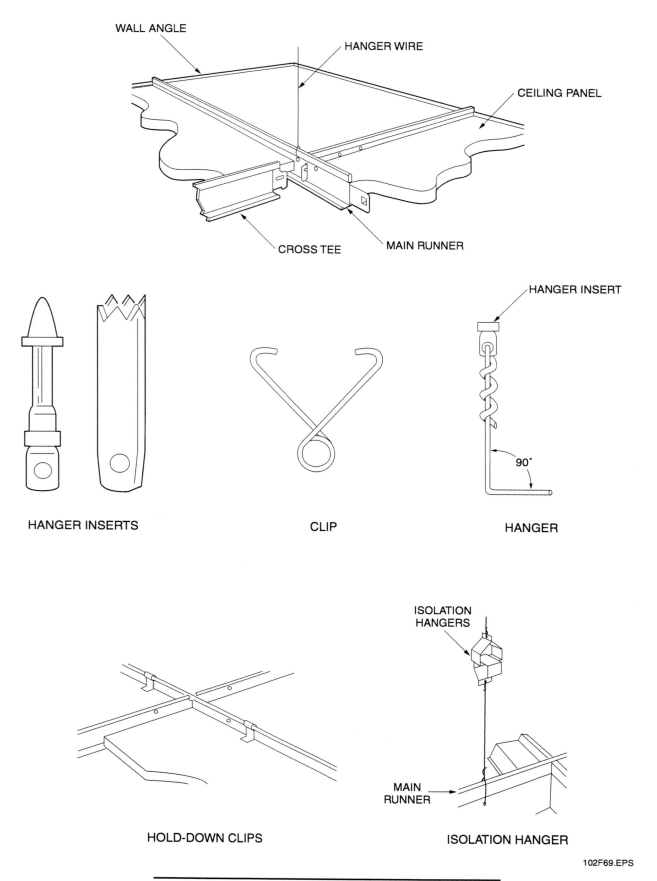

Figure 69. Typical Exposed Grid System Components

Note: It should be pointed out here that the terms *ceiling panel* and *ceiling tile* have specific meanings. Ceiling panels are typically any lay-in acoustical board that is designed for use with an exposed mounting system. They do not have finished edges or precise dimensional tolerances because the exposed grid system support members provide the trimout. Ceiling tiles are acoustical ceiling boards, usually 12" × 12" or 12" × 24", which are nailed, cemented, or suspended by a concealed grid system. The edges are often **kerfed** (notched) and cut back.

- *Hanger inserts and clips* – There are many types of fastening devices used to attach the grid system hangers or wires to the building's horizontal structure above the suspended ceiling. Screw eyes and star anchors are commonly used, and require an electric hammer for installation. Eye pin fasteners are also commonly used to fasten into reinforced concrete with a powder-actuated fastening tool. Clips are used where beams are available and are typically installed over the beam flanges, then the hanger wires are inserted through the loops in the clips and secured. These devices must be adequate to handle the load.
- *Hangers* – These are attached to the hanger inserts, pins, clips, etc., to support the suspended ceiling's main runners. The hangers can be made of No. 12 wire or heavier rod stock. Ceiling isolation hangers are also available that isolate ceilings from noise traveling through the building structure.
- *Hold-down clips* – Used in some systems to hold the ceiling panels in place.
- *Nails, screws, rawls, toppets, molly bolts* – Used to secure the wall angle to the wall. The specific item used depends on the wall construction and material.

4.4.2 Metal Pan Systems

The metal pan system is similar to the conventional suspended acoustical ceiling system, but metal tiles or pans are used in place of the conventional acoustical panel. See *Figure 70*.

The pans are made of steel or aluminum and are generally painted white; however, other colors are available by special order. Pans are also available in a variety of surface patterns. Tests have indicated that metal pan ceiling systems are effective for sound absorption. They are durable and easily cleaned and disinfected. In addition, the finished ceiling has little or no tendency to have sagging joint lines or drooping corners. The metal pans are die-stamped and have crimped edges that snap into the spring-locking main runner and provide a flush ceiling.

Take care in handling the pans if you have to remove them. Use white gloves or rub your hands with cornstarch to keep any perspiration marks from the surface of the pans. If care is not taken, fingerprints will be plainly visible when the units are reinstalled.

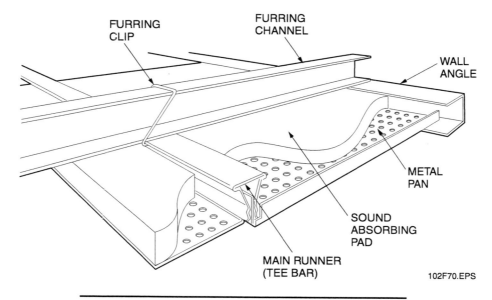

Figure 70. Typical Metal Pan Ceiling Components

When removing or installing pans, grasp the pan at its edge and force its crimp into the tee bar slots. Use the palms of your hands to seat the pan. After installing several of the pans as noted above, slide them along the tee bars into position. Use the side of your closed fist to bump the pan into level position if it does not seat readily. If metal pan hoods are required, slip them into position over the pans as they are installed. The purpose of the hood is to reduce the travel of sound through the ceiling into the room.

If a metal pan must be removed, a pan pulling device is available (*Figure 71*). To pull out a pan, insert the free ends of the device into two of the perforations at one corner of the pan and pull down sharply. Repeat this at each corner of the pan. By following this removal procedure, there is no danger of bending the pan out of shape.

An illustration of a metal pan ceiling is shown in *Figure 72*.

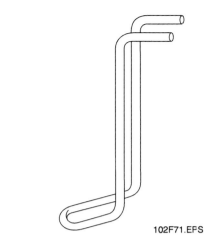

Figure 71. Pan Removal Tool

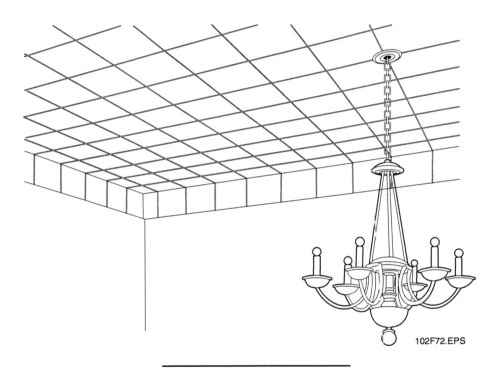

Figure 72. Metal Pan Ceiling

4.4.3 Direct-Hung Concealed Grid Systems

In this type of suspended acoustical ceiling system, the support runners are hidden from view, resulting in a patterned ceiling that is not broken by the pattern of the runners. See *Figure 73*.

The tiles used for this system are similar in composition to conventional acoustical tile but are manufactured with a kerf on all four edges. Kerfed and **rabbeted** 12" × 12" or 12" × 24" tiles are used with this system. Tiles of various colors and finishes are available. Refer to *Figure 74* for a diagram of a concealed grid system installation.

Figure 73. Typical Concealed Grid System

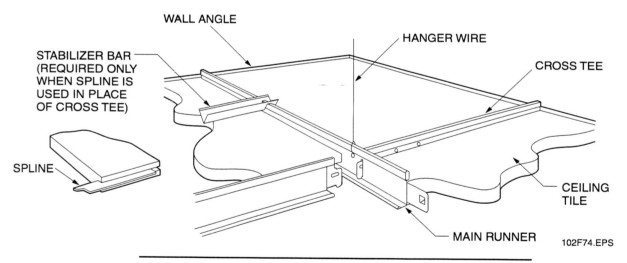

Figure 74. Direct-Hung Concealed Grid System Components

If regular access is needed to the area above the ceiling, special access systems are available which can be incorporated into the ceiling. See *Figure 75*.

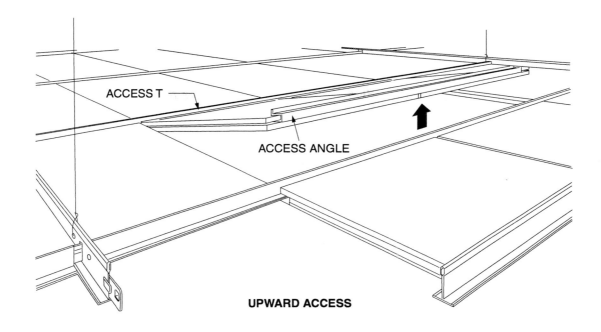

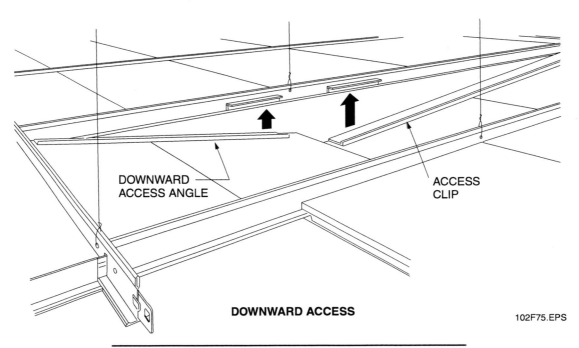

Figure 75. Typical Access For Concealed Grid Ceilings

4.4.4 Integrated Ceiling Systems

As noted by its name, the integrated ceiling system incorporates the lighting and/or air supply diffusers as part of the overall ceiling system, as shown in *Figures 76* and *77*.

This system is available in units called *modules*. The common sizes are 30" × 60" and 60" × 60". The dimensions refer to the spacing of the main runners and cross tees.

Figure 76. Typical Integrated Grid System

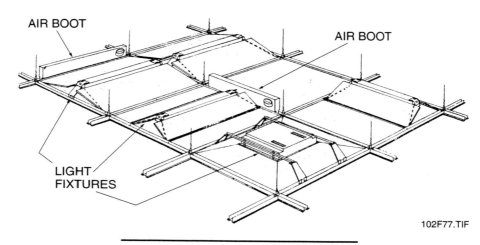

Figure 77. Integrated Ceiling Layout

4.4.5 Luminous Ceiling Systems

Luminous ceiling systems are available in many styles, such as exposed-grid systems with drop-in plastic light diffusers and aluminum or wood framework with translucent acrylic light diffusers. Refer to *Figure 78*.

Fluorescent fixtures are generally installed above the translucent diffusers. Standard modules of 2' × 2' up to sizes of 5' × 5' are available, as are custom sizes for special applications. There are two types of luminous ceilings—standard and nonstandard. Standard systems are, as their name indicates, those that are available in a series of standard sizes and patterns. Nonstandard systems differ in that they deviate from the normal spacing of main supports and/or have unusual sizes, shapes, and configurations of diffusing panels.

Figure 78. Typical Integrated Luminous Ceiling System

All surfaces in the luminous space, including pipes, ductwork, ceilings, and walls, are painted with a 75% to 90% reflectance matte white finish. Any surfaces in this area which might tend to flake, such as fireproofing and insulation, should receive an approved hard surface coating prior to painting to prevent flaking onto the ceiling below.

The installation of a standard luminous system is the same as for the exposed grid suspended system, with the exception of the border cuts. Luminous ceilings are placed into the grid members in full modules. Any remaining modules are filled in with acoustical material that has been cut to size.

With a 2' × 2' or 2' × 4' standard exposed grid system, luminous panels are used to provide the light diffusing element in the system. These panels are laid in between the runners. A variety of sizes and shapes of panels are available.

4.4.6 Suspended Drywall Furring Ceiling Systems

The suspended drywall furring system is used when it is desirable or specified to use a drywall finish or drywall backing for an acoustical tile ceiling.

When this type of ceiling is installed, the first step is to install a carrying channel, as shown in *Figure 79*. Furring channels are then installed at right angles to the carrying channels. *Figure 80* shows an example in which a hat-type metal furring channel is used.

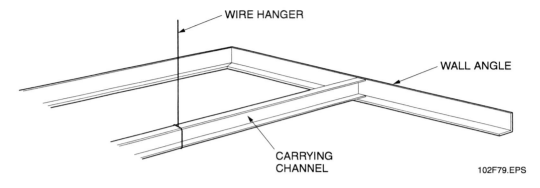

Figure 79. Carrying Channel Installed

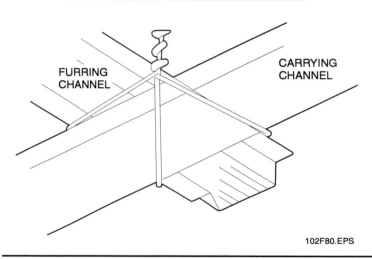

Figure 80. Furring Channel Attached To Carrying Channel

After the furring channels are in place, the drywall sheets are installed with drywall screws driven into the furring channel (*Figure 81*).

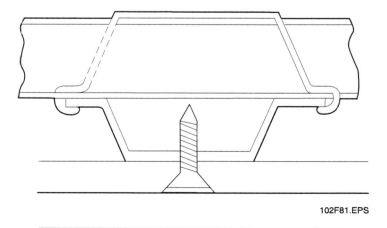

Figure 81. Drywall Secured To Furring Channel

In some cases, the furring channels are attached directly to structural members such as steel beams or wooden joists, instead of to suspended carrying channels. In other cases, ceiling tiles are attached to the drywall.

4.4.7 Special Ceiling Systems

There are numerous special ceiling systems that differ from those covered in this module. Some of these are the special metallic system, special pan system, planar system, mirrored (reflective) system, and translucent panel system (*Figures 82* through *86*).

Figure 82. Special Metallic System

Figure 83. Special Pan System

Figure 84. Planar System

Figure 85. Reflective Ceiling

Figure 86. Translucent Panels

4.4.8 General Guidelines For Accessing Suspended Ceilings

In new construction, the cable installer sometimes has the luxury of running cable before the suspended ceiling grid and panels are installed. More often, however, you will be faced with the retrofit of an existing system, or with a situation in which the ceiling has already been installed in the new building by the time you get there.

Keep in mind that the ceiling panels are delicate and the grid system is not capable of sustaining much weight. In addition, as you have already seen, there are many different types of ceiling systems, each of them with its own special requirements.

The following are some general guidelines for working with suspended ceilings:

- Contact building maintenance personnel to find out how the ceiling is constructed and how to obtain spare panels in case some of the existing panels get damaged. They should also have the special tools you will need to get access to some types of ceilings. One type of concealed grid system, for example, has a special hook that is used to reach under the panel and release the cross member. As previously discussed, pan ceilings require special procedures for removing and installing pans.
- Do not force ceiling panels. Some panels are clipped to the gridwork. If that is the case, you will need to find the panel that is not clipped and start there. A special tool may be needed to release the clips.
- As discussed earlier, some ceilings have hinged panels that can be raised or lowered to provide access to the area above.
- Keep your hands clean to avoid staining the ceiling panels. If a panel gets dirty, try cleaning it with a damp sponge or an art gum eraser. Vinyl-faced fiberglass and mylar-faced ceilings can be cleaned with mild detergents or germicidal cleaners.
- Pan ceiling panels require special handling. Wear gloves or rub cornstarch on your hands to prevent the transfer of fingerprints to the panels.

4.5.0 FIRESTOPPING

Firestopping means cutting off the air supply so that fire and smoke cannot readily move from one location to another. You will hear the term *firestop* used in two different ways.

In frame construction, a firestop is a piece of wood or fire-resistant material inserted into an opening such as the space between studs. This type of firestop acts as a barrier to block airflow that would allow the space to act as a chimney, carrying fire rapidly to the upper floors. It does not put out the fire, but it slows the fire's progress.

In commercial construction (and some residential uses), firestopping material is used to close wall penetrations such as those created to run conduit, piping, and air conditioning ducts. If such openings are not sealed, fire will travel through the openings in its search for oxygen.

In order to meet the fire rating standards established by the building and fire codes, the openings must be sealed. The firestopping methods used for this purpose are classified as mechanical and nonmechanical.

Mechanical firestops are devices such as the one shown in *Figure 87* that mechanically seal the opening. Nonmechanical firestops are fire-resistant materials such as caulks and putties that are used to fill the space around the conduit or piping.

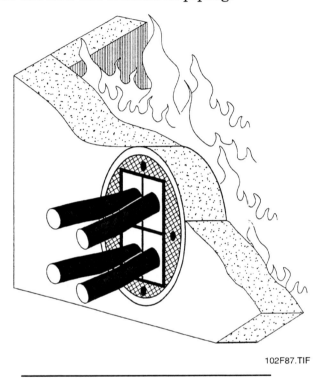

Figure 87. Mechanical Firestop Device

4.5.1 Nonmechanical Firestopping Materials

As an EST, you may be required to install firestopping materials when working with fire-rated walls and floors. Holes or gaps affect the fire rating of a floor or wall. Properly filling these penetrations with firestopping materials maintains the rating. Firestopping materials are typically applied around all types of piping, electrical conduit, ductwork, electrical and communications cables (*Figure 88*), and similar devices that run through openings in floor slabs, walls, and other fire-rated building partitions and assemblies.

Firestopping materials are classified as *intumescent* or *endothermic*. Both are formulated to help control the spread of fire before, during, and after exposure to open flames. When subjected to the extreme heat of a fire, intumescent materials expand (typically up to three times their original size) to form a strong insulating material that seals the opening for about three to four hours. Should the insulation on the cables, pipes, etc., passing through the penetration become consumed by the fire, the expansion of the firestopping material also acts to fill the void in the floor or wall in order to help stop the spread of smoke and toxic products of combustion.

CONSTRUCTION MATERIALS AND METHODS — TRAINEE TASK MODULE 33102

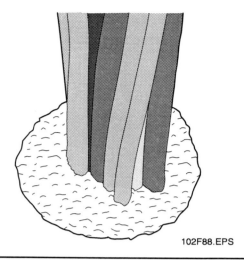

Figure 88. Moldable Putty Fire Barrier Installed Around Electrical Cables

Endothermic materials block heat by chemically releasing bound water, which causes them to absorb heat. Firestopping materials are formulated in such a way that when activated, they are free of corrosive gases, reducing the risks to building occupants and sensitive equipment.

Firestopping materials are made in a variety of forms including composite sheeting, caulks, silicone sealants, foams, moldable putty, wrap strips, and spray coatings. They come in both one-part and two-part formulations. The installation of these materials must always be done in accordance with the applicable building codes and the manufacturer's instructions for the product being used. Depending on the product, firestopping materials can be applied by spray, conventional caulking guns, pneumatic pumping equipment, or a putty knife.

Any firestopping materials used must meet the criteria of standard ASTM-E-814, *Fire Test*, as tested under positive pressure. They must also have an hourly rating that is equal to or greater than the hourly rating of the floor or wall being penetrated. Based on ASTM-E-814/UL-1479 tests, one of four ratings (measured in time) may be applied to firestopping materials and systems. These are explained below:

- *F rating* – A firestopping system meets the requirement of an F rating if it remains in the opening during the fire test for the rated period without permitting the passage of flame through the opening, or the occurrence of flaming on any element of the unexposed side of the assembly.

- *FT rating* – A firestopping system meets the requirement of an FT rating if it remains in the opening during the fire test within the limitations as specified for an F rating. In addition, the transmission of heat through the firestopping system during the rated period shall not have been such as to raise the temperature of any thermocouple on the unexposed surface of the firestopping system by more than 347.8°F (181°C) above its initial temperature.

- *FH rating* – A firestopping system meets the requirement of an FH rating if it remains in the opening during the fire test as well as a hose stream test within the limitations for an F rating. In addition, during the hose stream test, the firestopping system shall not develop any opening that would permit a projection of water from the stream beyond the unexposed side.
- *FTH rating* – A firestopping system shall be considered as meeting the requirements for an FTH rating if it remains in the opening during the fire test and hose stream test within the limitations as described for FT and FH ratings.

5.0.0 TOOLS USED FOR RUNNING CABLE

When you install low-voltage cable, it may be necessary to drill or cut through wood, steel, concrete, masonry, gypsum wallboard, and other building materials. It is important that you be able to select the correct tool and the correct bit or blade for each job.

The intent of this section is to familiarize you with each of the tools and the related safety rules that apply when using them. Before you will be allowed to operate a specific power tool, you must be able to show that you know the safety rules associated with it. As your training progresses, you will learn how to operate each of the tools while under the supervision of your instructor and/or supervisor. Specific operating procedures and safety rules for using a tool are provided in the Operator's/User's Manual supplied by the manufacturer with each tool. Before operating any power tool for the first time, you should always read this manual to familiarize yourself with the tool. If the manual is missing, you or your supervisor should contact the manufacturer for a replacement.

5.1.0 GUIDELINES FOR USING ALL POWER TOOLS

Before proceeding with our descriptions of power tools, it is important to review the general safety rules that apply when using all power tools, regardless of type. It is also important to overview the general guidelines that should be followed in order to properly care for power tools.

Note: Power tools may be operated by electricity (AC or DC current), air, combustion engines, or explosive powder.

5.1.1 Safety Rules Pertaining To All Power Tools

The rules for the safe use of all power tools are:

- Do not attempt to operate any power tool before being checked out by the instructor on that particular tool.
- Always wear eye protection, hearing protection, and a hard hat when operating all power tools.
- Wear face protection when necessary.

- Wear proper respirator equipment when necessary.
- Wear the appropriate clothing for the job being done. Never wear loose clothing that could become caught in the moving tool. Roll up long sleeves, tuck in shirttails, and tie back long hair.
- Do not distract others or let anyone distract you while operating a power tool.
- Do not engage in horseplay in the shop.
- Do not run or throw objects in the shop.
- Consider the safety of others, as well as yourself.
- Do not leave a power tool running while it is unattended.
- Assume a safe and comfortable position before using a power tool.
- Be sure that the power tool is properly grounded before using it.
- Be sure that the power tool is disconnected before performing maintenance or changing accessories.
- Do not use a dull or broken tool or accessory.
- Use a power tool only for its intended purpose.
- Keep your feet, fingers, and hair away from the blade and/or other moving parts of a power tool.
- Do not use a power tool with the guards or safety devices removed.
- Do not operate a power tool if your hands or feet are wet.
- Keep the work area clean at all times.
- Become familiar with the correct operation and adjustments of a power tool before attempting to use it.
- Keep a firm grip on the power tool at all times.
- Use electric extension cords of sufficient size to service the particular power tool you are using.
- Report unsafe conditions to your instructor or supervisor.

5.1.2 Guidelines Pertaining To The Care Of All Power Tools

Guidelines for the proper care of power tools are:

- Keep all tools clean and in good working order.
- Keep all machine surfaces clean and waxed.
- Follow the manufacturer's maintenance procedures.
- Protect cutting edges.
- Keep all tool accessories (such as blades and bits) sharp.
- Always use the appropriate blade for the arbor size.
- Report any unusual noises, sounds, or vibrations to your instructor or supervisor.
- Regularly inspect all tools and accessories.

5.2.0 DRILLING TOOLS

A wide variety of tools and drill bits are used to penetrate wood, metal, and masonry. Selection is based on the type of material, as well its thickness and hardness. *Figure 89* shows examples of how drills are used by cable installers.

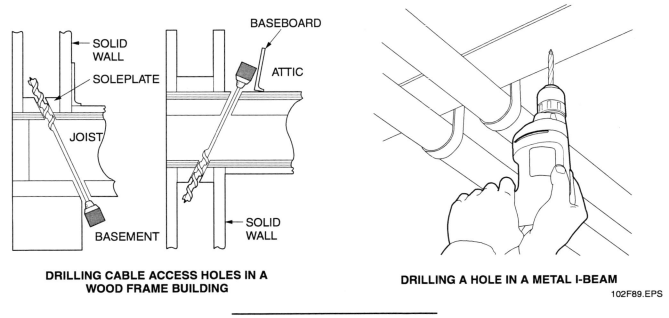

Figure 89. Using An Electric Drill

5.2.1 Portable Drills And Screwguns

Portable drills (*Figure 90*) are made in a great number of types and sizes. Light-duty drills generally have a pistol grip. Heavy-duty drills may have a spade-shaped or D-shaped handle and a side handle to provide a secure grip for better control on large drilling jobs. Drill sizes are based on the diameter of the largest drill shank that will fit into the chuck of the drill; ¼", ⅜", and ½" are common. Most drills have variable speed and reversible controls. Twist drill bits are used in electric drills to make holes in wood and metal. For boring larger holes in wood, hole saws and spade and power bore bits are used. Drills can be used for many operations, including:

- Boring and drilling
- Cutting holes with hole saws
- Mixing materials
- Driving screws

Figure 91 shows examples of common drill bits.

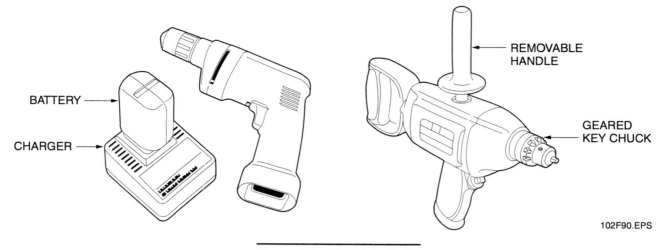

Figure 90. Portable Drills

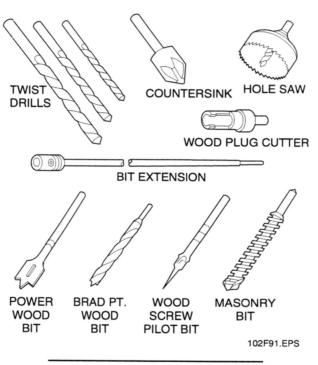

Figure 91. Examples Of Drill Bits

Power-driven screwdrivers (*Figure 92*), also called *screwguns*, are used for the rapid and efficient driving or removal of all types of screws, including wood, machine, and thread-cutting screws. Heavy-duty types can be used for driving and removing lag bolts, flooring screws, etc., Electric screwdrivers normally have an adjustable depth control to prevent over-driving the screws. Many have a clutch mechanism that disengages when the screw has been driven to a preset depth. Some power screwdrivers are designed to perform specific fastening jobs, such as fastening drywall to walls and ceilings.

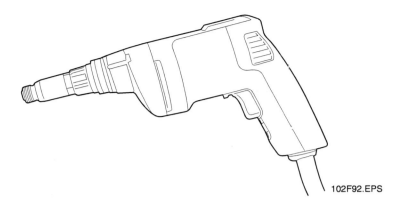

Figure 92. Power-Driven Screwdriver

5.2.2 Hammer Drills And Rotary Hammers

Hammer drills (*Figure 93*) and rotary hammers are special types of power drills. When equipped with a carbide-tipped percussion bit, they are used mainly to drill holes in concrete and other masonry materials. With different bits or cutters (*Figure 94*), they can also be used for:

- Drilling holes in wood and other materials
- Setting anchors
- Performing light chipping work
- Mixing materials

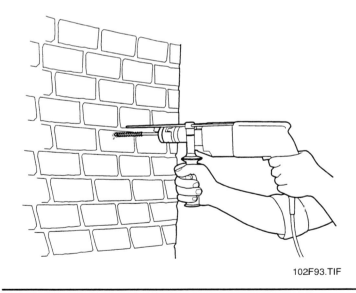

Figure 93. Hammer Drill Used To Drill Holes In Masonry

CONSTRUCTION MATERIALS AND METHODS — TRAINEE TASK MODULE 33102

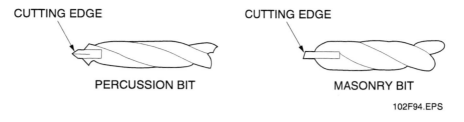

Figure 94. Hammer Drill Bits

The hammer drill is a lighter-duty tool used for drilling smaller holes; the rotary hammer is a heavier-duty tool used to drill larger holes. Both the hammer drill and rotary hammer operate with a dual action. They rotate and hammer at the same time, enabling them to drill holes much faster than can be done with a standard drill equipped with a masonry bit. Most models are reversible and can be easily switched to a standard rotary-action drill. They also have a depth gauge that can be set to control the depth of the hole being drilled. A cold chisel bit can be used with some models of rotary hammers for chipping and edging concrete. Note that chipping hammers, similar to rotary hammers, are also made specifically for use in chipping masonry and concrete.

A heavy-duty rotary hammer, such as the one shown in *Figure 95*, can accommodate a 6" core bit, as well as a variety of chisels used to penetrate or break up concrete and stone.

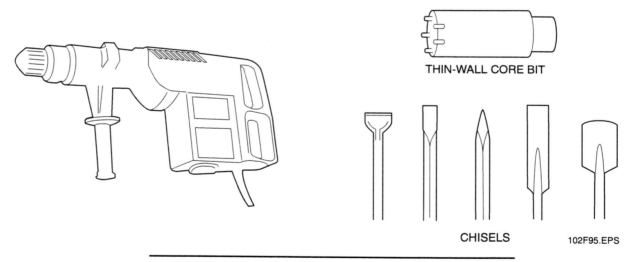

Figure 95. Heavy-Duty Rotary Hammer And Accessories

Rules for the safe use of portable drills and hammer drills are:

- Hold the tool firmly.
- Always remove the chuck key before starting the drill.
- Be sure the drill or tool is secure in the chuck before starting the drill.
- Never attempt to stop the drill by taking hold of the chuck.

- Do not force the drill into the material.
- Be sure the material is properly secured before drilling.
- Never point the drill at anyone.

Rules for the safe use of an electric screwdriver (screwgun) are:

- Hold the tool firmly.
- Use the correct type and size bit for the screws being used.
- Set the gun for the proper depth.
- Never place the screw point against any part of your body.
- Never hold your hand behind the material into which you are driving the screw.

Rules for the safe use of rotary hammers and chipping hammers are:

- Do not force or overload the hammer.
- Use the correct tool for the job being done.
- Keep your hands and feet clear of the tool.
- Never point the hammer at anyone while it is running.
- Do not use the point or chisel for prying.
- Be sure the tools are properly secured in the hammer.

5.2.3 Special Drilling Equipment

Some special drilling tools are used by electronic systems technicians to assist in running cable. *Figure 96* shows a system used to run cable after the finish work has been done. This system combines a flexible steel bit with a spring steel shaft. Once the bit is through to the other side, a hole in the bit is used to connect the cable, which can then be fished through the hole by reversing the drill through the hole in the bottom plate, then pulling it through the opening in the wall.

Other drill bits especially designed for wiring work contain a hole that can be used to fish a cable back through the drilled opening.

In some tight spots, there is not enough room for a standard drill. For that reason, tool manufacturers have developed the 90° drill (*Figure 97*). Another drill designed for working in close quarters has the head positioned at a 55° angle.

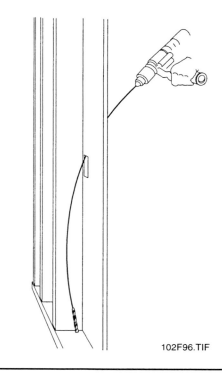

Figure 96. Flexible Steel Drilling System

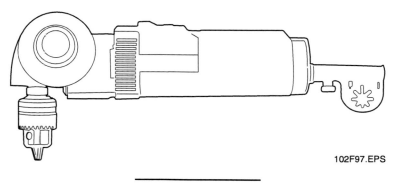

Figure 97. 90° Drill

Because of the wide variety of applications and tools used to drill or cut through various types of materials, two tables are provided for reference. *Table 4* relates the type of material to the tools that can be used to bore or cut through it. *Table 5* relates various types and sizes of bits and blades to the tools with which they are used.

TYPE & TYPICAL SIZES	MATERIAL										
	Wood	Steel	Metal Studs	Vinyl Siding	Alum. Siding	Drywall	Plaster	Ceiling Tile	Formica & Laminates	Asphalt Roofing	Metal Siding
Twist Bits, Standard Shank from 1/16" – 1/2"	•	•	•	•	•	•		•	•	•	•
Twist Bits, 1/2" – 1"	•		•	•	•	•		•	•	•	•
Brad Point Wood Bit (same as twist bit size)	•			•		•		•	•	•	•
Knockout Punch, Various Sizes & Shapes		•	•								
Metal Stud Punch, 7/8" – 1-3/8"			•								
Stepper Bits, 1/8" – 1-3/8"		•	•		•						•
Spade Bits, 1/4" – 1-1/2"	•					•		•	•		
Easy Bore Bits, 1/2" – 4-5/8"	•					•				•	
Bell Hanger Bits, 1/4" – 3/8"	•			•	•	•		•	•	•	•
Hole Saws, 9/16" – 6"	•			•		•		•	•	•	
Bimetal Hole Saws, 9/16" – 6"	•	•	•	•		•		•	•	•	•
Forstner Bits (Precision Boring), 1/4" – 2-1/8"	•			•		•		•	•		
Carbide Easy Bore Bits, 3/4" – 2-1/2"	•			•	•	•	•	•		•	
Auger Type, 9/16" – 1-1/2"	•					•			•		
Around the Corner Bits (Cuts Curved Holes), 3/4" – 1-1/8"	•			•		•		•			
Type "C" Combination Flex Bit, 4' – 6' Long 1/4" – 1" Dia.	•		•	•		•		•		•	
Type "B" Combination Flex Bit, 4' – 6' Long 3/8" – 9/16"	•					•					
Auger Flex Bit, 4' – 6' Long 3/8" – 1" Dia.	•										
Wallboard Saw						•	•				
Jab Saw with Reciprocating Saw Blade	•			•		•	•	•			
Utility Knife						•					
Reciprocating Saw Blades 3" – 12"	•	•	•	•	•	•	•	•	•	•	•

MASONRY	Concrete	Brick	Stone	Ceramic & Mosaic Floor & Wall Tile	Plaster with or without Wire Mesh	Corian & Granite	Block
Masonry Twist Bit from 3/16" to 1/2"	•	•		•	•	•	•
Percussion Carbide Bit from 3/16" to 1/2"	•	•	•	•	•	•	•
Type "M" Flex Bits from 1/4" to 3/4"	•			•	•		•
Hammer Core Bits from 2" to 6"	•	•	•				•

Table 4. Drill Bits And Blades Used For Various Materials (1 of 2)

TYPE & COMMON SIZE	SUGGESTED TOOL FOR CUT OR HOLE							
	Standard 3/8" & 1/2" Drill	Cordless 3/8" & 1/2" Drill	Hammer Drill	Close Quarter Drill	1/2" Hammer Drill	1/2" Hole Hawg Drill	Hand Auger	1/2" 90° Angle Drill
Twist Bits, Standard Shank from 1/16" – 1/2"	●	●	●	●	●			
Twist Bits, 1/2" – 1"	●	●	●		●			
Brad Point Wood Bit (same as twist bit size)	●	●	●	●	●			
Stepper Bits, 1/8" – 1-3/8"	●	●	●	●	●			
Spade Bits, 1/4" – 1-1/2"	●	●	●	●	●			
Easy Bore Bits, 1/2" – 4-5/8"	1/2" recommended	●	●		●			●
Bell Hanger Bits, 1/4" – 3/8"	●	●	●	●	●		●	
Hole Saws, 9/16" – 6"	1/2" required	●	●		●	●		●
Bimetal Hole Saws, 9/16" – 6"	1/2" required	●	●		●			●
Forstner Bits (Precision Boring), 1/4" – 2-1/8"	●	●	●		●			●
Carbide Easy Bore Bits, 3/4" – 2-1/2"						●		●
Auger Type, 9/16" – 1-1/2"	●				●	●		●
Around the Corner Bits (Cuts Curved Holes), 3/4" – 1-1/8"	●	●			●	●		●
Type "C" Combination Flex Bit, 4' – 6' Long 1/4" – 1" Dia.	1/2" HD					●		●
Type "B" Combination Flex Bit, 4' – 6' Long 3/8" – 9/16"	1/2" HD	●			●	●		●
Auger Flex Bit, 4' – 6' Long 3/8" – 1" Dia.	1/2" HD				●	●		●

MASONRY

	Standard 3/8" & 1/2" Drill	Cordless Drill	Hammer Drill	Rotary Hammer	Demolition Hammer
Masonry Twist Bit from 3/16" to 1/2"	●	●	●		
Percussion Carbide Bit from 3/16" to 1/2"			●		
Type "M" Flex Bits from 1/4" to 3/4"			●		
Hammer Core Bits from 2" to 6"				●	●

Table 4. Drill Bits And Blades Used With Various Tools (2 of 2)

5.3.0 CUTTING TOOLS

The reciprocating saw and the jig saw are the most commonly used tools for cutting openings in wood, metal, wallboard, and other building materials. Other specialty tools are also used to make openings in gypsum wallboard and metal studs.

5.3.1 Reciprocating Saws

Reciprocating saws (*Figures 98* and *99*) are heavy-duty saws with a horizontal back and forth movement of the blade. They are an all-purpose saw that can be used for cutting or notching wood and metal.

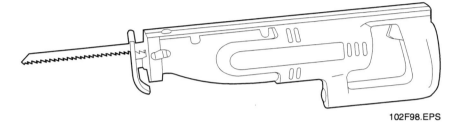

Figure 98. Reciprocating Saw

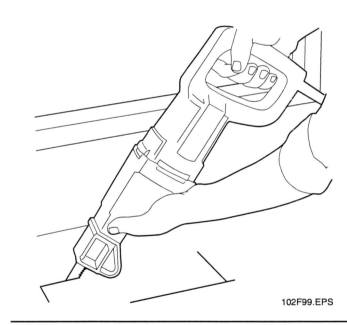

Figure 99. Reciprocating Saw Cutting Through A Floor

Reciprocating saw models with variable speeds ranging from 0 to 2,400 strokes per minute (SPM) are common. Speed selection is made with a variable control at the trigger on/off switch. Greater horsepower and slower speeds are generally needed when cutting through metals or when cutting along a curved or angled line. The typical length of the horizontal sawing stroke is 1⅛". A multipositioned foot at the front of the saw can be put in three different positions for use in flush cutting, **ripping**, and **crosscutting**.

A wide variety of blades are made for use with reciprocating saws. Each type of blade is designed to make an optimum cut in a different kind of material. The blade length determines the thickness of the material that can be cut. Use the shortest blade that will do the job. Reciprocating saw blade lengths range from 3½" to 12". The number of teeth range from 3½ to 32 per inch. Blades with 3½ to 6 teeth per inch are generally used for sawing wood, while blades with 6 to 10 teeth per inch are used for general-purpose sawing. Blades with 10 to 18 teeth per inch are used for cutting metal. Always use the blade recommended by the blade manufacturer for the type of material being cut.

5.3.2 Jig Saws

Jig saws (*Figure 100*), also commonly called *saber saws* or *bayonet saws*, are lighter-duty saws than reciprocating saws.

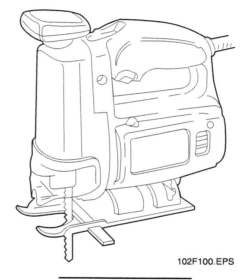

Figure 100. Jig Saw

With the proper blade, jig saws can cut wood, metal, plastic, and other materials. Variable-speed models with speeds ranging from 0 to 3,200 SPM are common. The typical length of the vertical sawing stroke is 1". Jig saws have a baseplate or shoe. The broad surface of the shoe helps keep the blade aligned. It also helps prevent the work from vibrating and allows the teeth to bite into the material. The baseplate can be tilted for making bevel cuts. Many jig saws have a large scrolling knob which can be unlocked from a stationary position, then used to rotate the blade while sawing the material. This makes it easier to cut tight curves, corners, and patterns.

Wood-cutting, metal-cutting, and special-purpose blades are available. Use the shortest blade that will do the job. Blades for cutting wood have as few as 6 teeth per inch for fast, coarse cutting, and as many as 14 teeth per inch for fine work. Tapered-ground blades are made that produce splinter-free cuts in plywood. Blades made for cutting both metals and plastic laminates typically have between 12 and 32 teeth per inch. Always use the blade recommended by the blade manufacturer for the type of material being cut.

5.3.3 Power Cutout Tool

This power tool (*Figure 101*), also known as a *wall router*, is used to cut openings in gypsum drywall. It uses a special cutting drill bit to penetrate and cut through the wallboard. Other bits are available to cut wall tile and cement board, which is commonly used as both a tile backer in showers and in subflooring.

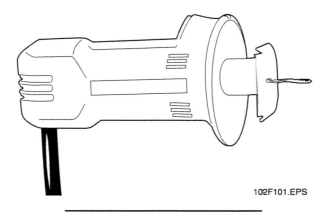

Figure 101. Power Cutout Tool

5.3.4 Light Box Cutter

This is a special tool used to cut exact openings for various types of electrical boxes (*Figure 102*). It will not damage the surrounding gypsum core or paper facings.

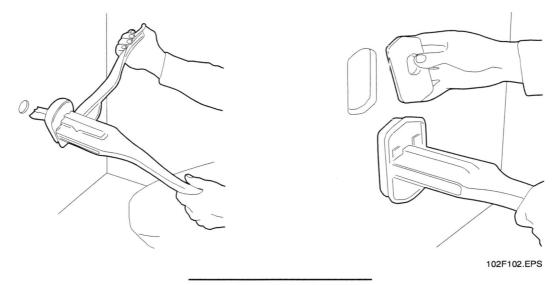

Figure 102. Light Box Cutter

5.3.5 Metal Stud Punches

As stated earlier in the module, metal studs normally come from the factory with holes pre-punched for routing cabling and other services. If it is necessary to punch holes on site, a metal stud punch (*Figure 103*) is used. In such cases, a special bushing is inserted into the hole to protect the insulation from sharp edges.

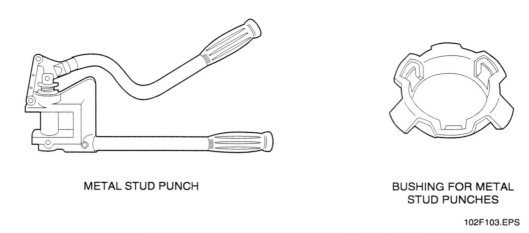

Figure 103. Metal Stud Punch And Bushing

Pneumatic nailers and staplers (*Figure 104*) are fastening tools powered by compressed air which is fed to the tool through an air hose connected to an air compressor. These tools, known as *guns*, are widely used for quick, efficient fastening of framing, subflooring, sheathing, etc. Nailers and staplers are made in a variety of sizes to serve different purposes. Under some conditions, staples have some advantages over nails. For example, staples do not split wood as easily as a nail when driven near the end of the board. Staples are also excellent for fastening sheathing, shingles, building paper, and other materials because their two-legged design covers more surface area. However, both fasteners are sometimes used to accomplish the same fastening jobs.

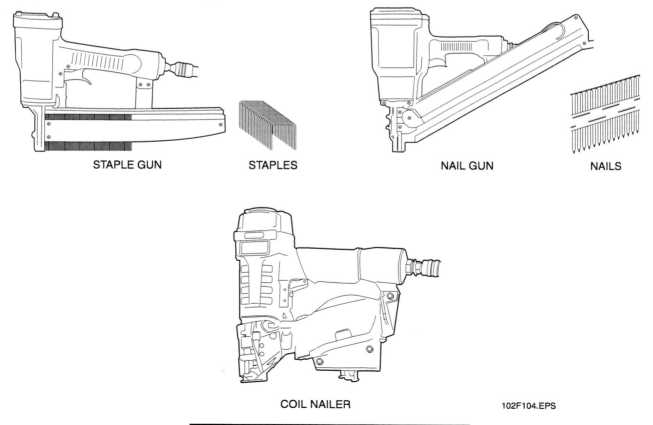

Figure 104. Pneumatic Stapler And Nailers

For some models of fasteners, the nails or staples come in strips and are loaded into a magazine, which typically holds 100 or more fasteners. Some tools have an angled magazine, which makes it easier for the tool to fit into tight places. Coil-fed models typically use a coil of 300 nails loaded into a circular magazine. Lightweight nailing guns can handle tiny finishing nails. Larger framing nailers can shoot smooth-shank nails up to 3¼" in length.

Rules for the safe use of pneumatic fasteners are:

- Be sure all safety devices are in good shape and are functioning properly.
- Use the pressure specified by the manufacturer.
- Always assume that the fasteners are loaded.
- Never point a pneumatic fastener at yourself or anyone else.
- Be sure the fastener is disconnected from the power source before making adjustments or repairs.
- Use caution when attaching the fastener to the air supply because the fastener may discharge.
- Never leave a pneumatic fastener unattended while it is still connected to the power source.
- Use nailers and staplers only for the type of work for which they were intended.
- Use only nails and staples designed for the fastener being used.
- Never use fasteners on soft or thin materials that nails may completely penetrate.

Nailers and staplers are also made in cordless models (*Figure 105*). These tools use a tiny internal combustion engine powered by a disposable fuel cell and a rechargeable battery. The action of the piston drives the fastener. A cordless stapler can drive about 2,500 staples with one fuel cell. A cordless framing nailer can drive about 1,200 nails on one fuel cell. The battery on a cordless tool must be periodically recharged. It pays to have a spare battery to use while one is being charged. The rules for the safe operation of a cordless nailer or stapler are basically the same as those described above for pneumatic nailers and staplers.

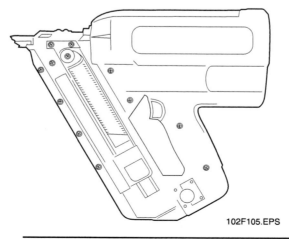

Figure 105. Gas-Driven Cordless Nail Gun

5.4.0 POWDER-ACTUATED FASTENING TOOLS

A powder-actuated fastening tool (*Figure 106*) is a low-velocity fastening system powered by gunpowder cartridges, commonly called *boosters*. Powder-actuated tools are used to drive specially-designed fasteners into masonry and steel.

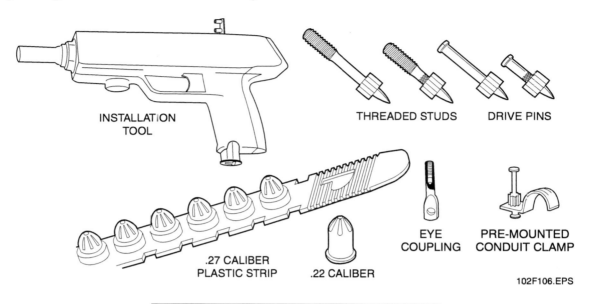

Figure 106. Powder-Actuated Fastening Tool

Manufacturers use color-coding schemes to identify the strength of a powder load charge. It is extremely important to select the right charge for the job, so learn the color-coding system that applies to the tool you are using. *Table 6* shows an example of a color-coding system.

Power Level*	Color
1	Gray
2	Brown
3	Green
4	Yellow
5	Red
6	Purple
*From the least powerful (1) to the most powerful (6).	

Table 6. Powder Charge Color-Coding System

> ***WARNING!*** OSHA requires that all operators of powder-actuated tools be qualified and certified by the manufacturer of the tool. Certification cards must be carried whenever using the tool. If the gun does not fire, hold it against the work surface for at least 30 seconds. Follow the manufacturer's instructions for removing the cartridge. Do not try to pry it out, as some cartridges are rim-fired and could explode.

Other rules for safely operating a powder-actuated tool are:

- Do not use a powder-actuated tool unless you are certified.
- Follow all safety precautions in the manufacturer's instruction manual.
- Always wear safety goggles and a hard hat when operating a powder-actuated tool.
- Use the proper size pin for the job you are performing.
- When loading the driver, put the pin in before the charge.
- Use the correct booster (powder load) according to the manufacturer's instructions.
- Never hold the end of the barrel against any part of your body or cock the tool against your hand.
- Never hold your hand behind the material you are fastening.
- Do not shoot close to the edge of concrete.
- Never attempt to pry the booster out of the magazine with a sharp instrument.
- Always wear ear protection.

5.5.0 STUD FINDERS

When you are running cable in a finished building, it is essential to find the studs behind the wallboard or other finish panels in order to mount connection boxes. There are two types of stud finders. One is magnetic and reacts to the nails or screws used to secure the panels to the studs. Another type of stud finder senses density. As the device is passed over a panel, it will detect the existence of a mass in an otherwise hollow area. Some very sophisticated types of stud finders are capable of differentiating between steel studs and copper piping and can also locate electrical wires. They can also be used to locate the steel rebars and other material embedded in concrete up to 6" thick.

The cost of stud finders ranges from a few dollars for the most basic type to more than a hundred dollars for the multifunction type.

5.6.0 FISH TAPES

Special devices known as *fish tapes* are used to fish wire and cable through walls, conduit, and other openings. *Figure 107* shows a basic fish tape and its use.

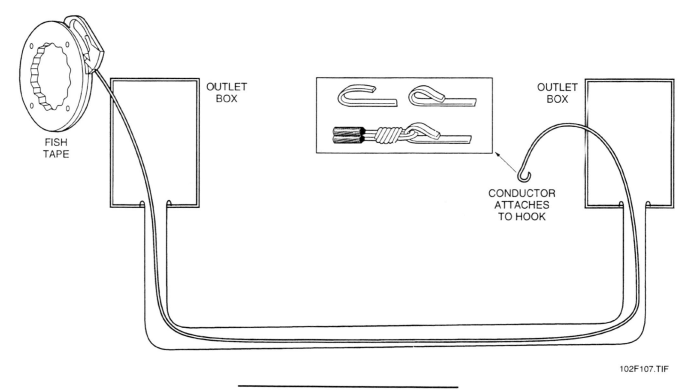

Figure 107. Fish Tape Installation

The tape is fed through an opening, then the wire is connected to the hook on the end of the tape and carefully pulled back through the opening.

There are several types of fishing tools, and some of them are designed to pull heavy bundles of cables. We will discuss various fishing tools in more detail in the conductor installation module.

6.0.0 PROJECT SCHEDULES

A construction project requires a lot of planning and scheduling because different trades, equipment, and materials are needed at different times in the process. Electrical and communications cabling is normally installed when the building has been dried-in (i.e., the exterior siding and roofing are applied so the building remains dry, but the framing is exposed in the interior of the building).

Project planning and scheduling will be covered in more detail later in your training. For now, *Figures 108* and *109* provide an overview of where each trade fits into the construction process for residential and commercial projects, respectively.

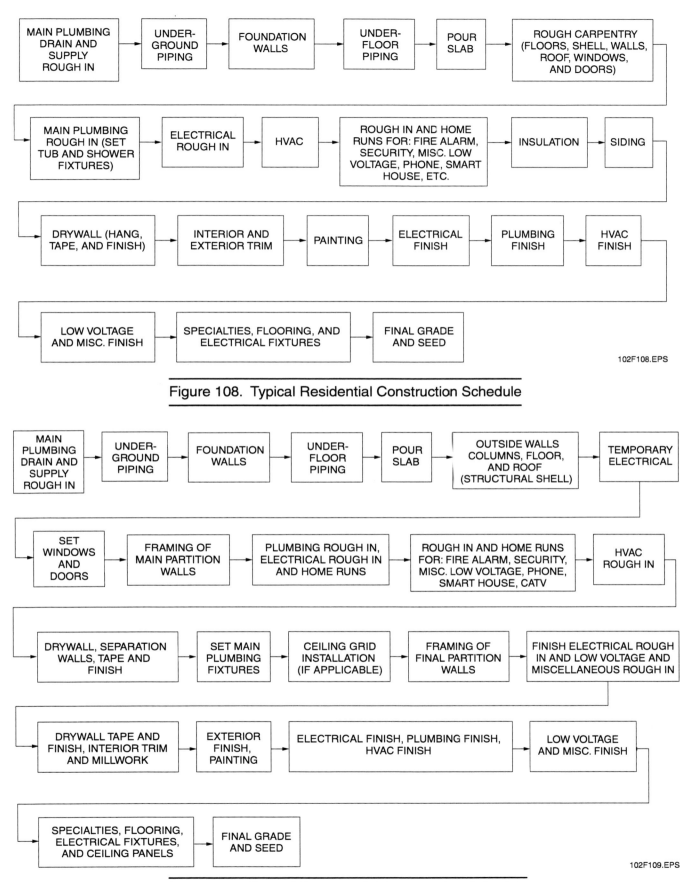

Figure 108. Typical Residential Construction Schedule

Figure 109. Typical Commercial Construction Schedule

SUMMARY

When you are called upon to run cable or install equipment in any kind of environment, whether it is residential or commercial, new construction or modification, it will be important for you to know how the building is put together. With that knowledge, you can anticipate potential problems. You will also be able to select the right drilling or cutting tools for the job.

As you now realize, there are many different types of construction. You also know that you will not get very far by guessing what might be behind a wall. It could be anything. Armed with a basic knowledge of construction principles, however, you will be able to face a new project with confidence.

References

For advanced study of topics covered in this Task Module, the following books are suggested:

Carpentry, Latest Edition, American Technical Publishers, Homewood, IL.

Carpentry, Latest Edition, Delmar Publishers, Albany, NY.

Modern Carpentry, Latest Edition, The Goodheart-Willcox Co., Inc., Tinley Park, IL.

ACKNOWLEDGMENTS

Figures 6 and 52 courtesy of Economy Form Corporation (EFCO)
Figure 47 courtesy of Bethlehem Steel
Figures 51 and 87 courtesy of BICSI
Figures 60, 61, and 62 courtesy of the Florida Chapter of the Associated Builders and Contractors
Figures 73, 76, and 77 courtesy of Armstrong Company
Figures 82 and 83 courtesy of Integrated Ceilings, Inc.
Figure 84 courtesy of Chicago Metallic
Figure 85 courtesy of Alcan Building Products
Figure 86 courtesy of Ceilings Plus

REVIEW/PRACTICE QUESTIONS

1. Which of the following materials must be handled with special care because it is treated with hazardous chemicals?
   a. Gypsum drywall
   b. Pressure-treated lumber
   c. Brick
   d. Plywood

2. Gypsum wallboard with foil backing is used as a _____.
   a. backer for tile
   b. decorator panel
   c. vapor barrier
   d. liner in elevator shafts

3. The typical concrete block (CMU) used in loadbearing construction is about _____ deep.
   a. 2"
   b. 4"
   c. 6"
   d. 8"

4. If you needed to drill an opening in an exterior wall built of brick veneer, wood framing, and gypsum wallboard, you should have a drill bit that is at least _____ long.
   a. 4"
   b. 6"
   c. 10"
   d. 16"

5. The main horizontal framing member of a floor is known as a _____.
   a. girder
   b. joist
   c. lally column
   d. rafter

6. If you are required to drill through a 2" × 12" floor joist, the diameter of the hole may not exceed _____.
   a. 1"
   b. 2"
   c. 3"
   d. 4"

CONSTRUCTION MATERIALS AND METHODS — TRAINEE TASK MODULE 33102

7. The purpose of a strongback is to _____.
   a. lift heavy objects
   b. hold ceiling joists in line
   c. support ceiling joists
   d. brace walls

8. In plank-and-beam construction, framing members are typically installed _____ on center.
   a. 12"
   b. 16"
   c. 24"
   d. 48"

9. Floors in large office buildings are typically constructed of _____.
   a. concrete
   b. metal joists and wood sheathing
   c. steel beams
   d. galvanized aluminum

10. Concrete tilt-up panels are typically _____ thick.
    a. 1" to 2"
    b. 2" to 3"
    c. 3" to 4"
    d. 5" to 8"

11. The wall shown in the illustration above is most likely located between _____.
    a. a doctor's office and the waiting room
    b. the garage and kitchen of a home
    c. a manufacturing area and an office
    d. a bedroom and a bathroom

12. The metal studs used in the construction of a nonloadbearing partition are usually _____ gauge.
    a. 25
    b. 20
    c. 18
    d. 16

13. When drywall ceilings are used in large commercial construction projects, the drywall sheets are typically attached _____.
    a. directly to the underside of the concrete floor above using powder-actuated fasteners
    b. to the wood joists
    c. to suspended furring channels
    d. to steel trusses

14. When drilling a 3" diameter hole in gypsum drywall, you would probably use a _____.
    a. twist bit
    b. hole saw
    c. percussion bit
    d. core bit

15. Which of the following tools would you use to drill a hole through a concrete wall?
    a. Rotary hammer with a carbide-tipped percussion bit
    b. Hammering drill with a hole saw
    c. Reciprocating saw with a concrete blade
    d. Rotary hammer with an auger bit

ANSWERS TO REVIEW/PRACTICE QUESTIONS

Answer	Section Reference
1. b	2.1.1
2. c	2.5.1
3. d	2.6.2
4. c	2.6.3/Figure 8
5. b	3.1.2
6. d	3.1.5
7. c	3.3.0/Figure 33
8. d	3.5.0/Figure 42
9. a	4.0.0
10. d	4.0.0
11. c	4.3.0/Figure 54
12. a	4.3.1
13. c	4.4.6
14. b	5.2.1
15. a	5.2.2

NCCER CRAFT TRAINING USER UPDATES

The NCCER makes every effort to keep these manuals up-to-date and free of technical errors. We appreciate your help in this process. If you have an idea for improving this manual, or if you find an error, a typographical mistake, or an inaccuracy in the NCCER's Craft Training Manuals, please write us, using this form or a photocopy. Be sure to include the exact module number, page number, a description of the problem, and the correction, if possible. Your input will be brought to the attention of the Technical Review Committee. Thank you for your assistance.

Instructors – If you found that additional materials were necessary in order to teach this module effectively, please let us know so that we may include them in the Equipment/Materials list in the Instructor's Guide.

Write: Curriculum Revision and Development Department
National Center for Construction Education and Research
P.O. Box 141104
Gainesville, FL 32614-1104
Fax: 352-334-0932

Craft _____ Module Name _____

Copyright Date _____ Module Number _____ Page Number(s) _____

Description of Problem
_____
_____
_____
_____

(Optional) Correction of Problem
_____
_____
_____

(Optional) Your Name and Address
_____
_____
_____

notes

Pathways and Spaces
Module 33103

Electronic Systems Technician Trainee Task Module 33103

PATHWAYS AND SPACES

NATIONAL
CENTER FOR
CONSTRUCTION
EDUCATION AND
RESEARCH

OBJECTIVES

Upon completion of this module, the trainee will be able to:

1. Describe various types of cable trays and raceways.
2. Identify and select various types and sizes of raceways and fittings.
3. Identify and select various types and sizes of cable trays.
4. Identify various methods used to install raceways.
5. Demonstrate knowledge of NEC raceway requirements.
6. Describe procedures for installing raceways and boxes on various surfaces.
7. Install and properly level D-rings and mushrooms.
8. Make a conduit-to-box connection.
9. Select cable support hardware for various applications.
10. Install an outlet box in drywall.

Prerequisites

Successful completion of the following Task Modules is recommended before beginning study of this Task Module: Core Curricula; Electronic Systems Technician Level One, Modules 33101 and 33102.

Required Trainee Materials

1. Trainee Task Module
2. Copy of the latest edition of the *National Electrical Code*
3. Appropriate Personal Protective Equipment

Note: The designations "National Electrical Code," "NE Code," and "NEC," where used in this document, refer to the *National Electrical Code®*, which is a registered trademark of the National Fire Protection Association, Quincy, MA. *All National Electrical Code (NEC) references in this module refer to the 1999 edition of the NEC.*

COURSE MAP

This course map shows all of the modules in the first level of the Electronic Systems Technician curricula. The suggested training order begins at the bottom and proceeds up. Skill levels increase as a trainee advances on the course map. The training order may be adjusted by the local Training Program Sponsor.

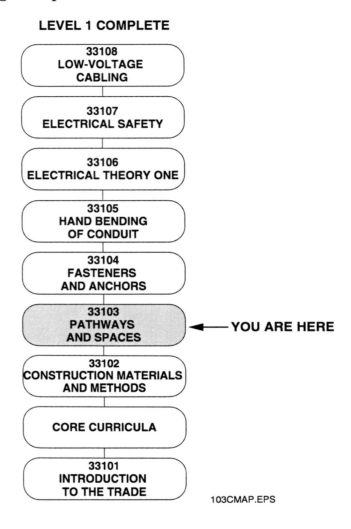

TABLE OF CONTENTS

Section	Topic	Page
	Course Map	2
1.0.0	Introduction	6
2.0.0	Raceways	7
3.0.0	Conduit	7
3.1.0	Types Of Conduit	7
3.1.1	Electrical Nonmetallic Tubing	7
3.1.2	Inner Duct	8
3.1.3	Electrical Metallic Tubing	9
3.1.4	Rigid Metal Conduit	12
3.1.5	Galvanized Rigid Steel Conduit	12
3.1.6	Plastic-Coated GRC	12
3.1.7	Aluminum Conduit	13
3.1.8	Intermediate Metal Conduit	13
3.1.9	Rigid Nonmetallic Conduit	14
3.1.10	Liquidtight Flexible Nonmetallic Conduit	16
3.1.11	Flexible Metal Conduit	17
4.0.0	Metal Conduit Fittings	18
4.1.0	Couplings	19
4.2.0	Conduit Bodies	19
4.2.1	Type C Conduit Bodies	20
4.2.2	Type L Conduit Bodies	20
4.2.3	Type T Conduit Bodies	21
4.2.4	Type X Conduit Bodies	21
4.2.5	Threaded Weatherproof Hubs	22
4.3.0	Insulating Bushings	22
4.3.1	Nongrounding Insulating Bushings	22
4.3.2	Grounding Insulating Bushings	23
4.4.0	Offset Nipples	23
5.0.0	Boxes	23
5.1.0	Metal Boxes	24
5.1.1	Pryouts	24
5.1.2	Knockouts	24
5.2.0	Nonmetallic Boxes	25
5.3.0	Low-Voltage Boxes	25
6.0.0	Bushings And Locknuts	27
7.0.0	Sealing Fittings	27
8.0.0	Cable And Raceway Supports	28
8.1.0	Straps	28
8.2.0	Standoff Supports	29
8.3.0	Electrical Framing Channels	30
8.4.0	Beam Clamps	30
8.5.0	Cable Supports	31
8.5.1	Cable Ties	31
8.5.2	Cable Hangers	31
9.0.0	Wireways	33
9.1.0	Types Of Wireways	34
9.2.0	Wireway Fittings	36
9.2.1	Connectors	36

TABLE OF CONTENTS (Continued)

Section	Topic	Page
9.2.2	End Plates	37
9.2.3	Tees	37
9.2.4	Crosses	37
9.2.5	Elbows	37
9.2.6	Telescopic Fittings	38
9.3.0	Wireway Supports	38
9.3.1	Suspended Hangers	38
9.3.2	Gusset Brackets	39
9.3.3	Standard Hangers	39
9.3.4	Wireway Hangers	40
9.4.0	Other Types Of Raceways	40
9.4.1	Surface Metal And Nonmetallic Raceways	41
9.4.2	Pole Systems	44
9.4.3	Underfloor Systems	44
9.4.4	Cellular Metal Floor Raceways	45
9.4.5	Cellular Concrete Floor Raceways	46
10.0.0	Cable Trays	46
10.1.0	Cable Tray Fittings	48
10.2.0	Cable Tray Supports	48
10.2.1	Direct Rod Suspension	48
10.2.2	Trapeze Mounting And Center Hung Support	48
10.2.3	Wall Mounting	49
10.2.4	Pipe Rack Mounting	49
11.0.0	Storing Raceways	50
12.0.0	Handling Raceways	50
13.0.0	Ducting	51
14.0.0	Underground Systems	51
14.1.0	Duct Materials	53
14.2.0	Plastic Conduit	53
14.3.0	Monolithic Concrete Duct	53
14.4.0	Controlled Environment Vaults	54
14.5.0	Pedestals And Cabinets	54
15.0.0	Making A Conduit-To-Box Connection	55
16.0.0	Various Construction Procedures	56
16.1.0	Masonry And Concrete Flush-Mount Construction	56
16.2.0	Metal Stud Environment	58
16.3.0	Wood-Frame Environment	59
16.4.0	Steel Environment	61
16.5.0	Suspended Ceilings	62
17.0.0	Overview Of Cable Distribution	62
17.1.0	Pathways	64
17.2.0	Spaces	65
	Summary	66
	Review/Practice Questions	67
	Answers To Review/Practice Questions	70

Trade Terms Introduced In This Module

Accessible: Able to be reached, as for service or repair.

Approved: Meeting the requirements of an appropriate regulatory agency.

Bonding wire: A wire used to make a continuous grounding path between equipment and ground.

Cable trays: Rigid structures used to support wiring and cabling.

Conductors: Wires or cables used to carry current.

Conduit: A round raceway, similar to pipe, that houses conductors.

Continuity: The existence of an uninterrupted circuit.

Exposed location: Not permanently closed in by the structure or finish of a building, and able to be installed or removed without damage to the structure.

Fire-rated: Constructed to meet code requirements for fire resistance in hours or fractions of an hour.

Ground: A conducting connection between an electrical circuit or equipment and the earth or another conducting body.

HVAC: Heating, ventilation, and air conditioning.

Kick: A bend in a piece of conduit, usually less than 45°, made to change the direction of the conduit.

Pathway: A device or group of devices that allow the installation of cables and conductors between building spaces. Also, the vertical and horizontal route of the cable.

Plenum: A designated area, open or closed, used for transporting environmental air.

Purlin: A horizontal framing member supporting the rafters of a roof.

Raceways: Enclosed channels designed expressly for holding wires, cables, or busbars, with additional functions as permitted in the NEC.

Splice: The connection of two or more conductors.

Tap: An intermediate point on a main circuit where another wire is connected to supply electrical current to another circuit.

Trough: A long, narrow box used to protect electrical connections from the environment.

Underwriters' Laboratories (UL): An agency that evaluates and approves electrical components and equipment.

Wireways: Steel troughs designed to carry electrical wire and cable.

1.0.0 INTRODUCTION

As you have already learned, the proper selection, installation, and termination of wiring and cables associated with low-voltage systems are important elements of an Electrical Systems Technician's job. In this module, you will learn about the **conduit**, **wireways**, and other types of **raceways** used in managing the hundreds, and often thousands, of cables and wires used to connect telecommunications and security equipment within a building.

Some of the knowledge and skills you gained in the *Construction Materials and Methods* module will be helpful as you learn the proper mounting techniques for raceways and electronic equipment.

Along with the study of this module, the following NEC articles should be referenced:

- *NEC Article 250* – *Grounding*
- *NEC Article 318* – *Cable Trays*
- *NEC Article 331* – *Electrical Nonmetallic Tubing*
- *NEC Article 345* – *Intermediate Metal Conduit*
- *NEC Article 346* – *Rigid Metal Conduit*
- *NEC Article 347* – *Rigid Nonmetallic Conduit*
- *NEC Article 348* – *Electrical Metallic Tubing*
- *NEC Article 350* – *Flexible Metal Conduit*
- *NEC Article 351* – *Liquidtight Flexible Metal Conduit and Flexible Nonmetallic Conduit*
- *NEC Article 362* – *Wireways*
- *NEC Article 725* – *Class 1, Class 2, and Class 3 Remote-Control, Signaling, and Power-Limited Circuits*
- *NEC Article 760* – *Fire Alarm Systems*
- *NEC Article 770* – *Optical Fiber Cables and Raceways*
- ***NEC Articles 800, 810, 820, and 830*** – *Communications Systems*

Note: Mandatory rules in the NEC are characterized by the use of the word *shall*. Explanatory material is in the form of Fine Print Notes (FPNs). When referencing specific sections of the NEC, always check to see if any exceptions apply.

Some of the NEC articles referenced in this module are intended to establish safe practices for high-voltage applications. Although electrical safety may not be an issue in some low-voltage applications, the NEC requirements represent good practices that have been established and refined over many years of use. It is therefore in your interest to follow these practices, regardless of the voltage level. Moreover, there will be situations in which high voltage and low voltage share a raceway. In these situations, the high-voltage requirements must be followed.

2.0.0 RACEWAYS

The term *raceway* refers to a wide range of circular and rectangular enclosed channels used to house electrical wiring. Raceways can be metallic or nonmetallic and come in different shapes. Depending on the particular purpose for which they are intended, raceways include enclosures such as underfloor raceways, flexible metal conduit, wireways, surface metal raceways, and surface nonmetallic raceways.

3.0.0 CONDUIT

Conduit is a raceway with a circular cross section, similar to pipe, that contains wires or cables. Conduit is used to provide protection for **conductors** and route them from one place to another. In addition, conduit makes it easier to replace or add wires to existing structures.

3.1.0 TYPES OF CONDUIT

There are many types of conduit used in the construction industry. The size of conduit to be used is determined by engineering specifications, local codes, and the NEC. Refer to ***NEC Chapter 9, Tables 1 through 8 and Appendix C*** for information on the allowable conduit fill with various conductors. We will examine several common types of conduit in this section.

3.1.1 Electrical Nonmetallic Tubing

Electrical nonmetallic tubing (ENT) is a corrugated raceway that can be bent by hand (*Figure 1*). It is made of polyvinyl chloride (PVC) plastic and comes in diameters ranging from ½" to 2". It is available in coils of 100' to 200' and in reels ranging from 500' to 1,500'. ENT is available in a color-coded form that conforms to the following code:

- *Yellow* – Communications and signal cable
- *Red* – Fire alarm circuits
- *Blue* – Power circuits

ENT is also known in the industry as *smurf tube* because the blue color used for power conductors is reminiscent of the Smurfs® cartoon characters.

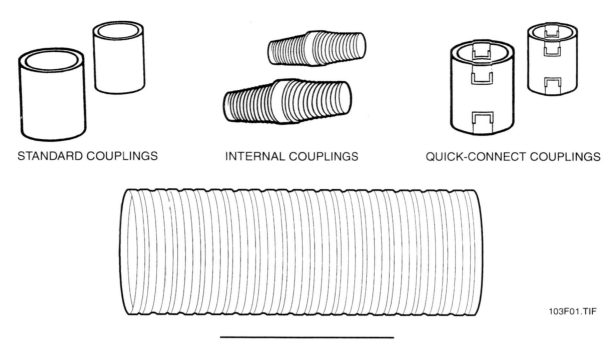

Figure 1. Example Of ENT

NEC Section 331-3 permits ENT to be exposed when used in a building not exceeding three floors. In buildings greater than three floors, the construction must be **fire-rated** and the conduit must be concealed in walls, floors, or ceilings. (The conduit must be fire-rated if it is installed in a **plenum** ceiling of any building.)

Note: Refer to ***NEC Section 331-4*** for restrictions that apply to ENT.

If it is necessary to **splice** sections of ENT, standard and quick-connect couplings are available.

ENT is also available as a pre-wired assembly. The conductors are installed at the factory under conditions that will prevent damage to the insulation. A special tool is required when cutting pre-wired ENT in order to prevent the conductor insulation from being cut.

3.1.2 Inner Duct

Inner duct is a flexible, nonmetallic tubing for use in walls, floors, slabs, and ceilings. It is intended to be pulled through an existing conduit system (*Figure 2*). Inner duct is very similar in construction to ENT. Like ENT, inner duct comes in colors; however, the color scheme is slightly different. The following color code has been established:

- *Orange* – Communications and signal cable (note the difference from ENT)
- *Red* – Fire alarm circuits
- *Blue* – Power circuits

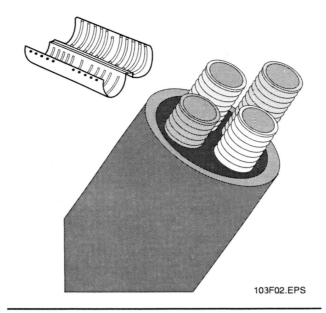

Figure 2. Conduit Containing Four 1" Inner Ducts

Inner duct is available in rolls of 100' and 200', as well as fixed lengths of 10'. Inside diameters range from ½" to 1¼". It normally comes with a pull rope installed, but is available without it.

Some of the advantages of inner duct include its relatively low cost and the fact that its use tends to reduce pulling tension, especially when multiple inner ducts are installed in conduit. A disadvantage is that because it is flexible, it must be carefully secured to the supporting structure to prevent kinking while the cable is being pulled through it.

Various connectors and terminations are available, as are adapters to transition from inner duct to rigid PVC.

Inner duct is often used for running optical fiber cables.

3.1.3 Electrical Metallic Tubing

Electrical metallic tubing (EMT) is the lightest duty tubing available for enclosing and protecting wiring. EMT is widely used for residential, commercial, and industrial wiring systems. It is lightweight, easily bent and/or cut to shape, and is the least costly type of metallic conduit. Because the wall thickness of EMT is less than that of rigid conduit, it is often referred to as *thinwall conduit*. A comparison of inside and outside diameters of EMT to rigid metal conduit and intermediate metal conduit (IMC) is shown in *Figure 3*.

NEC Section 348-4 permits the installation of EMT for either exposed or concealed work where it will not be subject to severe physical damage during installation or after construction. The installation of EMT using waterproof fittings is permitted in wet locations such as outdoors or indoors in dairies, laundries, and canneries.

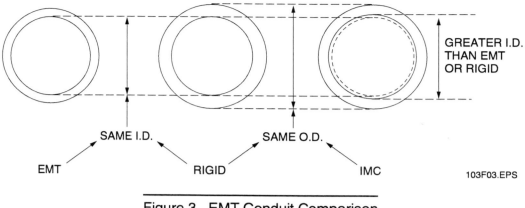

Figure 3. EMT Conduit Comparison

Note: Refer to **NEC Section 348-5** for restrictions that apply to the use of EMT.

EMT shall not be used (1) where, during installation or afterward, it will be subject to severe physical damage; (2) where protected from corrosion solely by enamel; (3) in cinder concrete or cinder fill where subject to permanent moisture, unless protected on all sides by a layer of noncinder concrete at least 2" thick or unless the tubing is at least 18" under the fill; (4) in any hazardous (classified) locations, except as permitted by **NEC Sections 502-4, 503-3, and 504-20**; or (5) for the support of fixtures or other equipment.

In a wet area, EMT and other conduit must be installed to prevent water from entering the conduit system. In locations where walls are subject to regular wash-down (see **NEC Section 300-6**), the entire conduit system must be installed to provide a ¼" air space between it and the wall or supporting surface. The entire conduit system is considered to include the conduit, boxes, and fittings. To ensure resistance to corrosion caused by wet environments, EMT is galvanized. The term *galvanized* is used to describe the procedure in which both the interior and exterior surfaces of the conduit are coated with a corrosion-resistant zinc compound.

Because EMT is a good conductor of electricity, it may be used as an equipment **grounding** conductor. In order to qualify as an equipment grounding conductor [see **NEC Section 250-68(b)**], the conduit system must be tightly connected at each joint and provide a continuous grounding path. The connectors used in an EMT system ensure electrical and mechanical **continuity** throughout the system.

Note: Support requirements for EMT are covered in **NEC Section 348-13**. Various types of supports will be discussed later in this module.

EMT fittings are manufactured in two basic types. One type of fitting is the compression coupling (see *Figure 4*).

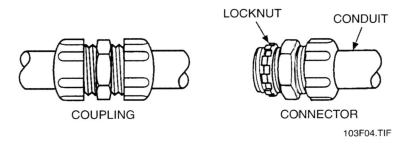

Figure 4. Compression Fittings

Because EMT is too thin for threads, special fittings must be used. For wet or damp locations, compression fittings such as those shown in *Figure 4* are used. These fittings contain a compression ring made of metal that forms a watertight seal.

When EMT compression couplings are used, they must be securely tightened, and when installed in masonry concrete, they must be of the concrete-tight type. If installed in a wet location, they must be of the raintight type. Refer to **NEC Article 348**.

EMT fittings for dry locations can be either the setscrew type or the indenting type. To use the setscrew type, the ends of the EMT are inserted into the sleeve and the setscrews are tightened to make the connection. Various types of setscrew couplings are shown in *Figure 5*.

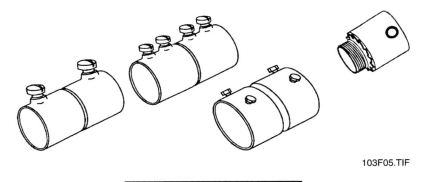

Figure 5. Setscrew Fittings

EMT sizes of 2½" and larger have the same outside diameter as corresponding sizes of galvanized rigid steel conduit (GRC). GRC threadless connectors may be used to connect EMT.

Note: EMT connectors, although they are the same size as GRC threadless connectors, may not be used to connect GRC.

Both setscrew and compression couplings are available in die-cast or steel construction. Steel couplings are stronger than die-cast couplings and have superior quality.

Support requirements for EMT are presented in **NEC Section 348-13**. As with most other metal conduit, EMT must be supported at least every 10' and within 3' of each outlet box, junction box, cabinet, fitting, or terminating end of the conduit. An exception to **NEC Section**

348-13 allows the fastening of unbroken lengths of EMT to be increased to a distance of 5' (1.52m) where structural members do not readily permit fastening within 3' (914mm).

3.1.4 Rigid Metal Conduit

Rigid metal conduit is conduit that is constructed of metal having sufficient thickness to permit the cutting of pipe threads at each end. Rigid metal conduit provides the best physical protection for conductors of any of the various types of conduit. Rigid metal conduit is supplied in 10' lengths including a threaded coupling on one end.

Note: Specific information on rigid metal conduit may be found in **NEC Article 346**.

Rigid metal conduit may be fabricated from steel or aluminum. Rigid metal steel conduit may either be galvanized or enamel-coated inside and out. Because of its threaded fittings, rigid metal conduit provides an excellent equipment grounding conductor, as defined in **NEC Section 250-118(2)**. A piece of rigid metal conduit is shown in *Figure 6*. The support requirements for rigid metal conduit are presented in **NEC Section 346-12**.

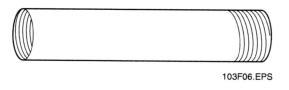

Figure 6. Rigid Metal Conduit

Rigid metal conduit may be used in underground installations if proper corrosion protection is used.

3.1.5 Galvanized Rigid Steel Conduit

Galvanized rigid steel conduit (GRC) is mostly used in industrial applications. GRC is heavier than EMT and IMC. It is more difficult to cut and bend, requires threading of each end, and has a higher purchase price than EMT and IMC. As a result, the cost of installing GRC is generally higher than the cost of installing EMT and IMC.

3.1.6 Plastic-Coated GRC

Plastic-coated GRC has a thin coating of PVC over the GRC. This combination is useful when an environment calls for the ruggedness of GRC along with the corrosion resistance of PVC. Typical installations where plastic-coated GRC may be required are:

- Chemical plants
- Food plants
- Refineries

- Fertilizer plants
- Paper mills
- Wastewater treatment plants

Plastic-coated GRC requires special threading and bending techniques.

3.1.7 Aluminum Conduit

Aluminum conduit has several characteristics that distinguish it from steel conduit. Because it has better resistance to wet environments and some chemical environments, aluminum conduit generally requires less maintenance in installations such as sewage treatment plants.

Direct burial of aluminum conduit results in a self-stopping chemical reaction on the conduit surface that forms a coating on the conduit. This coating acts to prevent further corrosion, increasing the life of the installation.

Note: Caution must be exercised to avoid burial of aluminum conduit in soil or concrete that contains calcium chloride. Calcium chloride may interfere with the corrosion resistance of aluminum conduit. Calcium chloride and similar materials are often added to concrete to speed concrete setting. It is important to determine if chlorides are to be used in the concrete prior to installing aluminum conduit. If chlorides are to be used, aluminum conduit must be avoided. Check with local authorities regarding this type of usage.

Since aluminum conduit is lighter than steel conduit, there are some installation advantages to using aluminum. For example, a 10' section of 3" aluminum conduit weighs about 23 pounds, compared to the 68-pound weight of its steel counterpart.

Because of its cost, aluminum conduit is not widely used. In addition, its use is limited by code. Since it is nonmagnetic, however, aluminum conduit is used with the magnetic resonance imaging (MRI) equipment used for medical diagnostic purposes.

3.1.8 Intermediate Metal Conduit

Intermediate metal conduit (IMC) has a wall thickness that is less than that of rigid metal conduit but greater than that of EMT. The weight of IMC is approximately $\frac{2}{3}$ that of rigid metal conduit. Because of its lower purchase price, lighter weight, and thinner walls, IMC installations are generally less expensive than comparable rigid metal conduit installations. However, IMC installations still have high strength ratings.

With the proper corrosion protection, IMC is suitable for use in underground applications.

Note: Additional information on intermediate metal conduit may be found in *NEC Article 345*.

The outside diameter of a given size of IMC is the same as that of the comparable size of rigid metal conduit. Therefore, rigid metal conduit fittings may be used with IMC. Since the threads on IMC and rigid metal conduit are the same size, no special threading tools are needed to thread IMC. Some technicians feel that threading IMC is more difficult than threading rigid metal conduit because IMC is somewhat harder.

The internal diameter of a given size of IMC is somewhat larger than the internal diameter of the same size of rigid metal conduit because of the difference in wall thickness. Bending IMC is considered easier than bending rigid metal conduit because of the reduced wall thickness. However, bending is sometimes complicated by kinking, which may be caused by the increased hardness of IMC.

The NEC requires that intermediate metal conduit be identified along its length at 5' intervals with the letters *IMC*. **NEC Sections 110-21 and 345-16(c)** describe this marking requirement.

Intermediate metal conduit, like rigid metal conduit, is permitted to act as an equipment grounding conductor, as defined in **NEC Section 250-118(3)**.

The use of IMC may be restricted in some jurisdictions. It is important to investigate the requirements of each jurisdiction before selecting any materials.

3.1.9 Rigid Nonmetallic Conduit

The most common type of rigid nonmetallic conduit is manufactured from polyvinyl chloride (PVC). Because PVC conduit is noncorrosive, chemically inert, and non-aging, it is often used for installation in wet or corrosive environments and for most underground applications. Corrosion problems found with steel and aluminum rigid metal conduit do not occur with PVC. However, PVC conduit may deteriorate under some conditions, such as long periods of direct sunlight.

All PVC conduit is marked according to standards established by the National Electrical Manufacturers' Association (NEMA) or **Underwriters' Laboratories (UL)**. A section of PVC conduit is shown in *Figure 7*.

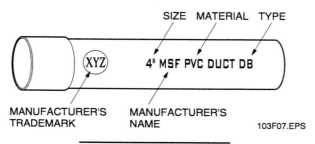

Figure 7. PVC Conduit

Since PVC conduit is lighter than steel or aluminum rigid conduit, IMC, or EMT, it is considered easier to handle. PVC conduit can usually be installed much faster than other types of conduit because the joints are made up with cement and require no threading.

PVC conduit contains no metal. This characteristic reduces the voltage drop of conductors carrying alternating current in PVC compared to identical conductors in steel conduit.

Because PVC is nonconducting, it cannot be used as an equipment grounding conductor. An equipment grounding conductor sized in accordance with **NEC Table 250-122** must be pulled in each PVC conductor run (except for underground service-entrance conductors).

Note: PVC is available in lengths up to 20', but some local codes require PVC to be cut to 10' lengths prior to installation because it expands and contracts significantly when exposed to changing temperatures. This expansion and contraction characteristic is worse in long runs. To avoid damage to PVC conduit caused by temperature changes, expansion couplings such as the one shown in *Figure 8* are used. The inside of the coupling is sealed with one or more O-rings. This type of coupling may allow up to 6" of movement. Check the requirements of the local codes prior to installing PVC.

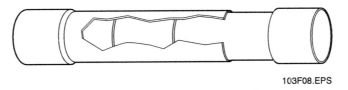

Figure 8. PVC Expansion Coupling

PVC conduit is manufactured in two types:

- *Type EB* – Thin wall for underground use only when encased in concrete. Also referred to as *Type I*.
- *Type DB* – Thick wall for underground use without encasement in concrete. Also referred to as *Type II*.

Type DB is available in two wall thicknesses: Schedule 40 and Schedule 80.
- Schedule 40 is heavy wall for direct burial in the earth and above-ground installations.
- Schedule 80 is extra heavy wall for direct burial in the earth, above-ground installations for general applications, and installations where the conduit is subject to physical damage.

PVC conduit is affected by higher-than-usual ambient temperatures. Support requirements for PVC are found in **NEC Table 347-8**. As with other conduit, it must be supported within 3' of each termination, but the maximum spacing between supports depends upon the size of the conduit.

Some of the regulations for the maximum spacing of supports are:

- ½" to 1" conduit – every 3'
- 1¼" to 2" conduit – every 5'
- 2½" to 3" conduit – every 6'
- 3½" to 5" conduit – every 7'
- 6" conduit – every 8'

3.1.10 Liquidtight Flexible Nonmetallic Conduit

Liquidtight flexible nonmetallic conduit (LFNC) was developed as a raceway for industrial equipment where flexibility was required and protection of conductors from liquids was also necessary. This conduit is covered in **NEC Article 351, Part B**. The use of LFNC has been expanded from industrial applications to outside and direct burial usage where listed and marked.

Several varieties of LFNC have been introduced. The first product (LFNC-A) was commonly referred to as *hose*. It consisted of an inner and outer layer of neoprene with a nylon reinforcing web between the layers. A second-generation product (LFNC-B) consisted of a smooth wall, flexible PVC with a rigid PVC integral reinforcement rod. The third product (LFNC-C) was a nylon corrugated shape without any integral reinforcements. These three permitted LFNC raceway designs must be flame-resistant with fittings **approved** for installation of electrical conductors. Nonmetallic connectors are listed for use and some liquidtight metallic flexible conduit connectors are dual-listed for both metallic and nonmetallic liquidtight flexible conduit.

LFNC is sunlight-resistant and suitable for use at conduit temperatures of 80°C dry and 60°C wet. It is available in ⅜" through 4" sizes. **NEC Section 351-23(b)** states that LFNC cannot be used where subject to physical damage or in lengths longer than 6', except where properly secured, where flexibility is required, or as permitted by **NEC Section 351-23(a)(5)**.

Liquidtight flexible metal conduit is a raceway with a circular cross section having an outer liquidtight, nonmetallic, sunlight-resistant jacket over an inner flexible metal core with associated couplings and connectors.

Flex connectors are used to connect flexible conduit to boxes or equipment. They are available in straight, 45°, and 90° configurations (*Figure 9*).

STRAIGHT CONNECTOR   45° CONNECTOR   90° CONNECTOR

Figure 9. Flex Connectors

3.1.11 Flexible Metal Conduit

Flexible metal conduit, sometimes called *flex*, may be used for many kinds of wiring systems. Flexible metal conduit is made from a single strip of steel or aluminum, wound and interlocked. It is typically available in sizes from ⅜" to 4" in diameter. An illustration of flexible metal conduit is shown in *Figure 10*.

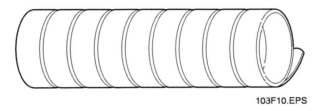

Figure 10. Flexible Metal Conduit

Flexible metal conduit is often used to connect equipment or machines that vibrate or move slightly during operation. It may also be used in the final connection to equipment having an electrical connection point that is difficult to access.

Flexible metal conduit is easily bent, but the minimum bending radius is the same as for other types of conduit. It should not be bent more than the equivalent of four quarter bends (360° total) between pull points (e.g., conduit bodies and boxes). It can be connected to boxes with a flexible conduit connector and to rigid conduit or EMT by using a combination coupling.

Note: When using a combination coupling, be sure the flexible conduit is pushed as far as possible into the coupling. This covers the end and protects the conductors from damage.

Two types of combination couplings are shown in *Figure 11*.

Flexible metal conduit is generally available in two types: nonliquidtight and liquidtight. **NEC Articles 350 and 351** cover the uses of flexible metal conduit.

 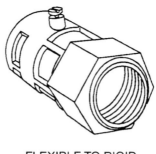

FLEXIBLE TO EMT   FLEXIBLE TO RIGID

Figure 11. Combination Couplings

Liquidtight flexible metal conduit has an outer covering of liquidtight, sunlight-resistant, flexible material that acts as a moisture seal. It is intended for use in wet locations. It is used primarily for equipment and motor connections when movement of the equipment is likely to occur. The number of bends, size, and support requirements for liquidtight conduit are the same as for all flexible conduit. Fittings used with liquidtight conduit must also be of the liquidtight type.

Support requirements for flexible metal conduit are found in **NEC Sections 350-18 and 351-8**. Straps or other means of securing the flexible metal conduit must be spaced every 4½' and within 12" of each end. (This spacing is closer together than for rigid conduit.) However, at terminals where flexibility is necessary, lengths of up to 36" without support are permitted. Failure to provide proper support for flexible conduit can make pulling conductors difficult.

Note: The NEC specifies that conduit bends must be made so that the conduit is not damaged and the radius of the conduit is not reduced. **NEC Table 346-10** provides specific minimum bend radii for various diameters of conduit.

The NEC also specifies that there shall not be more than the equivalent of four quarter-bends (360°) between pull points (e.g., a conduit body and a box).

There are raceways specifically designed to maintain the correct bend radii for Category 5 and fiber optic installations.

4.0.0 METAL CONDUIT FITTINGS

Manufacturers design and construct conduit fittings to permit a multitude of applications. The type of conduit fitting used in a particular application depends upon the size and type of conduit, the type of fitting needed for the application, the location of the fitting, and the installation method. The requirements and proper applications of boxes and fittings (conduit bodies) are found in **NEC Section 300-15**. Some of the more common types of fittings are examined in the following sections.

4.1.0 COUPLINGS

Couplings are sleeve-like fittings that are typically threaded inside to join two male threaded pieces of rigid conduit or IMC. A piece of conduit with a coupling is shown in *Figure 12*.

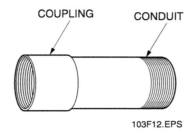

Figure 12. Conduit And Coupling

Other types of couplings may be used, depending upon the location and type of conduit. Several types are shown in *Figure 13*.

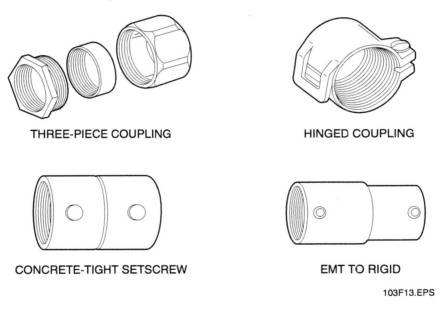

Figure 13. Metal Conduit Couplings

4.2.0 CONDUIT BODIES

Conduit bodies are a separate portion of a conduit or tubing system that provide access through a removable cover to the interior of the system. They are used at a junction of two or more sections of the system or at a terminal point of the system. They are usually cast and are significantly higher in cost than the stamped steel boxes permitted with EMT. However, there are situations in which conduit bodies are preferable, such as in outdoor locations, for appearance's sake in an **exposed location**, or to change types or sizes of raceways. Also, conduit bodies do not have to be supported in the same manner as stamped steel boxes. They are also used when elbows or bends would not be appropriate or to provide wire pulling locations in longer conduit runs.

NEC Section 370-16(c) states that conduit bodies cannot contain splices, **taps**, or devices unless they are durably and legibly marked by the manufacturer with their cubic inch capacity. The free space required for various conductors can be found using *Table 1*.

Size of Conductor	Free Space Within Box for Each Conductor
No. 18	1.5 cubic inches
No. 16	1.75 cubic inches
No. 14	2.0 cubic inches
No. 12	2.25 cubic inches
No. 10	2.5 cubic inches
No. 8	3.0 cubic inches
No. 6	5.0 cubic inches

Table 1. Volume Required Per Conductor

4.2.1 Type C Conduit Bodies

Type C conduit bodies may be used to provide a pull point in a long conduit run or a conduit run that has bends totaling more than 360°. A Type C conduit body is shown in *Figure 14*.

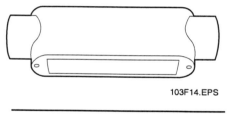

Figure 14. Type C Conduit Body

4.2.2 Type L Conduit Bodies

When referring to conduit bodies, the letter *L* represents an *elbow*. A Type L conduit body is used as a pulling point for conduit that requires a 90° change in direction. Because of the 90° bend, Type L conduit bodies are not acceptable for use with high-performance cable. The cover is removed, then the wire is pulled out, coiled on the ground or floor, reinserted into the other conduit body's opening, and pulled. The cover and its associated gasket are then replaced. Type L conduit bodies are available with the cover on the back (Type LB), on the sides (Type LL or LR), or on both sides (Type LRL). Several Type L conduit bodies are shown in *Figure 15*.

Note: The cover and gasket must be ordered separately. Do not assume that these parts come with conduit bodies when they are ordered.

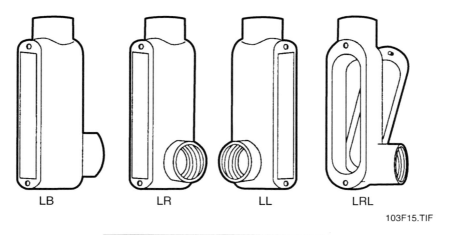

Figure 15. Type L Conduit Bodies

To identify Type L conduit bodies, use the following method:

Step 1 Hold the body like a pistol.

Step 2 Locate the opening on the body:
- If the opening is to the left, it is a Type LL.
- If the opening is to the right, it is a Type LR.
- If the opening is on the top (back), it is a Type LB.
- If there are openings on both the left and the right, it is a Type LRL.

4.2.3 Type T Conduit Bodies

Type T conduit bodies are used to provide a junction point for three intersecting conduits and are used extensively in conduit systems. A Type T conduit body is shown in *Figure 16*.

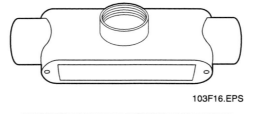

Figure 16. Type T Conduit Body

4.2.4 Type X Conduit Bodies

Type X conduit bodies are used to provide a junction point for four intersecting conduits. The removable cover provides access to the interior of the X so that wire pulling and splicing may be performed. A Type X conduit body is shown in *Figure 17*.

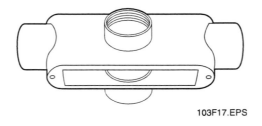

Figure 17. Type X Conduit Body

4.2.5 Threaded Weatherproof Hubs

Threaded weatherproof hubs are used for conduit entering a box in a wet location. *Figure 18* shows a typical threaded weatherproof hub.

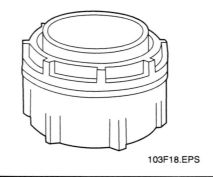

Figure 18. Threaded Weatherproof Hub

4.3.0 INSULATING BUSHINGS

An insulating bushing is either nonmetallic or has an insulated throat. Insulating bushings are installed on the threaded end of conduit that enters a sheet metal enclosure.

4.3.1 Nongrounding Insulating Bushings

The purpose of a nongrounding insulating bushing is to protect the conductors from being damaged by the sharp edges of the threaded conduit end. ***NEC Sections 345-15 and 347-12*** state that where a conduit enters a box, fitting, or other enclosure, a bushing must be provided to protect the wire from abrasion unless the design of the box, fitting, or enclosure is such as to afford equivalent protection. ***NEC Section 373-6(c)*** references ***NEC Section 300-4(f)*** that states where ungrounded conductors (No. 4 or larger) enter a raceway in a cabinet or box enclosure, the conductors shall be protected by a substantial fitting providing a smoothly rounded insulating surface, unless the conductors are separated from the raceway fitting by substantial insulating material securely fastened in place. An exception is where threaded hubs or bosses that are an integral part of a cabinet, box enclosure, or raceway provide a smoothly rounded or flared entry for conductors. An insulating bushing is shown in *Figure 19*.

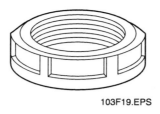

Figure 19. Insulating Bushing

4.3.2 Grounding Insulating Bushings

Grounded insulating bushings, usually called *grounding bushings*, are used to protect conductors and also have provisions for connection of an equipment grounding conductor. The ground wire, once connected to the grounding bushing, may be connected to the enclosure to which the conduit is connected. A grounding insulating bushing is shown in *Figure 20*.

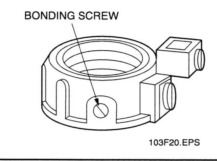

Figure 20. Grounding Insulating Bushing

4.4.0 OFFSET NIPPLES

Offset nipples are used to connect two pieces of electrical equipment in close proximity where a slight offset is required. They come in sizes ranging from ½" to 2" in diameter. See *Figure 21*.

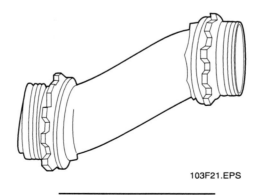

Figure 21. Offset Nipple

5.0.0 BOXES

Boxes are made from either metallic or nonmetallic material. *Figure 22* shows various boxes used in raceway systems.

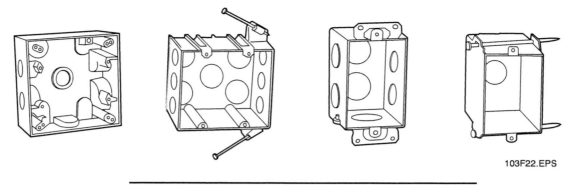

Figure 22. Various Boxes Used In Raceway Systems

5.1.0 METAL BOXES

Metal boxes are made from sheet steel. The surface is galvanized to resist corrosion and provide a continuous ground. Refer to **NEC Section 370-40** for information on thickness and grounding provisions. Metal boxes are made with removable circular sections called *pryouts* or *knockouts*. These circular sections are removed to make openings for conduit or cable connections.

5.1.1 Pryouts

In a pryout, a section is cut completely through the metal but only part of the way around, leaving solid metal tabs at two points. A slot is cut in the center of the pryout. To remove the pryout, a screwdriver is inserted into the slot and twisted to break the solid tabs (*Figure 23*).

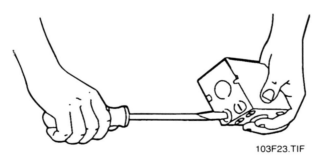

Figure 23. Pryout Removal

5.1.2 Knockouts

Knockouts are pre-punched circular sections that do not include a pryout slot. The knockout is easily removed when sharply hit by a hammer and punch, as shown in *Figure 24*.

Conduit must often enter boxes, cabinets, or panels that do not have pre-cut knockouts. In these cases, a knockout punch can be used to make a hole for the conduit connection. A knockout punch is shown in *Figure 25*.

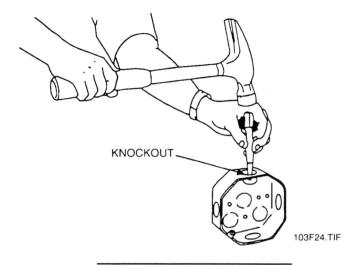

Figure 24. Knockout Removal

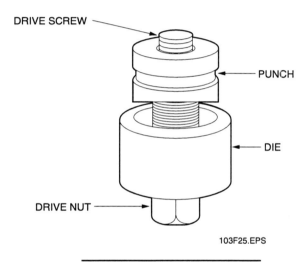

Figure 25. Knockout Punch

5.2.0 NONMETALLIC BOXES

Nonmetallic boxes are made of PVC or fiber-reinforced Bakelite. Nonmetallic boxes are often used in corrosive environments. **NEC Section 370-3** covers the use of nonmetallic boxes and the types of conduit, fittings, and grounding requirements for specific applications.

5.3.0 LOW-VOLTAGE BOXES

Low-voltage systems encompass computer networks, telecommunications lines, low-voltage wiring, audio, and video. For that reason, many different types of termination boxes and outlets are required. Several examples are shown in *Figure 26*.

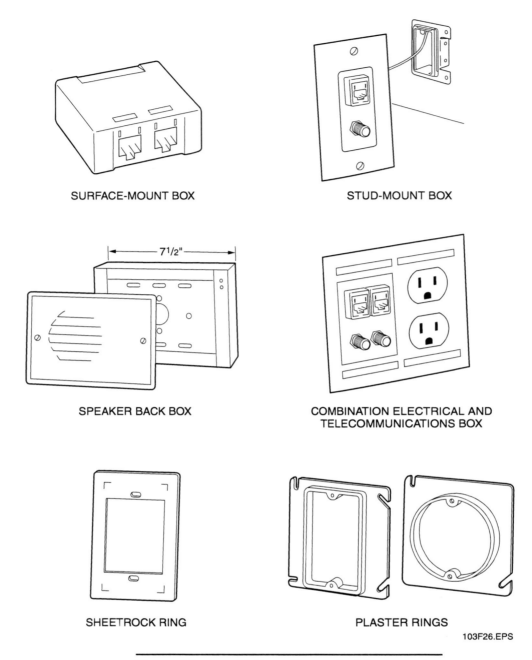

Figure 26. Low-Voltage Boxes And Accessories

The speaker back box is much larger than the standard outlet box. It is used to mount a speaker and microphone combination such as an intercom.

As shown in *Figure 26*, electrical and telecommunications outlets can reside in the same box, but there must be a partition separating them.

Several types of trim rings are also used. You should note that the term *ring* does not necessarily mean that the component is circular. The sheetrock (gypsum board) ring fits behind the cover plate and is used to cover up the opening when it is cut too big for the cover plate. The plaster ring is used to build out the box in cases where plaster or similar material is applied over the sheetrock.

6.0.0 BUSHINGS AND LOCKNUTS

Conduit is joined to boxes by connectors, adapters, threaded hubs, or locknuts.

Bushings protect the wires from the sharp edges of the conduit. Bushings are usually made of plastic or metal. Some metal bushings have a grounding screw to permit a **bonding wire** to be installed. Some different types of plastic and metal bushings are shown in *Figure 27*.

Figure 27. Bushings

Locknuts are used on the inside and outside walls of the box to which the conduit is connected. A grounding locknut may be needed if a bonding wire is to be installed. Special sealing locknuts are also used in wet locations. Several types of locknuts are shown in *Figure 28*.

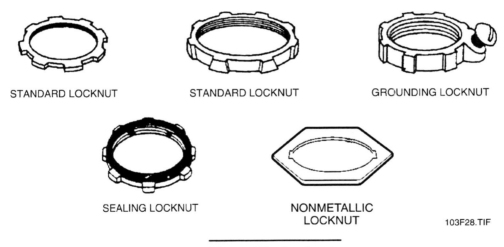

Figure 28. Locknuts

7.0.0 SEALING FITTINGS

Hazardous locations in manufacturing plants and other industrial facilities involve a wide variety of flammable gases and vapors and ignitable dusts. These hazardous substances have widely different flash points, ignition temperatures, and flammable limits requiring fittings that can be sealed. Sealing fittings are installed in conduit runs to minimize the passage of gases, vapors, or flames through the conduit and reduce the accumulation of moisture. They are required by **NEC Article 500** in hazardous locations where explosions may occur. They are also required where conduit passes from a hazardous location of one classification to another classification or to an unclassified location. Several types of sealing fittings are shown in *Figure 29*.

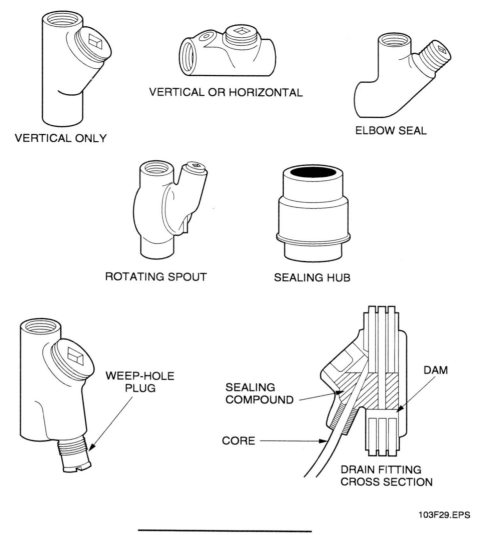

Figure 29. Sealing Fittings

8.0.0 CABLE AND RACEWAY SUPPORTS

Cable and raceway supports are available in many types and configurations. This section discusses the most common cable and conduit supports found in low-voltage cable installations. ***NEC Section 300-11*** discusses the requirements for branch circuit wiring that is supported from above. Equipment and raceways must have their own supporting methods and may not be supported by the supporting hardware of a fire-rated roof/ceiling assembly.

8.1.0 STRAPS

Straps are used to support conduit to a surface (see *Figure 30*). The spacing of these supports must conform to the minimum support spacing requirements for each type of conduit. One-hole and two-hole straps are used for all types of conduit: EMT, GRC, IMC, PVC, and flex. The straps can be flexible or rigid. Two-part straps are used to secure conduit to electrical framing channels (struts). Parallel and right angle beam clamps are also used to support conduit to structural members.

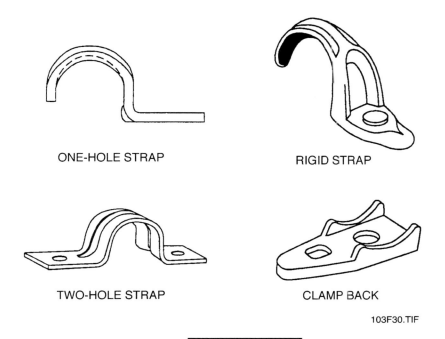

Figure 30. Straps

Clamp back straps can also be used with a backplate to maintain the ¼" spacing from the surface required for installations in wet locations.

8.2.0 STANDOFF SUPPORTS

The standoff support, often referred to as a *Minerallac* (the name of a manufacturer of this type of support), is used to support conduit away from the supporting structure. In the case of the one-hole and two-hole straps, the conduit must be kicked up wherever a fitting occurs. If standoff supports are used, the conduit is held away from the supporting surface and no offsets (**kicks**) are required in the conduit at the fittings. Standoff supports may be used to support all types of conduit including GRC, IMC, EMT, PVC, and flex, as well as tubing installations. A standoff support is shown in *Figure 31*.

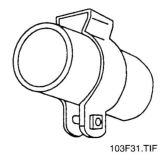

Figure 31. Standoff Support

8.3.0 ELECTRICAL FRAMING CHANNELS

Electrical framing channels or other similar framing materials are used together with Unistrut-type conduit clamps to support conduit (see *Figure 32*). They may be attached to a ceiling, wall, or other surface or be supported from a trapeze hanger.

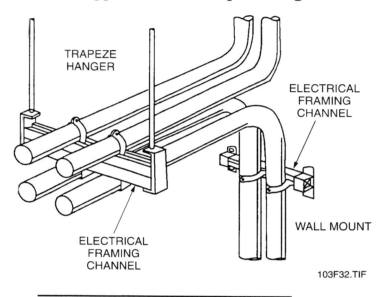

Figure 32. Electrical Framing Channels

8.4.0 BEAM CLAMPS

Beam clamps are used with suspended hangers. The raceway is attached to or laid in the hanger. The hanger is suspended by a threaded rod. One end of the threaded rod is attached to the hanger and the other end is attached to a beam clamp. The beam clamp is then attached to a beam. A beam clamp is shown in *Figure 33*.

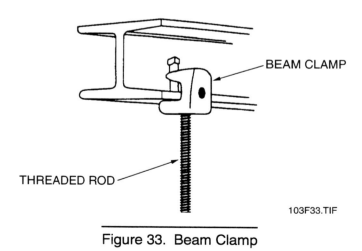

Figure 33. Beam Clamp

8.5.0 CABLE SUPPORTS

All cable runs must be securely fastened to the building structure. Several types of mechanical devices are used to support cable.

8.5.1 Cable Ties

Ties are used for a variety of purposes (*Figure 34*). The basic tie can be wrapped around a cable bundle to keep it neatly dressed. The lashing tie or velcro tie can be used to temporarily keep cables in place while additional cables are being pulled. The mounting tie and the tie with the mounting base can be used to secure small cable bundles to a flat surface. The mounting bases are available with adhesive backing or a screw hole. With the exception of the velcro tie, these ties are generally made of nylon.

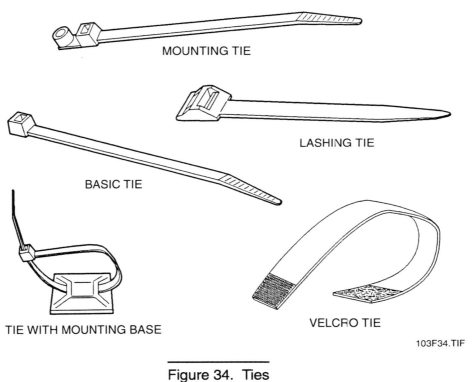

Figure 34. Ties

When installing ties, keep the following precautions in mind:

- Fire-rated cable ties must be used in cases where cables run through a ceiling plenum.
- Ties should be loosely connected to prevent pinching of the cable; otherwise, the performance of the cable will be affected.

8.5.2 Cable Hangers

J-clamps (J-hooks) and bridle rings are among the most common types of cable hangers (*Figure 35*). These devices can be mounted in a variety of ways.

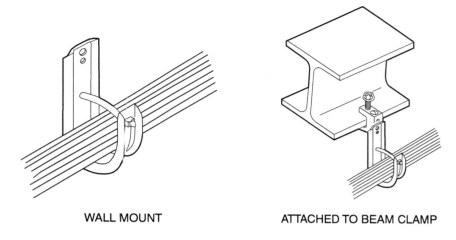

WALL MOUNT   ATTACHED TO BEAM CLAMP

J-HOOKS

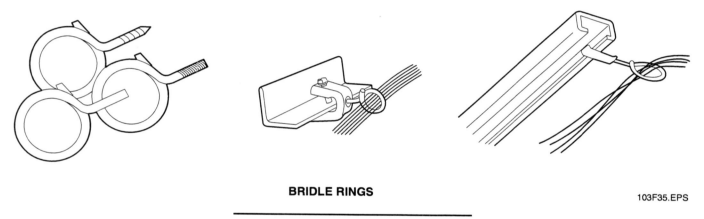

BRIDLE RINGS

Figure 35. J-Hooks And Bridle Rings

When high-performance cable such as Category 5, 6, or 7 telecommunications cabling is involved, bridle rings and ties are not suitable because they can crimp the cable and thereby reduce its performance. In such cases, J-hooks spaced every four or five feet are generally recommended. J-hooks have a wider base and are therefore less likely to crimp the cable. Wide strap-type hangers are also available to handle bundles from 4" to 6" in diameter.

Category 5, 6, and 7 cables may not be laid over a suspended ceiling. They must be routed through suitable hangers or raceways.

Other types of cable supports are D-rings, half D-rings, and mushroom posts (*Figure 36*). D-rings come in a variety of sizes and colors, and are available in aluminum or plastic.

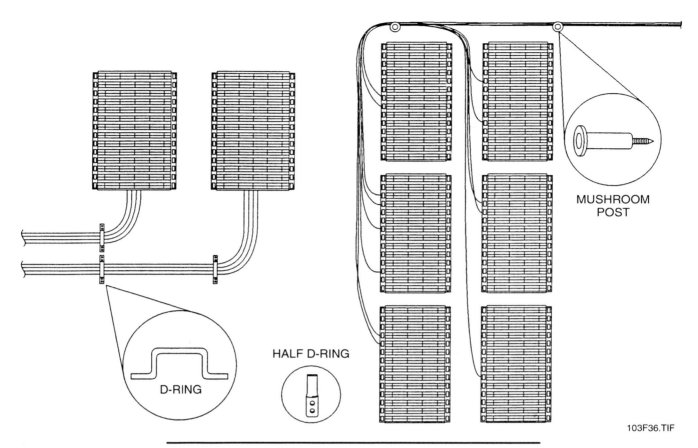

Figure 36. Hardware Used To Manage Cables And Wiring

9.0.0 WIREWAYS

Wireways are sheet metal **troughs** provided with hinged or screw-on removable covers. Like other types of raceways, wireways are used to house, support, and protect electric wires and cables. Wireways are available in various lengths, including 1', 2', 3', 4', 5', and 10'. The availability of various lengths allows runs of any exact number of feet to be made without cutting the wireway ducts.

It is noted in ***NEC Section 362-7*** that conductors, together with splices and taps, must not fill the wireway to more than 75% of its cross-sectional area. Each wireway is also rated for a maximum permitted conductor size for any single conductor. No conductor larger than the rated size is to be installed in any wireway.

In many situations, it is necessary to make extensions from the wireways to wall receptacles and control devices. In these cases, ***NEC Section 362-11*** specifies that these extensions be made using any wiring method presented in ***NEC Chapter 3*** that includes a means for equipment grounding. Finally, as required in ***NEC Section 362-12***, wireways must be marked in such a way that their manufacturer's name or trademark will be visible.

As you can see in *Figure 37*, a wide range of fittings is required for connecting wireways to one another and to fixtures.

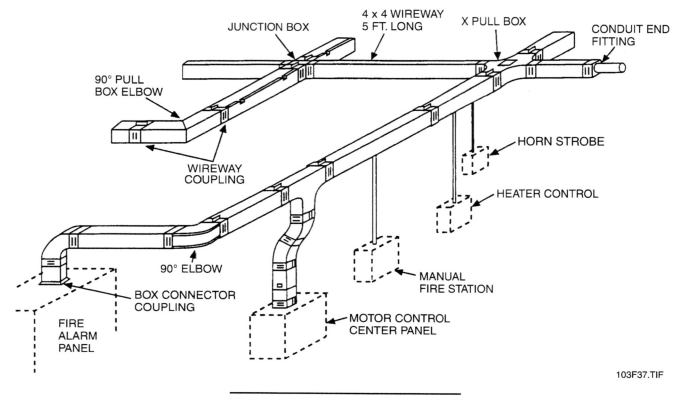

Figure 37. Wireway System Layout

9.1.0 TYPES OF WIREWAYS

Rectangular duct-type wireways come as either hinged-cover or screw-cover troughs. Typical lengths are 1', 2', 3', 4', 5', and 10'. Shorter lengths are also available. Raintight troughs are used in environments where moisture is not permitted within the raceway. However, the raintight trough should not be confused with the raintight lay-in wireway, which has a hinged cover. *Figure 38* shows a raintight trough with a removable side cover.

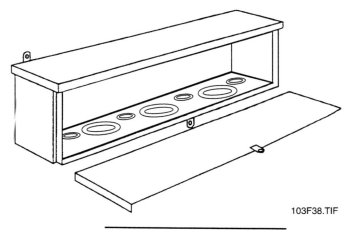

Figure 38. Raintight Trough

Wireway troughs are exposed when first installed. Whenever possible, they are mounted on the ceilings or walls, although they may sometimes be suspended from the ceiling. Note that in *Figure 39*, the trough has knockouts similar to those found on junction boxes. The conduit is joined to the wireway at the most convenient knockout possible.

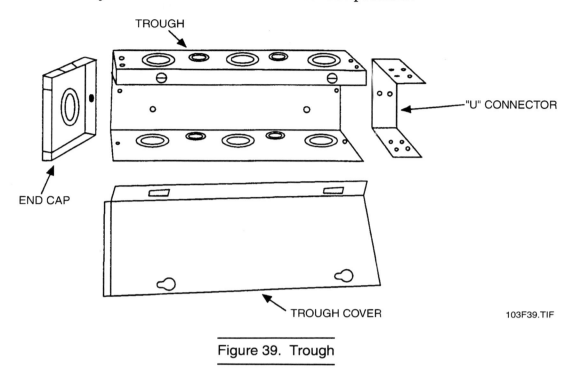

Figure 39. Trough

Wireway components such as trough crosses, 90° internal elbows, and tee connectors serve the same function as fittings on other types of raceways. The fittings are attached to the duct using slip-on connectors. All attachments are made with nuts and bolts or screws. When assembling wireways, always place the head of the bolt on the inside and the nut on the outside so that the conductors will not be resting against a sharp edge. It is usually best to assemble sections of the wireway system on the floor, and then raise the sections into position. An exploded view of a section of wireway is shown in *Figure 40*. Both the wireway fittings and the duct come with screw-on, hinged, or snap-on covers to permit conductors to be laid in or pulled through.

The NEC specifies that wireways may be used only for exposed work. Therefore, they cannot be used in underfloor installations. If they are used for outdoor work, they must be of an approved raintight construction. It is important to note that wireways must not be installed where they are subject to severe physical damage, corrosive vapors, or hazardous locations.

Wireway troughs must be installed so that they are supported at distances not exceeding 5'. When specially approved supports are used, the distance between supports must not exceed 10'.

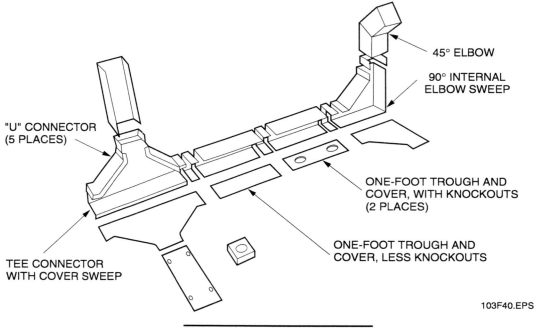

Figure 40. Wireway Sections

9.2.0 WIREWAY FITTINGS

Many different types of fittings are available for wireways, especially for use in exposed, dry locations.

9.2.1 Connectors

Connectors are used to join wireway sections and fittings. Connectors are slipped inside the end of a wireway section and are held in place by small bolts and nuts. Alignment slots allow the connector to be moved until it is flush with the inside surface of the wireway. After the connector is in position, it can be bolted to the wireway. This helps to ensure a strong, rigid connection. Connectors have a friction hinge that helps hold the wireway cover open when needed. A connector is shown in *Figure 41*.

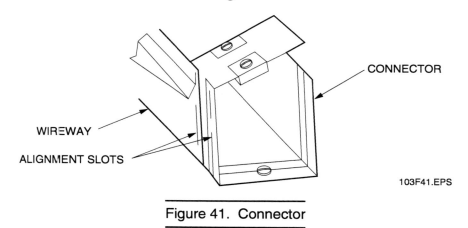

Figure 41. Connector

9.2.2 End Plates

End plates, or closing plates, are used to seal the ends of wireways. They are inserted into the end of the wireway and fastened by screws and bolts. End plates contain knockouts so that conduit or cable may be extended from the wireway. An end plate is shown in *Figure 42*.

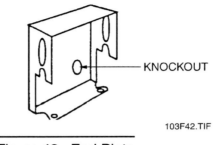

Figure 42. End Plate

9.2.3 Tees

Tee fittings or tees are used when a tee connection is needed in a wireway system. A tee connection is used where conductors may branch in different directions. The tee fitting cover and sides can be removed for access to splices and taps. Tee fittings are attached to other wireway sections using standard connectors. A tee is shown in *Figure 43*.

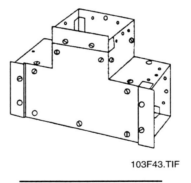

Figure 43. Tee Fitting

9.2.4 Crosses

Crosses have four openings and are attached to other wireway sections with standard connectors. The cover is held in place by screws and can be easily removed for laying in wires or for making connections. A cross is shown in *Figure 44*.

9.2.5 Elbows

Elbows are used to make a bend in the wireway. They are available in angles of 22½°, 45°, or 90°, and are either internal or external. They are attached to wireway sections with standard connectors. Covers and sides can be removed for wire installation. The inside corners of elbows are rounded to prevent damage to conductor insulation. An inside elbow is shown in *Figure 45*.

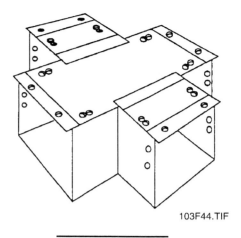

Figure 44. Cross

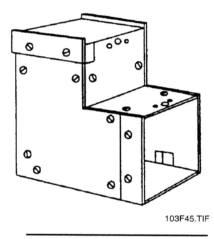

Figure 45. 90° Inside Elbow

9.2.6 Telescopic Fittings

Telescopic or slip fittings may be used between lengths of wireway. Slip fittings are attached to standard lengths by setscrews and usually adjust from ½" to 11½". Slip fittings have a removable cover for installing wires and are similar in appearance to a nipple.

9.3.0 WIREWAY SUPPORTS

Horizontal wireways should be securely supported at each end and at intervals of no more than 5' or for individual lengths greater than 5' at each end or joint, unless listed for other support intervals. In no case shall the support distance be greater than 10', in accordance with *NEC Section 362-8*. If possible, wireways can be mounted directly to a surface. Otherwise, wireways are supported by hangers or brackets.

9.3.1 Suspended Hangers

In many cases, the wireway is supported from a ceiling, beam, or other structural member. In such installations, a suspended hanger (*Figure 46*) may be used to support the wireway.

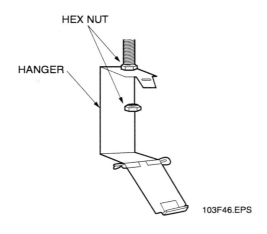

Figure 46. Suspended Hanger

The wireway is attached to or laid in the hanger. The hanger is suspended by a threaded rod. One end of the rod is attached to the hanger with hex nuts. The other end of the rod is attached to a beam clamp or anchor.

9.3.2 Gusset Brackets

Another type of support used to mount wireways is a gusset bracket. This is an L-type bracket that is mounted to a wall. The wireway rests on the bracket and is attached by screws or bolts. A gusset bracket is shown in *Figure 47*.

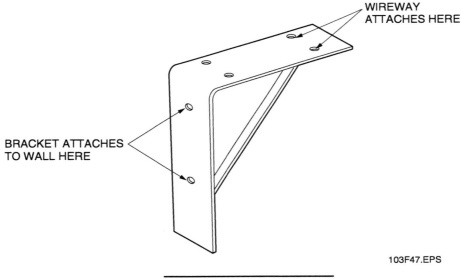

Figure 47. Gusset Bracket

9.3.3 Standard Hangers

Standard hangers are made in two pieces. The two pieces are combined in different ways for different installation requirements. The wireway is attached to the hanger by bolts and nuts. A standard hanger is shown in *Figure 48*.

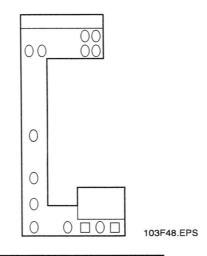

Figure 48. Standard Hanger

9.3.4 Wireway Hangers

When a larger wireway must be suspended, a wireway hanger may be used. A wireway hanger is made by suspending a piece of strut from a ceiling, beam, or other structural member. The strut is suspended by threaded rods attached to beam clamps or other ceiling anchors, as shown in *Figure 49*.

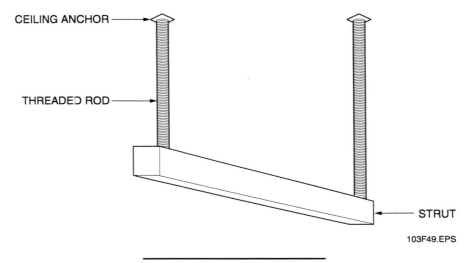

Figure 49. Wireway Hanger

9.4.0 OTHER TYPES OF RACEWAYS

In this section, other types of raceways will be discussed. Depending on the particular purpose for which they are intended, raceways include enclosures such as surface metal and nonmetallic raceways, and underfloor raceways.

9.4.1 Surface Metal And Nonmetallic Raceways

Surface metal raceways consist of a wide variety of special raceways designed primarily to carry power and communications wiring to locations on the surface of ceilings or walls of building interiors.

Low-voltage perimeter raceway (*Figure 50*) is specifically designed to route, protect, and seal data, voice, video, fiber optic, or power cabling.

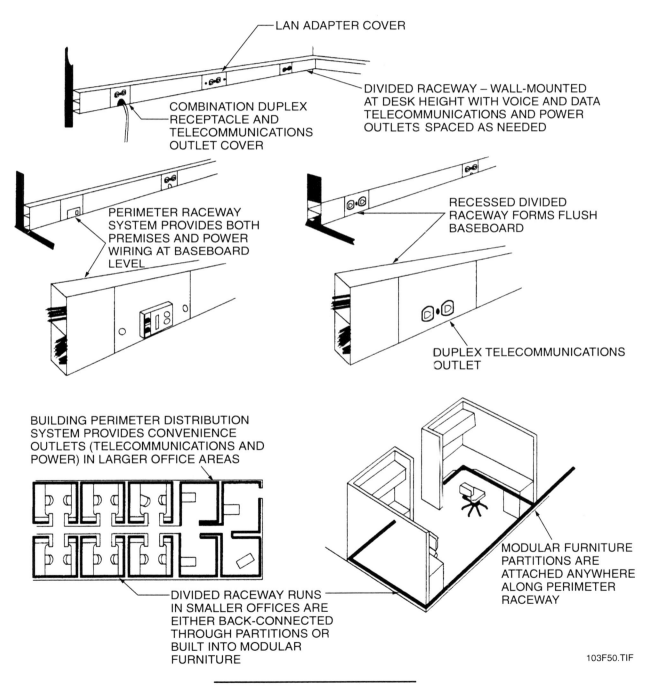

Figure 50. Perimeter Raceway

There are three types of surface raceway (*Figure 51*). One is a single-channel, one-piece type. Another is a one-piece, self-latching model designed for low-capacity applications (e.g., one- and two-channel). This type is available with a peel-and-stick adhesive system.

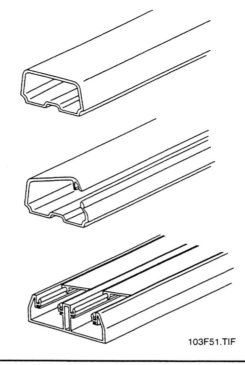

Figure 51. Examples Of Surface Raceway

The third type of low-voltage raceway is a two-piece model with a base that can be screwed or nailed to a wall, floor, or ceiling. A compatible raceway cover fits over the base and locks into place. This type of raceway is used for multi-channel applications and has available divider walls to separate power and low-voltage cables.

The perimeter raceways shown in *Figure 51* are available in widths ranging from ¾" to over 7".

Accessories are available to connect the 8' or 10' sections of raceway, create junctions, and make corners. Corners are compatible with the minimum 1" bend radius required by TIA/EIA 568-A. A selection of termination boxes and faceplates is also available. See *Figure 52* for an example.

Installation specifications of both surface metal raceways and surface nonmetallic raceways are listed in detail in **NEC Article 352**. All these raceways must be installed in dry, interior locations. The number of conductors, their amperage, and the allowable cross-sectional area of the conductors are specified in **NEC Tables 310-16 through 310-19**.

One use of surface metal raceways is to protect conductors that run to nonaccessible outlets.

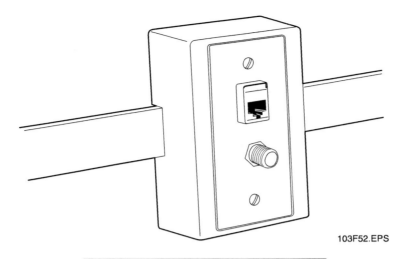

Figure 52. Low-Voltage Raceway Box

Other surface metal raceway designs are referred to as *pancake raceways* because of their flat cross sections. Their primary use is to extend power, lighting, telephone, or signal wire to locations away from the walls of a room without embedding them under the floor. A pancake raceway is shown in *Figure 53*.

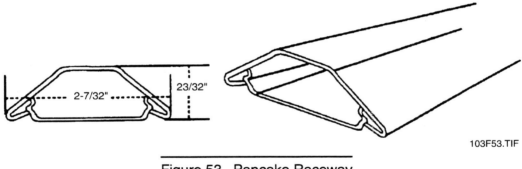

Figure 53. Pancake Raceway

There are also surface metal raceways available that house two or three different conductor raceways. These are referred to as *twin duct* or *triple duct*. These raceways permit different circuits, such as power and signal circuits, to be placed within the same raceway.

NEC Section 800-52 specifies minimum separations between power and communications cable.

The reason for separating these cables is to limit the effect that electromagnetic interference (EMI) has on communications systems. EMI is basically electronic noise that is generated by high-voltage systems and by appliances such as microwave ovens. This noise can interfere with, and even interrupt, communications lines.

While communications cable can share pathways with other low-voltage cables in most instances, communications and power cables must be separated by a barrier if they are run in the same raceway. If the cables are not run in a partitioned raceway, there must be a 2" separation.

The number and types of conductors permitted to be installed and the capacity of a particular surface raceway must be calculated and matched with NEC requirements, as discussed previously. **_NEC Tables 310-16 through 310-19_** are used for surface raceways in the same manner in which they are used for wireways. For surface raceway installations with more than three conductors in each raceway, refer to **_NEC Table 310-15(b)(2)(a)_**.

9.4.2 Pole Systems

There are many situations in which cables and wires have to be carried from overhead wiring systems to devices that are not located near existing wall outlets or control circuits. This type of wiring is typically used in open office spaces where cubicles are provided by temporary dividers. Poles are used to accomplish this. Some common manufacturers' names for these poles include _Tele-Power poles, Quick-E poles,_ and _Walkerpoles._ The poles usually come in lengths suitable for 10', 12', or 15' ceilings. _Figure 54_ shows a pole base. There must be a divider between electrical and low-voltage cabling.

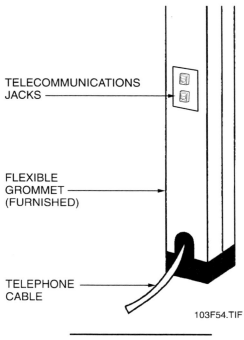

Figure 54. Pole Base

9.4.3 Underfloor Systems

Underfloor raceway systems were developed to provide a practical means of bringing conductors for lighting, power, and signaling to cabinets and consoles. Underfloor raceways are available in 10' lengths and widths of 4" and 8". The sections are made with inserts spaced every 24". The inserts can be removed for outlet installation. These are explained in **_NEC Article 354_**.

Note: Inserts must be installed so that they are flush with the finished grade of the floor.

Junction boxes are used to join sections of underfloor raceways. Conduit is also used with underfloor raceways by using a raceway-to-conduit connector (conduit adapter). A typical underfloor raceway duct is shown in *Figure 55*.

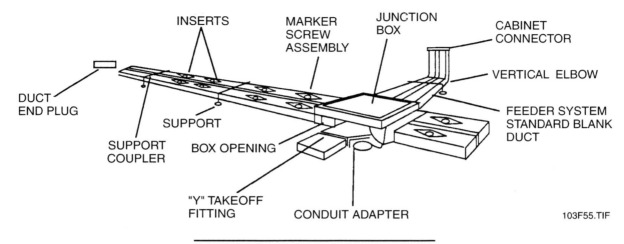

Figure 55. Underfloor Raceway Duct

This wiring method makes it possible to place a desk or table in any location where it will always be over, or very near to, a duct. The wiring method for lighting and power between cabinets and the raceway junction boxes may be conduit, underfloor raceway, wall elbows, and cabinet connectors. Optional dividers are available with some junction boxes to allow power and low-voltage conductors to share the same box. Special collars and flanges are also available to elevate the surface of the junction box so it is even with the floor.

9.4.4 Cellular Metal Floor Raceways

A cellular metal floor raceway is a type of floor construction designed for use in steel-frame buildings. In these buildings, the members supporting the floor between the beams consist of sheet steel rolled into shapes. These shapes are combined to form cells, or closed passageways, which extend across the building. The cells are of various shapes and sizes, depending upon the structural strength required. The cells of this type of floor construction form the raceways, as shown in *Figure 56*.

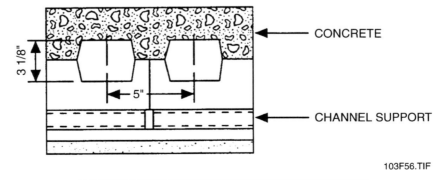

Figure 56. Cross Section Of A Cellular Floor

Connections to the cells are made using headers which extend across the cells. A header connects only to those cells which are to be used as raceways for conductors. A junction box or access fitting is necessary at each joint where a header connects to a cell. Two or three separate headers, connecting to different sets of cells, may be used for different systems. For example, light and power, signaling systems, and public telephones would each have a separate header. A special elbow fitting is used to extend the headers up to the distribution equipment on a wall or column.

9.4.5 Cellular Concrete Floor Raceways

The term *precast cellular concrete floor* refers to a type of floor used in steel-frame, concrete-frame, and wall-bearing construction. In this type of system, the floor members are precast with hollow voids which form smooth, round cells. The cells form raceways which can be adapted for use as underfloor raceways by using special fittings. A precast cellular concrete floor is fire-resistant and requires no further fireproofing. The precast reinforced concrete floor members form the structural floor and are supported by beams or bearing walls. Connections to the cells are made with headers which are secured to the precast concrete floor.

10.0.0 CABLE TRAYS

Cable trays function as a support for conductors and tubing (see **NEC Article 318**). A cable tray has the advantage of easy access to conductors, and thus lends itself to installations where the addition or removal of conductors is a common practice. Cable trays are fabricated from aluminum, steel, and fiberglass.

Cable trays are available in two basic forms: ladder and trough. Ladder tray, as the name implies, consists of two parallel channels connected by rungs. Trough-type trays consist of two parallel channels (side rails) having a corrugated, ventilated bottom, or a corrugated, solid bottom. There is also a special center rail cable tray available for use in telephone and sound wiring.

Cable trays are commonly available in 12' and 24' lengths. They are usually available in widths of 6", 9", 12", 18", 24", 30", and 3', and load depths of 4", 6", and 8".

Cable trays may be used in most installations. Cable trays may be used in air handling (plenum) ceiling spaces but only to support the wiring methods permitted in such spaces by **NEC Section 300-22(c)**. Also, cable trays may be used in Class 1, Division 2 locations according to **NEC Section 501-4(b)**. Cable trays may also be used above a suspended ceiling that is not used as an air handling space. Some manufacturers offer an aluminum cable tray that is coated with PVC for installation in caustic environments. A typical cable tray system with fittings is shown in *Figure 57*.

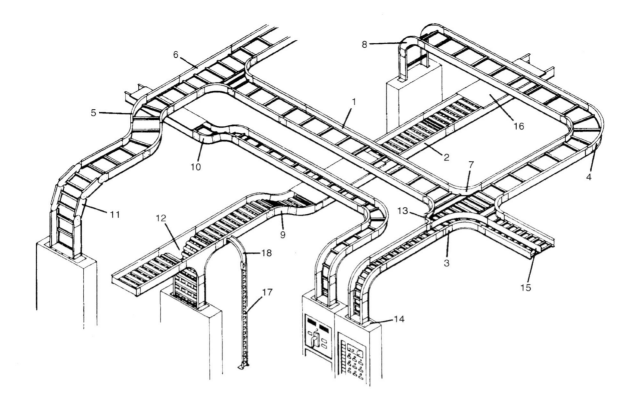

1. LADDER CABLE TRAY
2. VENTILATED TROUGH CABLE TRAY
3. SPLICE PLATES
4. 90° HORIZONTAL BEND (LADDER TYPE)
5. 45° HORIZONTAL BEND (LADDER TYPE)
6. HORIZONTAL TEE (LADDER TYPE)
7. HORIZONTAL CROSS (LADDER TYPE)
8. 90° VERTICAL OUTSIDE BEND (LADDER TYPE)
9. 45° VERTICAL OUTSIDE BEND (LADDER TYPE)
10. 30° VERTICAL OUTSIDE BEND (LADDER TYPE)
11. VBS-2 VERTICAL BEND SEGMENT
12. VERTICAL TEE (VENTILATED TROUGH TYPE)
13. LEFT-HAND REDUCER (LADDER TYPE)
14. BOX CONNECTOR
15. BARRIER STRIP
16. SOLID FLANGED TRAY COVER
17. CABLE CHANNEL, NON-VENTILATED
18. CABLE CHANNEL, 90° VERTICAL

Figure 57. Cable Tray System

Wire and cable installation in cable trays is defined by the NEC. Read **NEC Article 318** to become familiar with the requirements and restrictions made by the NEC for safe installation of wire and cable in a cable tray.

Metallic cable trays that support electrical conductors must be grounded as required by **NEC Article 250**. Where steel and aluminum cable tray systems are used as an equipment grounding conductor, all of the provisions of **NEC Section 318-7(b)** must be complied with.

WARNING!   Do not stand on, climb in, or walk on a cable tray.

PATHWAYS AND SPACES — TRAINEE TASK MODULE 33103

10.1.0 CABLE TRAY FITTINGS

Cable tray fittings are part of the cable tray system and provide a means of changing the direction or dimension of the different trays. Some of the uses of horizontal and vertical tees, horizontal and vertical bends, horizontal crosses, reducers, barrier strips, covers, and box connectors are shown in *Figure 57*.

10.2.0 CABLE TRAY SUPPORTS

Cable trays are usually supported in one of five ways: direct rod suspension, trapeze mounting, center hung, wall mounting, and pipe rack mounting.

10.2.1 Direct Rod Suspension

The direct rod suspension method of supporting cable tray uses threaded rods and hanger clamps. One end of the threaded rod is connected to the ceiling or other overhead structure. The other end is connected to hanger clamps that are attached to the cable tray side rails. A direct rod suspension assembly is shown in *Figure 58*.

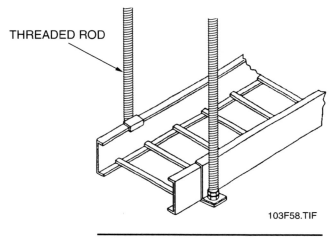

Figure 58. Direct Rod Suspension

10.2.2 Trapeze Mounting And Center Hung Support

Trapeze mounting of cable tray is similar to direct rod suspension mounting. The difference is in the method of attaching the cable tray to the threaded rods. A structural member, usually a steel channel or strut, is connected to the vertical supports to provide an appearance similar to a swing or trapeze. The cable tray is mounted to the structural member. Often, the underside of the channel or strut is used to support conduit. A trapeze mounting assembly is shown in *Figure 59*.

A method that is similar to trapeze mounting is a center hung tray support. In this case, only one rod is used and it is centered between the cable tray side rails.

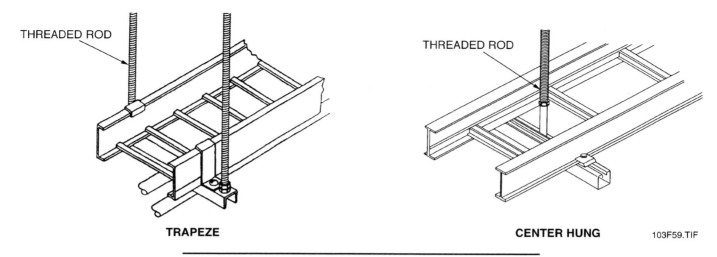

TRAPEZE   **CENTER HUNG**

Figure 59. Trapeze Mounting And Center Hung Support

10.2.3 Wall Mounting

Wall mounting is accomplished by supporting the cable tray with structural members attached to the wall. This method of support is often used in tunnels and other underground or sheltered installations where large numbers of conductors interconnect equipment that is separated by long distances. A wall mounting assembly is shown in *Figure 60*.

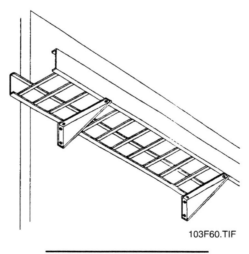

Figure 60. Wall Mounting

10.2.4 Pipe Rack Mounting

Pipe racks are structural frames used to support piping that interconnects equipment in outdoor industrial facilities. Usually, some space on the rack is reserved for conduit and cable tray. Pipe rack mounting of cable tray is often used when power distribution and electrical wiring is routed over a large area.

11.0.0 STORING RACEWAYS

Proper and safe methods of storing conduit, wireways, raceways, and cable trays may sound like a simple task, but improper storage techniques can result in wasted time and damage to the raceways, as well as personal injury. There are correct ways to store raceways that will help avoid costly damage, save time in identifying stored raceways, and reduce the chance of personal injury.

Pipe racks are commonly used for storing conduit. The racks provide support to prevent bending, sagging, distorting, scratching, or marring of conduit surfaces. Most racks have compartments where different types and sizes of conduit can be separated for ease of identification and selection. The storage compartments in racks are usually elevated to help avoid damage that might occur at floor level. Conduit that is stored at floor level is easily damaged by people and other materials or equipment in the area.

The ends of stored conduit should be sealed to help prevent contamination and damage. Conduit ends can be capped, taped, or plugged.

Always inspect raceway before storing it to make sure that it is clean and is not damaged. Also, make sure that the raceway is stored securely so that when someone comes to get it for a job, it will not fall in any way that could cause injury.

To prevent contamination and corrosion of stored raceway, it should be covered with a tarpaulin or other suitable covering. It should also be separated from noncompatible materials such as hazardous chemicals.

Wireways, surface metal raceways, and cable trays should always be stored off the ground on boards in an area where people will not step on them and equipment will not run over them. Stepping on or running over raceway bends the metal and makes it unusable.

12.0.0 HANDLING RACEWAYS

Raceway is made to strict specifications. It can easily be damaged by careless handling. From the time raceway is delivered to a job site until the installation is complete, use proper and safe handling techniques. These are a few basic guidelines for handling raceway that will help avoid damaging or contaminating it:

- Never drag raceway off a delivery truck or off other lengths of raceway, and never drag it on the ground or floor. Dragging raceway can cause damage to the ends.
- Keep the thread protection caps on when handling or transporting conduit raceway.
- Keep raceway away from any material that might contaminate it during handling.
- Flag the ends of long lengths of raceway when transporting it to the job site.
- Never drop or throw raceway when handling it.
- Never hit raceway against other objects when transporting it.

- Always use two people when carrying long pieces of raceway. Make sure that you both stay on the same side and that the load is balanced. Each person should be about ¼ of the length of the raceway from the end. Lift and put down the raceway at the same time.

13.0.0 DUCTING

A duct is a single enclosed raceway through which conductors or cables can be led. Underground duct systems may include maintenance holes, vaults, and risers.

There are several reasons for running cabling underground rather than overhead. Architectural plans may require buried lines throughout a subdivision or a planned community. Tunnels may already exist, or be planned, for carrying steam or water lines. In any of these situations, underground installations are appropriate. Underground cables may be buried directly in the ground or run through appropriate conduit.

In underground construction, a duct system provides a safe passageway for communications cables. In buildings, underfloor raceways and cellular floor raceways are built to provide ducting so that cabling can be distributed throughout a large area. As an electronics system technician, you need to know the approved methods of constructing underground ducting. You also need to know how to avoid potential electrical hazards, both during construction and when performing system maintenance. Also, it is essential to understand the requirements and limitations imposed on running wires through underfloor and cellular floor raceways and ducts.

14.0.0 UNDERGROUND SYSTEMS

There are three different ways to install cable underground:
- Duct
- Conduit
- Direct burial

The method used will depend on the application, the materials available, and the number and types of conductors to be pulled.

A duct consists of at least one **pathway** placed in a trench and covered with earth. In some cases, conduit can be classified as a *pathway*. Pathways are available as either a single duct or multiple ducts (2, 3, 4, or 6 pathways per section). The depth at which the duct will be placed is determined using ***NEC Table 300-5***. The conduit pathways are encased in concrete or other materials. This provides good mechanical strength.

In underground cable installations, a duct is a buried conduit through which a cable passes. Maintenance holes are set at intervals in an underground duct run. Maintenance holes provide access through throats (sometimes called *chimneys*). At ground level, or street surface level, a maintenance hole cover closes off the maintenance hole area tightly. An individual cable length running underground normally terminates at a maintenance hole, where it is spliced to another length of cable. A duct may consist of a single conduit or several, each carrying a cable length from one maintenance hole to the next. See *Figure 61*.

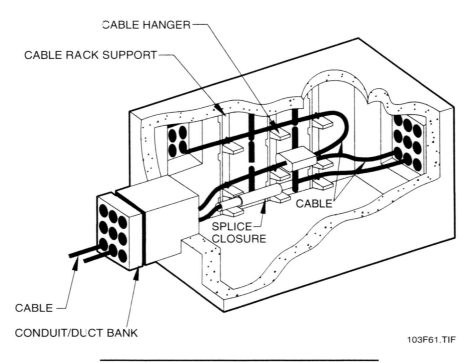

Figure 61. Typical Cable Maintenance Hole

Maintenance holes provide room for installing lengths of cable in conduit lines (*Figure 62*). They are also used for maintenance work and for performing tests. Workers enter a maintenance hole from above. In a two-way maintenance hole, cables enter and leave in two directions. There are also three-way and four-way maintenance holes. Maintenance holes are often located at the intersection of two streets so that they can be used for cables leaving in four directions. Maintenance holes are usually constructed of brick or concrete. Their design must provide room for drainage and for workers to move around inside them. A similar opening, known as a *handhole,* is sometimes provided for splicing on lateral two-way ducts.

For direct-buried cable, filled polyethylene insulated conductor (PIC) cable is the only type of copper cable recommended. It contains a gel which inhibits water penetration and migration, and therefore protects the integrity of the cable. Direct-buried cable may require an armored sheath to resist damage from rodents and other sources.

Note: The difference between direct-buried cable and underground cable is that underground cable is run through a conduit and/or maintenance hole.

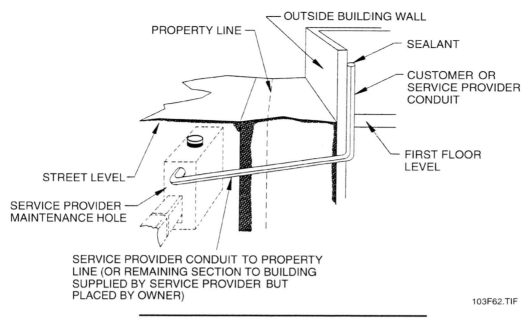

Figure 62. Installing Underground Entrances

14.1.0 DUCT MATERIALS

Underground ducts can be made of fiber, vitrified tile, iron conduit, plastic, or poured concrete. The inside diameter of the ducting for a specific job is determined by the size of the cable that will be drawn into the duct. Sizes from 2" to 6" (inside diameter) are available.

CAUTION: Be careful when working with unfamiliar duct materials. In older installations, asbestos/cement duct may have been used. You must be certified to remove or disturb asbestos.

14.2.0 PLASTIC CONDUIT

Plastic conduit may be made of PVC (polyvinyl chloride), PE (polyethylene), or styrene. Since this type of conduit is available in lengths up to 20', fewer couplings are needed than with other types of ducting. Plastic conduit is popular because it is easy to install, requires less labor than other types of conduit, and is low in cost.

14.3.0 MONOLITHIC CONCRETE DUCT

Monolithic concrete duct is poured at the job site. Multiple duct pathways can be formed using rubber tubing cores on spacers. The cores may be removed after the concrete has set. A die containing steel tubes, known as a *boat*, can also be used to form ducts. It is pulled slowly through the trench on a track as concrete is poured from the top. Poured concrete ducting made by either method is relatively expensive, but it offers the advantage of creating a very clean duct interior with no residue that can decay. The rubber core method is especially useful for forming curves or turns in duct systems.

14.4.0 CONTROLLED ENVIRONMENT VAULTS

The controlled environment vault (CEV) is a precast concrete structure consisting of top and bottom sections. It is available in 16' and 24' lengths and is normally placed near a maintenance hole. The vault is designed to house environmentally sensitive electronic equipment and is, therefore, sealed against gases or moisture that may enter from the outside. The CEV is generally equipped with electrical power, lights, sump pumps, ventilation blowers, heaters, and atmospheric monitors. It may also be air conditioned.

14.5.0 PEDESTALS AND CABINETS

Pedestals and cabinets are outdoor enclosures used as junction points to provide above-ground access to cables from underground, direct-buried, or aerial sources. There are many types of pedestals; some examples are shown in *Figure 63*.

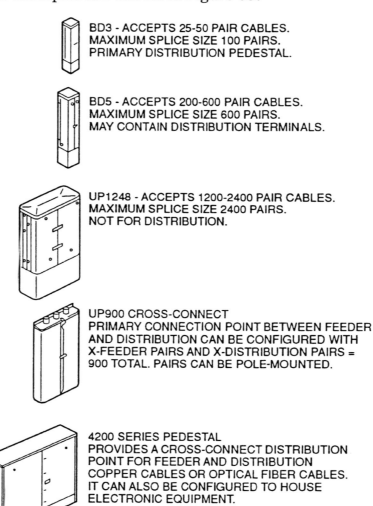

Figure 63. Examples Of Pedestals

Pedestals may be installed directly on the ground or they may be wall-mounted, pole-mounted, or pad-mounted. Depending on the application, the pedestal may house electronic or fiber optic equipment. Pedestals and cabinets containing electronic equipment may also have environmental control systems, including heating, cooling, and air circulation.

15.0.0 MAKING A CONDUIT-TO-BOX CONNECTION

A proper conduit-to-box connection is shown in *Figure 64*.

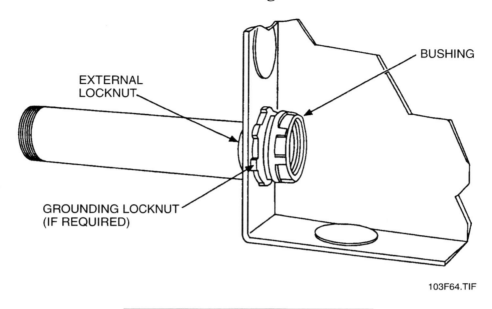

Figure 64. Conduit-To-Box Connection

In order to make a good connection, use the following procedure:

Step 1 Thread the external locknut onto the conduit. Run the locknut to the bottom of the threads.

Step 2 Insert the conduit into the box opening.

Step 3 If an inside locknut or grounding locknut is required, screw it onto the conduit inside the box opening.

Step 4 Screw the bushing onto the threads projecting into the box opening. Make sure the bushing is tightened as much as possible.

Step 5 Tighten the external locknut to secure the conduit to the box.

It is important that the bushings and locknuts fit tightly against the box. For this reason, the conduit must enter straight into the box (*Figure 65*). This may require that a box offset or kick be made in the conduit.

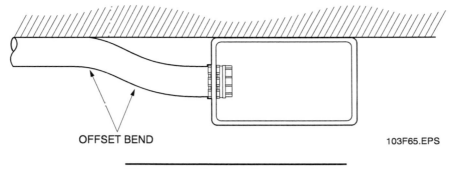

Figure 65. Correct Entrance Angle

16.0.0 VARIOUS CONSTRUCTION PROCEDURES

16.1.0 MASONRY AND CONCRETE FLUSH-MOUNT CONSTRUCTION

In a reinforced concrete construction environment, the conduit and boxes must be embedded in the concrete to achieve a flush surface. Ordinary boxes may be used, but special concrete boxes are preferred and are available in depths up to six inches. These boxes have special ears by which they are nailed to the wooden forms for the concrete. When installing them, stuff the boxes tightly with paper to prevent concrete from seeping in. *Figure 66* shows an installed box.

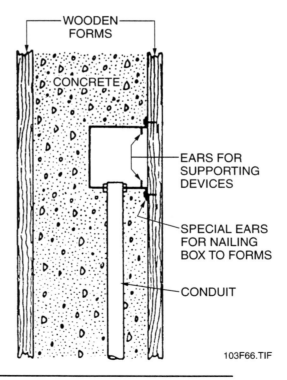

Figure 66. Concrete Flush-Mount Installation

Flush construction can also be done on existing concrete walls, but this requires chiseling a channel and box opening, anchoring the box and conduit, and then resealing the wall.

To achieve flush construction with masonry walls, the most acceptable method is to work closely with the mason laying the blocks. When the construction blocks reach the convenience outlet elevation, boxes are made up as shown in *Figure 67*. The figure shows a raised tile ring or box device cover.

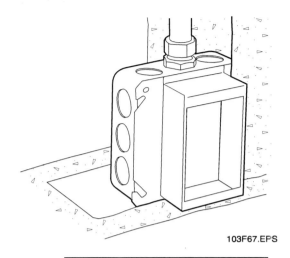

Figure 67. Concrete Outlet Box

Note: Ensure that the box is properly grouted and sealed.

Figure 68 shows the use of a 4-S extension ring installed to bring the box to the masonry surface.

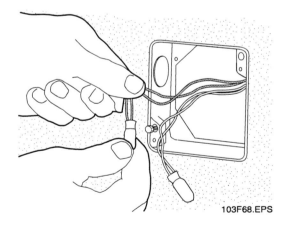

Figure 68. 4-S Extension Ring Used To Bring The Box To the Masonry Surface

Figure 69 shows a masonry box that needs no extension or deep plaster ring to bring it to the surface.

Note: EMT should be installed in the rear knockout of this masonry switch box.

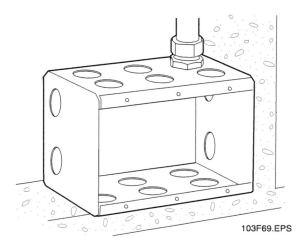

Figure 69. Three-Gang Switch Box

Sections of conduit are then coupled in short (4' or 5') lengths. This is done because it is impractical for the mason to maneuver blocks over 10' sections of conduit.

16.2.0 METAL STUD ENVIRONMENT

Metal stud walls are a popular method of construction for the interior walls of commercial buildings. Metal stud framing consists of relatively thin metal channel studs, usually constructed of galvanized steel and with the same overall dimensions as standard 2 × 4 wooden studs. Wiring in this type of construction is relatively easy when compared to masonry.

EMT conduit is the most common type of raceway specified for metal stud wiring. Metal studs usually have some number of pre-punched holes that can be used to route the conduit. If a pre-punched hole is not located where it needs to be, holes can easily be punched in the metal stud using a hole cutter or knockout punch.

WARNING! Cutting or punching metal studs can create sharp edges. Avoid contact with these edges, which can result in severe cuts.

Boxes can be secured to the metal stud using self-tapping screws or one of the many types of box supports available. EMT conduit is supported by the metal studs using conduit straps or other approved methods. It is important that the conduit be properly supported to facilitate pulling the conductors through the tubing. Boxes are mounted on the metal studs so that the box will be flush with the finished walls. You must know the finished wall thickness in order to properly secure the boxes to the metal studs. For example, if the finished wall will be $\frac{5}{8}$" drywall, then the box must be fastened so that it protrudes $\frac{5}{8}$" from the metal stud.

WARNING! When using a screw gun or cordless drill to mount boxes to studs, keep the hand holding the box away from the gun/drill to avoid injury.

Figure 70 shows several examples of clips, known as *caddy-fastening devices*, that are used in metal stud environments.

Figure 70. Caddy-Fastening Devices

16.3.0 WOOD-FRAME ENVIRONMENT

At one time, the use of rigid conduit in partitions and ceilings was a laborious and time consuming operation. Thinwall conduit makes an easier and far quicker job, largely because of the types of fittings that are specially adapted to it.

Figure 71 shows two methods of running thinwall conduit in these locations: boring timbers and notching them. When boring, holes must be drilled large enough for the tubing to be inserted between the studs. The tubing is cut rather short, calling for multiple couplings. EMT can be bowed quite a bit while threading through holes in studs. Boring is the preferred method.

WARNING! Always wear safety goggles when drilling wood.

NEC Section 300-4 addresses the requirements for preventing physical damage to conductors and cabling in wood members. By keeping the edge of the drilled hole 1¼" from the closest edge of the stud, nails are not likely to penetrate the stud far enough to damage the cables. The building codes provide maximum requirements for bored or notched holes in studs.

The exception in the NEC permits IMC, RMC, RNMC, and EMT to be installed through bored holes or laid-in notches less than 1¼" from the nearest edge without a steel plate or bushing. Also, riser-rated conduit must be used in some applications. Check the specifications for fire ratings when working in a wood-frame environment.

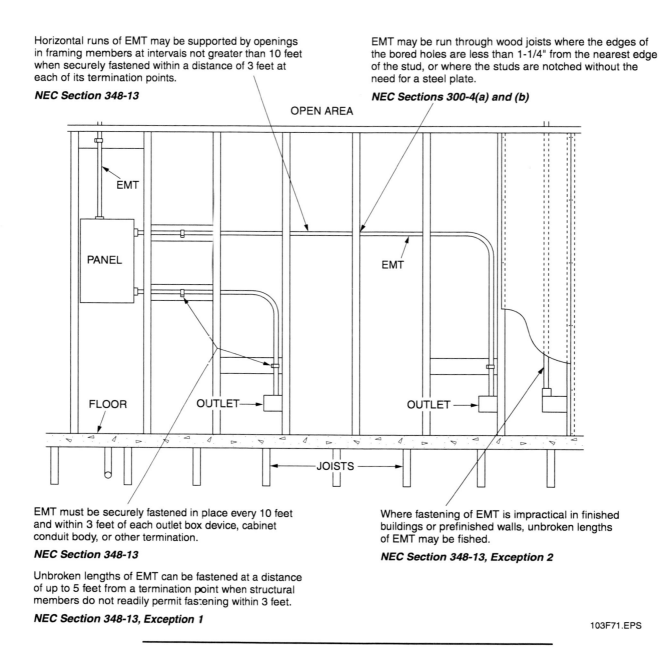

Figure 71. Installing Wire Or Conduit In A Wood-Frame Building

A bearing timber supports floor joists or other weight. Because of its weakening effect upon the structure, notching should be resorted to only where absolutely necessary. Notches should be as narrow as possible and in no case deeper than $\frac{1}{5}$ the stock of the bearing timber. An additional requirement is for the notch to be covered with a steel reinforcement bracket. This bracket aids in retaining the original strength of the timber.

Note: Always check with the architect before notching.

16.4.0 STEEL ENVIRONMENT

Installations in buildings where steel beams are the structural framework are most often found in industrial buildings and warehouses. This type of construction is typically found in pre-engineered buildings where beams and other supports are pre-cut and pre-drilled so that erection of the building is fast and simple.

The interior of the building will in most cases be unfinished, and the wiring will be supported by the metal beams and **purlins**. Beams and purlins should not be drilled through; consequently, the conduit is supported from the metal beams by anchoring devices designed especially for that purpose. The supports attach to the beams and have clamps to secure the conduit to the structure. All conduit runs should be plumb since they are exposed. Bends should be correct and have a neat and orderly appearance.

Since steel construction usually takes place in warehouses and industrial buildings where load handling and the movement of large and heavy items is common, rigid metal conduit is often required. If several runs of conduit are installed along the same path, strut-type systems are used. These systems are sometimes referred to as *Unistrut* systems (Unistrut is a manufacturer of these systems). Another manufacturer of strut systems is *B-Line*. They are very similar. These systems use a channel-type member that can support conduit from the ceiling by using threaded rod supports for the channel, as shown in *Figure 72*. Strut channel can also be secured to masonry walls to support vertical runs of conduit, wireways, and various types of boxes.

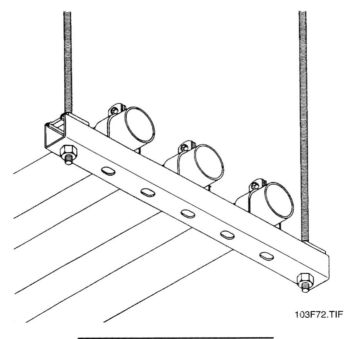

Figure 72. Steel Strut System

16.5.0 SUSPENDED CEILINGS

In commercial buildings, it is common to run cabling through the space between the suspended ceiling and the floor or roof above it. Cables may not be laid on the suspended ceiling tiles or its support grid; they must be carried in conduit or wireways approved for the purpose. *Figure 73* shows an example of a cable installation above a suspended ceiling.

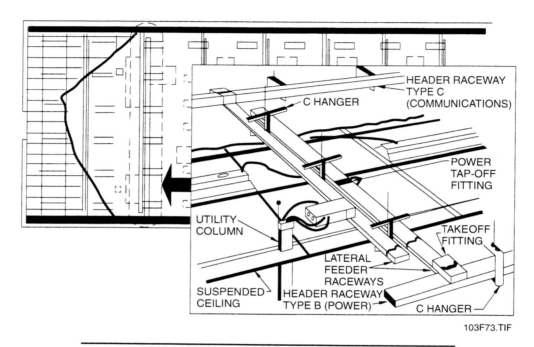

Figure 73. Raceway Installed Above A Suspended Ceiling

If the ceiling space is used as the return air plenum for the **HVAC** system, special requirements apply to the selection of the cable and raceway. For example, only fire-rated cable may be used in this application. These requirements are covered in ***NEC Section 300-22***.

For a review of the types and construction methods for suspended ceilings, refer to the module entitled *Construction Materials and Methods*.

17.0.0 OVERVIEW OF CABLE DISTRIBUTION

As previously discussed, the various types of power and low-voltage cable that enter a building commonly come from an underground source such as a maintenance hole or a CEV. Aerial sources, in which the cables are run from utility poles to the building entry point, are also used. *Figure 74* shows an overview of the cable entrance and distribution for a commercial building.

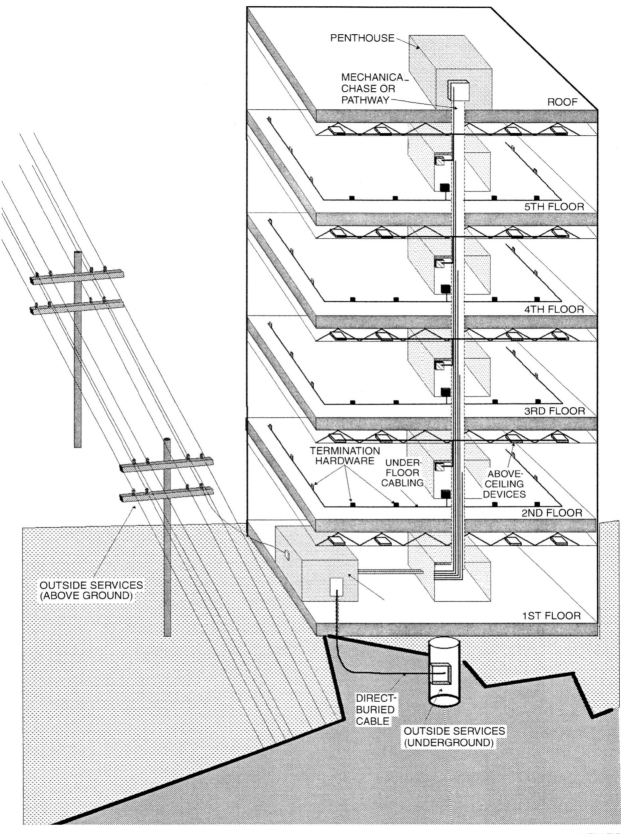

Figure 74. Example Of A Building Distribution System

The backbone cable will normally enter a building through pathways made in the exterior wall. In a large commercial building, it is common to have the entrance facility for power and low-voltage cabling in the equipment room (*Figure 75*). From there, it is distributed to the remainder of the building. Each floor of the building will have a telephone room or closet from which cabling is routed to the occupied spaces. Equipment rooms are designed to house large components such as equipment cabinets and relay racks, mainframe computers, uninterruptible power supplies, and/or video head-end equipment.

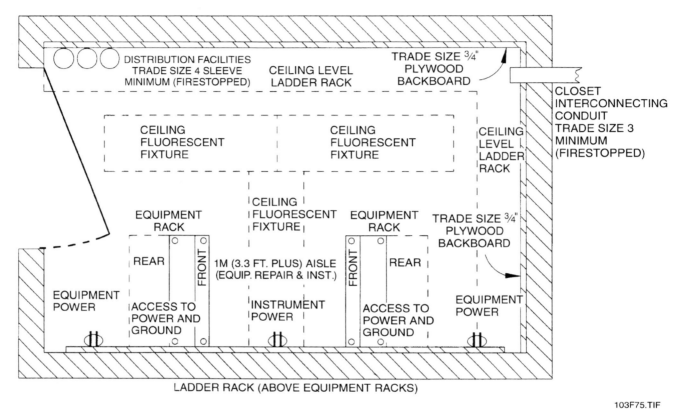

Figure 75. Example Of A Telecommunications Closet/Equipment Room Layout (Top View)

17.1.0 PATHWAYS

Bundles of cables leaving the equipment room are distributed using conduit or wireways, which may travel above a suspended ceiling, through a cellular concrete floor, or under a raised floor. Cables may also be distributed through underfloor or perimeter raceways.

The ultimate distribution of cabling and wiring depends on the application. Some cable and wiring is connected (hard-wired) directly to the equipment. In other cases, the cables terminate in boxes with wall plates to be used with various devices.

As your training progresses, you will learn how to select, pull, and terminate cables and wiring for each of the many types of electronic equipment and systems covered under the broad umbrella of low-voltage systems.

17.2.0 SPACES

The telecommunications closet located on each floor of the building is, in most cases, actually a room that houses the telecommunications equipment for the floor. Within the telecommunications closet, ¾" or 1" plywood boards will be mounted to provide an attachment base for equipment and cable management devices. Plywood with a grade of A/C is preferred. This grading means that the plywood surface on which the equipment is mounted is Grade A, which has no blemishes. The Grade C side, which faces the wall, may have some blemishes and/or small knotholes.

If the plywood is to be painted, fire-resistant plywood cannot be used because it is pressure-treated with a solution that resists paint. Standard A/C plywood may be painted with two coats of fire-resistant paint.

The plywood is installed in 4' × 8' sheets, with the long side vertical and the sheets butted together. The bottoms of the sheets are placed at floor level so that the tops are **accessible** without a ladder. Plywood sheets should be installed all the way around the closet, if possible.

There may be thousands of wires and cables coming into and going out of the telecommunications room. The hardware commonly used to manage this wiring includes D-rings and mushroom posts (*Figure 76*).

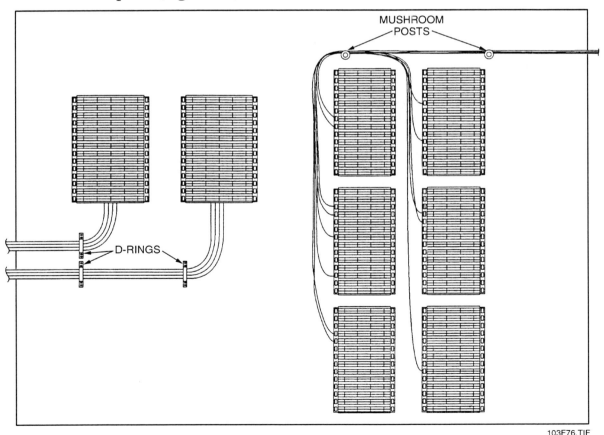

Figure 76. Hardware Used To Manage Cables And Wiring

SUMMARY

This module discussed the various types of raceways, boxes, and fittings, including their uses and procedures for installation. The primary purpose of raceways is to house wires and cables used for power distribution, communications, and electronic signal transmission. Raceways provide protection to the wiring and even a means of identifying one type of wire from another when they are run adjacent to each other. This process requires proper planning to allow for current needs, future expansion, and a neat and orderly appearance.

References

For advanced study of topics covered in this Task Module, the following books are suggested:

Benfield Conduit Bending Manual, Latest Edition, McGraw-Hill Publishing Company, New York, NY.

National Electrical Code Handbook, Latest Edition, National Fire Protection Association, Quincy, MA.

Telecommunications Cabling Installation Manual, Latest Edition, BICSI, Tampa, FL. www.bicsi.org

Telecommunications Distribution Methods Manual, Latest Edition, BICSI, Tampa, FL. www.bicsi.org

ACKNOWLEDGMENTS
Figures 2, 50, 61, 62, 63, 73, and 75 courtesy of BICSI.

REVIEW/PRACTICE QUESTIONS

1. In any building exceeding _____ floor(s), ENT must be fire-rated and concealed.
   a. one
   b. two
   c. three
   d. four

2. The lightest duty and most widely used nonflexible metal conduit available for enclosing and protecting cabling and wiring is _____.
   a. electrical metallic tubing
   b. rigid metal conduit
   c. aluminum conduit
   d. plastic-coated GRC

3. Because the wall thickness of _____ is less than that of rigid metal conduit, it is often referred to as *thinwall conduit*.
   a. intermediate metal conduit
   b. electrical metallic tubing
   c. rigid nonmetallic conduit
   d. galvanized rigid steel conduit

4. The type of conduit that provides the best physical protection for the wire inside is _____.
   a. flexible metal conduit
   b. rigid metal conduit
   c. electrical metallic tubing
   d. intermediate metal conduit

5. Bending IMC is often easier than bending the same size and quantity of rigid metal conduit because IMC has a slightly _____.
   a. larger outside diameter
   b. larger internal diameter
   c. smaller outside diameter
   d. smaller internal diameter

6. Flexible metal conduit _____.
   a. is made from a single strip of steel or aluminum
   b. can be used in wet locations if lead-covered conductors are installed
   c. can be used in underground locations
   d. is available in sizes up to six inches in diameter

7. A Type LB conduit body has the cover on _____.
   a. the left
   b. the right
   c. the back
   d. both sides

8. When conduit is joined to metal boxes, _____ protect the wires from the sharp edges of the conduit.
   a. washers
   b. locknuts
   c. bushings
   d. couplings

9. A J-clamp is used to attach _____.
   a. perimeter raceways to a wall
   b. cable to beams
   c. wireway hangers to beams
   d. pancake raceways to a floor

10. _____ are rigid rectangular raceways used for housing electric wires and cables.
    a. Troughs
    b. Gutters
    c. Pull boxes
    d. Conduits

11. Pancake raceways are a type of _____.
    a. rigid tubing
    b. flexible metal conduit
    c. surface metal raceway
    d. surface nonmetallic raceway

12. Surface metal raceway systems are _____.
    a. designed for concealed wiring within walls and partitions
    b. suited for use in wet or damp locations
    c. mainly used for making the transition from a masonry surface to a wood surface
    d. designed for exposed installations on the surface of walls and ceilings

13. Which of the following statements applies when running low-voltage and power conductors?
    a. They may not be run in the same raceway.
    b. They may be run in the same raceway but must be at least 2" apart.
    c. They may be run in the same raceway but must be separated by an approved barrier.
    d. If they are run outside a raceway, they must be at least 10' apart.

14. When transporting metal conduit, it is best to _____.
    a. remove your gloves so that you do not contaminate the conduit
    b. remove the thread protection caps so that they do not get lost
    c. place the metal conduit inside a larger raceway to keep it from bending
    d. keep the thread protection caps on to prevent damage to the threads

15. If you plan to use treated fire-resistant plywood as a backboard in a telecommunications closet, it should _____.
    a. be painted with fire-resistant paint
    b. not be painted
    c. be primed with shellac first
    d. be painted with two coats of latex paint

ANSWERS TO REVIEW/PRACTICE QUESTIONS

Answer	Section Reference
1. c	3.1.1
2. a	3.1.3
3. b	3.1.3
4. b	3.1.4
5. b	3.1.8
6. a	3.1.11
7. c	4.2.2
8. c	6.0.0
9. b	8.5.2/Figure 35
10. a	9.0.0
11. c	9.4.1
12. d	9.4.1
13. c	9.4.2
14. d	12.0.0
15. b	17.2.0

NCCER CRAFT TRAINING USER UPDATES

The NCCER makes every effort to keep these manuals up-to-date and free of technical errors. We appreciate your help in this process. If you have an idea for improving this manual, or if you find an error, a typographical mistake, or an inaccuracy in the NCCER's Craft Training Manuals, please write us, using this form or a photocopy. Be sure to include the exact module number, page number, a description of the problem, and the correction, if possible. Your input will be brought to the attention of the Technical Review Committee. Thank you for your assistance.

Instructors – If you found that additional materials were necessary in order to teach this module effectively, please let us know so that we may include them in the Equipment/Materials list in the Instructor's Guide.

Write: Curriculum Revision and Development Department
National Center for Construction Education and Research
P.O. Box 141104
Gainesville, FL 32614-1104

Fax: 352-334-0932

Craft _____ Module Name _____

Copyright Date _____ Module Number _____ Page Number(s) _____

Description of Problem

_____
_____
_____
_____
_____

(Optional) Correction of Problem

_____
_____
_____
_____

(Optional) Your Name and Address

_____
_____
_____

notes

Fasteners and Anchors
Module 33104

Electronic Systems Technician Trainee Task Module 33104

FASTENERS AND ANCHORS

NATIONAL
CENTER FOR
CONSTRUCTION
EDUCATION AND
RESEARCH

OBJECTIVES

Upon completion of this module, the trainee will be able to:

1. Identify and explain the use of threaded fasteners.
2. Identify and explain the use of non-threaded fasteners.
3. Identify and explain the use of anchors.
4. Demonstrate the correct applications for fasteners and anchors.
5. Install fasteners and anchors.

Prerequisites

Successful completion of the following Task Modules is recommended before beginning study of this Task Module: Core Curricula; Electronic Systems Technician Level One, Modules 33101 through 33103.

Required Trainee Materials

1. Trainee Task Module
2. Copy of the latest edition of the *National Electrical Code*
3. Appropriate Personal Protective Equipment

Note: The designations "National Electrical Code," "NE Code," and "NEC," where used in this document, refer to the *National Electrical Code®*, which is a registered trademark of the National Fire Protection Association, Quincy, MA. *All National Electrical Code (NEC) references in this module refer to the 1999 edition of the NEC.*

COURSE MAP

This course map shows all of the modules in the first level of the Electronic Systems Technician curricula. The suggested training order begins at the bottom and proceeds up. Skill levels increase as a trainee advances on the course map. The training order may be adjusted by the local Training Program Sponsor.

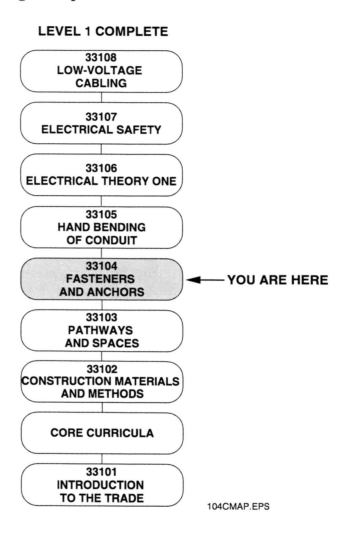

TABLE OF CONTENTS

Section	Topic	Page
	Course Map	2
1.0.0	Introduction	6
2.0.0	Threaded Fasteners	6
2.1.0	Thread Standards	6
2.1.1	Thread Series	7
2.1.2	Thread Classes	7
2.1.3	Thread Identification	7
2.1.4	Grade Markings	8
2.2.0	Bolt And Screw Types	8
2.2.1	Machine Screws	8
2.2.2	Machine Bolts	10
2.2.3	Cap Screws	10
2.2.4	Setscrews	11
2.2.5	Stud Bolts	12
2.3.0	Nuts	13
2.3.1	Jam Nuts	14
2.3.2	Castellated, Slotted, And Self-Locking Nuts	15
2.3.3	Acorn Nuts	15
2.3.4	Wing Nuts	16
2.4.0	Washers	16
2.4.1	Lock Washers	17
2.4.2	Flat And Fender Washers	17
2.5.0	Installing Fasteners	17
2.5.1	Torque Tightening	19
2.5.2	Installing Threaded Fasteners	20
3.0.0	Nonthreaded Fasteners	21
3.1.0	Retainer Fasteners	21
3.2.0	Keys	22
3.3.0	Pin Fasteners	23
3.3.1	Dowel Pins	23
3.3.2	Taper And Spring Pins	24
3.3.3	Cotter Pins	25
3.4.0	Blind Rivets	25
3.5.0	Tie Wraps	27
4.0.0	Special Threaded Fasteners	28
4.1.0	Eye Bolts	28
4.2.0	Anchor Bolts	29
4.3.0	J-Bolts	30
5.0.0	Screws	30
5.1.0	Wood Screws	31
5.2.0	Lag Screws And Shields	31
5.3.0	Concrete/Masonry Screws	32
5.4.0	Thread-Forming And Thread-Cutting Screws	33
5.5.0	Deck Screws	34
5.6.0	Drywall Screws	34
5.7.0	Drive Screws	35
6.0.0	Hammer-Driven Pins And Studs	35

TABLE OF CONTENTS (Continued)

Section	Topic	Page
7.0.0	Powder-Actuated Tools And Fasteners	36
8.0.0	Mechanical Anchors	38
8.1.0	One-Step Anchors	38
8.1.1	Wedge Anchors	39
8.1.2	Stud Bolt Anchors	39
8.1.3	Sleeve Anchors	39
8.1.4	One-Piece Anchors	40
8.1.5	Hammer-Set Anchors	40
8.2.0	Bolt Anchors	40
8.2.1	Drop-In Anchors	41
8.2.2	Single- And Double-Expansion Anchors	41
8.2.3	Lead (Caulk-In) Anchors	41
8.3.0	Screw Anchors	42
8.4.0	Self-Drilling Anchors	42
8.5.0	Guidelines For Drilling Anchor Holes In Hardened Concrete Or Masonry	43
8.6.0	Hollow-Wall Anchors	44
8.6.1	Toggle Bolts	45
8.6.2	Sleeve-Type Wall Anchors	46
8.6.3	Wallboard Anchors	47
8.6.4	Metal Drive-In Anchors	47
9.0.0	Epoxy Anchoring Systems	47
	Summary	49
	Review/Practice Questions	50
	Answers To Review/Practice Questions	52
	Appendix A	53

Trade Terms Introduced In This Module

American Society for Testing and Materials (ASTM): An organization that publishes specifications and standards relating to fasteners.

Clearance: The amount of space between the threads of bolts and their nuts.

Foot-pounds (ft. lbs.): The normal method used for measuring the amount of torque being applied to bolts or nuts.

Inch-pounds (in. lbs.): A method of measuring the amount of torque applied to small bolts or nuts that require measurement in smaller increments than foot-pounds.

Key: A machined metal part that fits into a keyway and prevents parts such as gears or pulleys from rotating on a shaft.

Keyway: A machined slot in a shaft and on parts such as gears and pulleys that accepts a key.

Nominal size: A means of expressing the size of a bolt or screw. It is the approximate diameter of a bolt or screw.

Society of Automotive Engineers (SAE): An organization that publishes specifications and standards relating to fasteners.

Thread classes: Threads are distinguished by three classifications according to the amount of tolerance the threads provide between the bolt and nut.

Thread identification: Standard symbols used to identify threads.

Thread standards: An established set of standards for machining threads.

Tolerance: The amount of difference allowed from a standard.

Torque: The turning force applied to a fastener.

Unified National Coarse (UNC) thread: A standard type of coarse thread.

Unified National Extra Fine (UNEF) thread: A standard type of extra-fine thread.

Unified National Fine (UNF) thread: A standard type of fine thread.

1.0.0 INTRODUCTION

Fasteners are used to assemble and install many different types of equipment, parts, and materials. Fasteners include screws, bolts, nuts, pins, clamps, retainers, tie wraps, rivets, and **keys**. You need to be familiar with the many different types of fasteners in order to identify, select, and properly install the correct fastener for a specific application.

Note: Staples may sometimes be used to secure cables and/or wiring to wood framing. Because of the potential for damage to the conductors, however, the use of staples is limited. Check the standards applicable to the type of wiring or cabling you are installing. In situations where staples are acceptable, a special staple gun is used, along with special staples designed to avoid crimping low-voltage cable.

The two primary categories of fasteners are:

- Threaded fasteners
- Nonthreaded fasteners

Within each of these two categories, there are numerous different types and sizes of fasteners. Each type of fastener is designed for a specific application. The kind of fastener used for a job may be listed in the project specifications, or you may have to select an appropriate fastener.

Failure of fasteners can result in a number of different problems. To perform quality work, it is important to use the correct type and size of fastener for the particular job. It is equally important that the fastener be installed properly.

In this module, you will be introduced to common fasteners.

2.0.0 THREADED FASTENERS

Threaded fasteners are the most commonly used type of fastener. Many threaded fasteners are assembled with nuts and washers. The following sections describe standard threads used on threaded fasteners, as well as different types of bolts, screws, nuts, and washers. *Figure 1* shows several types of threaded fasteners.

2.1.0 THREAD STANDARDS

There are many different types of threads used for manufacturing fasteners. The different types of threads are designed to be used for different jobs. Threads used on fasteners are manufactured to industry established standards for uniformity. The most common **thread standard** is the Unified standard, sometimes referred to as the *American standard*. Unified standards are used to establish thread series and classes.

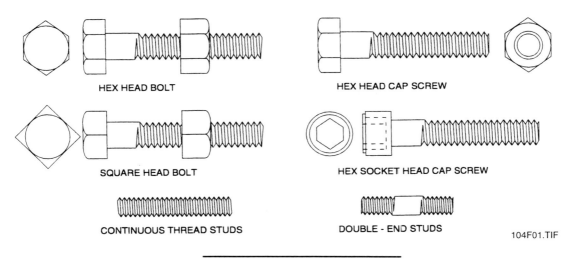

Figure 1. Threaded Fasteners

2.1.1 Thread Series

Unified standards are established for three series of threads, depending on the number of threads per inch for a certain diameter of fastener. These three series are:

- ***Unified National Coarse (UNC) thread*** – Used for bolts, screws, nuts, and other general purposes. Fasteners with UNC threads are commonly used for rapid assembly or disassembly of parts and where corrosion or slight damage may occur.
- ***Unified National Fine (UNF) thread*** – Used for bolts, screws, nuts, and other applications where a finer thread for a tighter fit is desired.
- ***Unified National Extra Fine (UNEF) thread*** – Used on thin-walled tubes, nuts, ferrules, and couplings.

2.1.2 Thread Classes

The Unified standards also establish **thread classes**. Classes 1A, 2A, and 3A apply to external threads only. Classes 1B, 2B, and 3B apply to internal threads only. Thread classes are distinguished from each other by the amounts of **tolerance** provided. Classes 3A and 3B provide a minimum **clearance** and classes 1A and 1B provide a maximum clearance.

Classes 2A and 2B are the most commonly used. Classes 3A and 3B are used when close tolerances are needed. Classes 1A and 1B are used where quick and easy assembly is needed and a large tolerance is acceptable.

2.1.3 Thread Identification

Thread identification is done using a standard method. *Figure 2* shows how screw threads are designated for a common fastener.

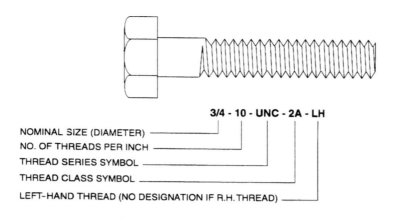

Figure 2. Screw Thread Designations

- *Nominal size* – The **nominal size** is the approximate diameter of the fastener.
- *Number of threads per inch (TPI)* – The TPI is standard for all diameters.
- *Thread series symbol* – The Unified standard thread type (UNC, UNF, or UNEF).
- *Thread class symbol* – The closeness of fit between the bolt threads and nut threads.
- *Left-hand thread symbol* – Specified by the symbol LH. Unless threads are specified with the LH symbol, the threads are right-hand threads.

2.1.4 Grade Markings

Special markings on the head of a bolt or screw can be used to determine the quality of the fastener. The **Society of Automotive Engineers (SAE)** and the **American Society for Testing and Materials (ASTM)** have developed the standards for these markings. These grade or line markings for steel bolts and screws are shown in *Figure 3*.

Generally, the higher-quality steel fasteners have a greater number of marks on the head. If the head is unmarked, the fastener is usually considered to be made of mild steel (having low carbon content).

2.2.0 BOLT AND SCREW TYPES

Bolts and screws are made in many different sizes and shapes and from a variety of materials. They are usually identified by the head type or other special characteristics. The following sections describe several different types of bolts and screws.

2.2.1 Machine Screws

Machine screws are used for general assembly work. They come in a variety of types with slotted or recessed heads. Machine screws are generally available in diameters ranging from 0 (0.060") to ½ (0.500"). The length of machine screws typically varies from ⅛" to 3".

ASTM AND SAE GRADE MARKINGS FOR STEEL BOLTS & SCREWS

GRADE MARKING	SPECIFICATION	MATERIAL
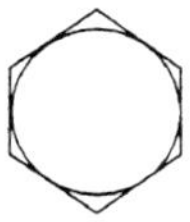 (no marks)	SAE-GRADE 0	STEEL
(no marks)	SAE-GRADE 1 ASTM-A 307	LOW CARBON STEEL
(no marks)	SAE-GRADE 2	LOW CARBON STEEL
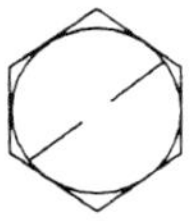 one line	SAE-GRADE 3	MEDIUM CARBON STEEL, COLD WORKED
 A 449	SAE-GRADE 5 ASTM-A 449	MEDIUM CARBON STEEL, QUENCHED AND TEMPERED
 A 325	ASTM-A 325	MEDIUM CARBON STEEL, QUENCHED AND TEMPERED
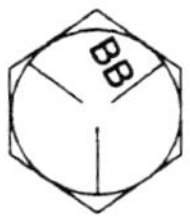 BB	ASTM-A 354 GRADE BB	LOW ALLOY STEEL, QUENCHED AND TEMPERED
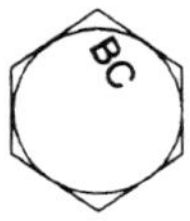 BC	ASTM-A 354 GRADE BC	LOW ALLOY STEEL, QUENCHED AND TEMPERED
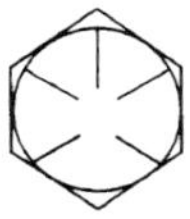 six lines	SAE-GRADE 7	MEDIUM CARBON ALLOY STEEL, QUENCHED AND TEMPERED ROLL THREADED AFTER HEAT TREATMENT
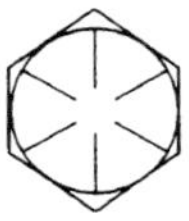 six lines	SAE-GRADE 8	MEDIUM CARBON ALLOY STEEL, QUENCHED AND TEMPERED
	ASTM-A 354 GRADE BD	ALLOY STEEL, QUENCHED AND TEMPERED
 A 490	ASTM-A 490	ALLOY STEEL, QUENCHED AND TEMPERED

ASTM SPECIFICATIONS
- A 307 - LOW CARBON STEEL EXTERNALLY AND INTERNALLY THREADED STANDARD FASTENERS.
- A 325 - HIGH STRENGTH STEEL BOLTS FOR STRUCTURAL STEEL JOINTS, INCLUDING SUITABLE NUTS AND PLAIN HARDENED WASHERS.
- A 449 - QUENCHED AND TEMPERED STEEL BOLTS AND STUDS.
- A 354 - QUENCHED AND TEMPERED ALLOY STEEL BOLTS AND STUDS WITH SUITABLE NUTS.
- A 490 - HIGH STRENGTH ALLOY STEEL BOLTS FOR STRUCTURAL STEEL JOINTS, INCLUDING SUITABLE NUTS AND PLAIN HARDENED WASHERS.

SAE SPECIFICATION
- J 429 - MECHANICAL AND QUALITY REQUIREMENTS FOR THREADED FASTENERS.

104F03.TIF

Figure 3. Grade Markings For Steel Bolts And Screws

Machine screws are also manufactured in metric sizes. *Figure 4* shows different types of machine screws.

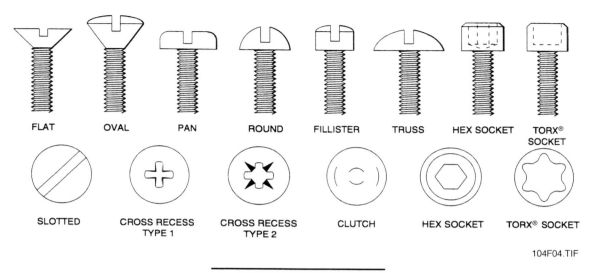

Figure 4. Machine Screws

As shown, the heads of machine screws are made in different shapes and with slots made to fit various kinds of manual and power tool screwdrivers. Flat head screws are used in a countersunk hole and tightened so that the head is flush with the surface. Oval head screws are also used in a countersunk hole in applications where a more decorative finish is desired. Pan and round head screws are general use fastening screws. Fillister, hex socket, and TORX® socket screws are typically used in confined space applications on machined assemblies that need a finished appearance. They are often installed in a recessed hole. Truss screws are a low-profile screw generally used without a washer. To prevent damage when tightening and removing machine screws (regardless of head type), make sure to use a screwdriver or power tool bit with the proper tip to drive them.

2.2.2 Machine Bolts

Machine bolts are generally used to assemble parts where close tolerances are not required. Machine bolts have square or hexagonal heads and are generally available in diameters ranging from ¼" to 3". The length of machine bolts typically varies from ½" to 30". Nuts used with machine bolts are similar in shape to the bolt heads. The nuts are usually purchased at the same time as the bolts. Figure 5 shows two different types of machine bolts.

2.2.3 Cap Screws

Cap screws are often used on high-quality assemblies requiring a finished appearance. The cap screw passes through a clearance hole in one of the assembly parts and screws into a threaded hole in the other part. The clamping action occurs by tightening the cap screw.

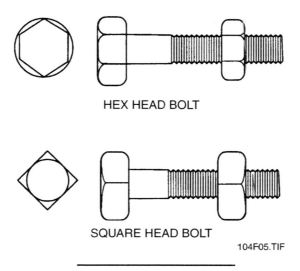

Figure 5. Machine Bolts

Cap screws are made to close tolerances and are provided with a machined or semi-finished bearing surface under the head. They are normally made in coarse and fine thread series and in diameters from ¼" to 2". Lengths may range from ⅜" to 10". Metric sizes are also available. *Figure 6* shows typical cap screws.

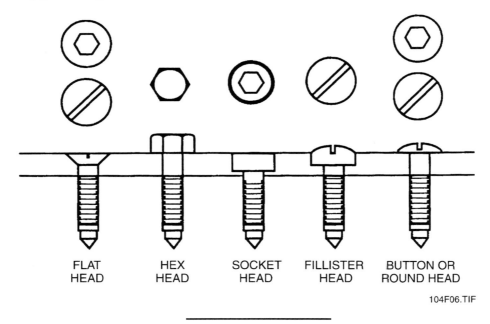

Figure 6. Cap Screws

2.2.4 Setscrews

Heat-treated steel is normally used to make setscrews. Common uses of setscrews include preventing pulleys from slipping on shafts, holding collars in place on shafts, and holding shafts in place. The head style and point style are typically used to classify setscrews. *Figure 7* shows several setscrew heads and point styles.

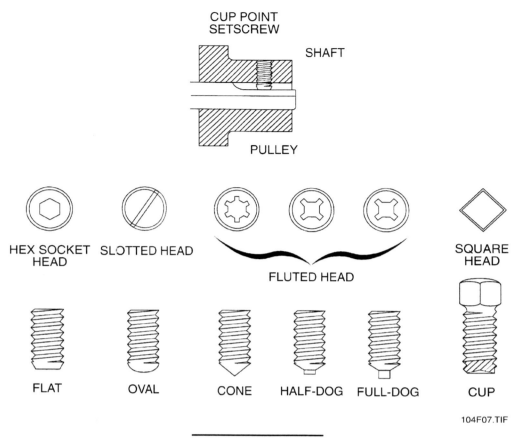

Figure 7. Setscrews

2.2.5 Stud Bolts

Stud bolts (*Figure 8*) are headless bolts that are threaded over the entire length of the bolt or for a length on both ends of the bolt. One end of the stud bolt is screwed into a tapped hole. The part to be clamped is placed over the remaining portion of the stud, and a nut and washer are screwed on to clamp the two parts together. Other stud bolts have machine-screw threads on one end and lag-screw threads on the other so that they can be screwed into wood.

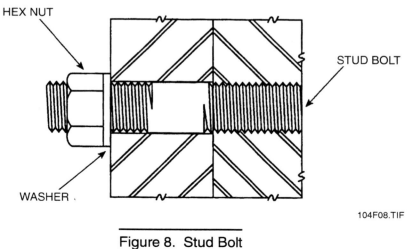

Figure 8. Stud Bolt

Stud bolts are used for several purposes, including holding together inspection covers on equipment and bearing caps.

2.3.0 NUTS

Most nuts used with threaded fasteners are hexagonal or square. They are usually used with bolts having the same shaped head. *Figure 9* shows several different types of nuts that are used with threaded fasteners.

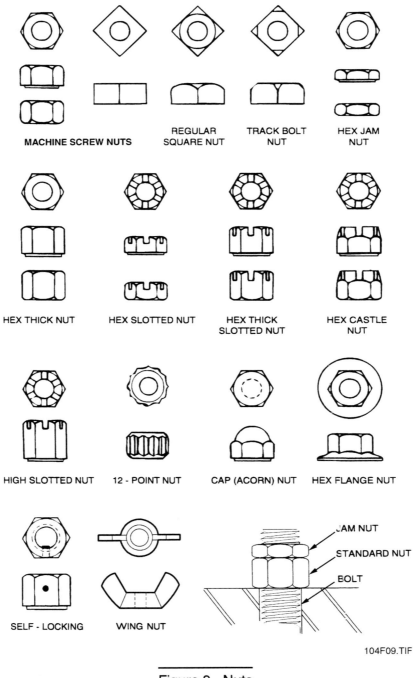

Figure 9. Nuts

FASTENERS AND ANCHORS — TRAINEE TASK MODULE 33104

Nuts are typically classified as regular, semi-finished, or finished. The only machining done on regular nuts is to the threads. In addition to the threads, semi-finished nuts are also machined on the bearing face. Machining the bearing face makes a truer surface for fitting the washer. The only difference between semi-finished and finished nuts is that finished nuts are made to closer tolerances.

The standard machine screw nut has a regular finish. Regular and semi-finished nuts are shown in *Figure 10*.

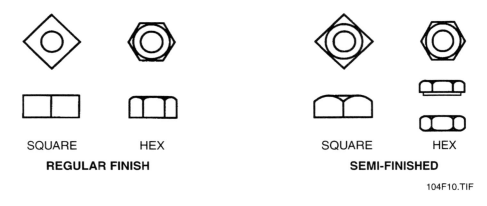

Figure 10. Nut Finishes

2.3.1 Jam Nuts

A jam nut is used to lock a standard nut in place. A jam nut is a thin nut installed on top of the standard nut. *Figure 11* shows an example of a jam nut installation. Note that a regular nut can also be used as a jam nut.

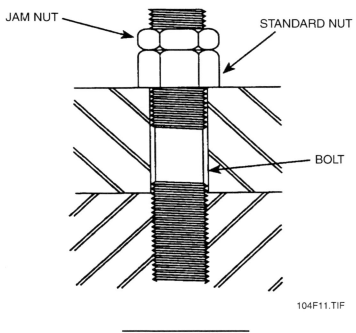

Figure 11. Jam Nut

2.3.2 Castellated, Slotted, And Self-Locking Nuts

Castellated (castle) and slotted nuts are slotted across the flat part of the nut. They are used with specially manufactured bolts in applications where little or no loosening of the fastener can be tolerated. After the nut has been tightened, a cotter pin is fitted in through a hole in the bolt and one set of slots in the nut. The cotter pin keeps the nut from loosening under working conditions.

Self-locking nuts are also used in many applications where loosening of the fastener cannot be tolerated. Self-locking nuts are designed with nylon inserts, or they are deliberately deformed in such a manner so they cannot work loose. An advantage of self-locking nuts is that no hole in the bolt is needed. *Figure 12* shows typical castellated, slotted, and self-locking nuts.

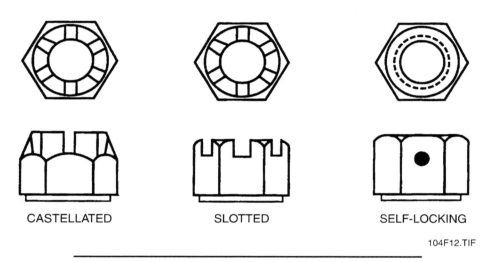

Figure 12. Castellated, Slotted, And Self-Locking Nuts

2.3.3 Acorn Nuts

When appearance is important or exposed, sharp thread edges on the fastener must be avoided, acorn (cap) nuts are used. The acorn nut tightens on the bolt and covers the ends of the threads. The tightening capability of an acorn nut is limited by the depth of the nut. *Figure 13* shows a typical acorn nut.

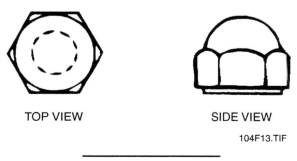

Figure 13. Acorn Nut

FASTENERS AND ANCHORS — TRAINEE TASK MODULE 33104

2.3.4 Wing Nuts

Wing nuts are designed to allow rapid loosening and tightening of the fastener without the need for a wrench. They are used in applications where limited **torque** is required and where frequent adjustments and service are necessary. *Figure 14* shows a typical wing nut.

Note: Wing nuts should be used for applications where hand tightening is sufficient.

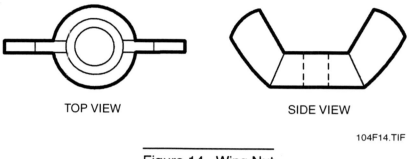

Figure 14. Wing Nut

2.4.0 WASHERS

There are several different types and sizes of washers. They fit over a bolt or screw to provide an enlarged surface for bolt heads and nuts. Washers also serve to distribute the fastener load over a larger area and to prevent marring of the surfaces. Standard washers are made in light, medium, heavy-duty, and extra heavy-duty series. *Figure 15* shows different types of washers.

Note: The threads of the bolt or screw should have minimal clearance from the hole in the washer.

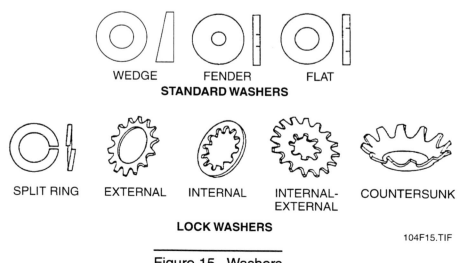

Figure 15. Washers

2.4.1 Lock Washers

Lock washers are designed to keep bolts or nuts from working loose. There are various types of lock washers for different applications.

- *Split-ring* – Commonly used with bolts and cap screws.
- *External* – Used for the greatest resistance.
- *Internal* – Used with small screws.
- *Internal-external* – Used for oversized mounting holes.
- *Countersunk* – Used with flat or oval-head screws.

2.4.2 Flat And Fender Washers

Flat washers are used under bolts or nuts to spread the load over a larger area and protect the surface. Common flat washers are made to fit bolt or screw sizes ranging from No. 6 up to 1" with outside diameters ranging from $\frac{3}{8}$" to 2", respectively.

Fender washers are wide-surfaced washers made to bridge oversized holes or other wide clearances to keep bolts or nuts from pulling through the material being fastened. They are flat washers that have a larger diameter and surface area than regular washers. They may also be thinner than a regular washer. Fender washers are typically made to fit bolt or screw sizes ranging from $\frac{3}{16}$" to $\frac{1}{2}$" with outside diameters ranging from $\frac{3}{4}$" to 2", respectively.

2.5.0 INSTALLING FASTENERS

Different types of fasteners require different installation techniques. However, all installations require knowing the proper installation methods, tightening sequence, and torque specifications for the type of fastener being used. Some bolts and nuts require that special safety wires or pins be installed to keep them from working loose.

Most fastener manufacturers provide charts that specify the size hole that should be drilled into the base material for use with each of their products (*Figure 16*). The charts typically show the proper size drill bit to use if it is necessary to first drill and tap holes for use with machine bolts, screws, or other threaded fasteners. They also show the proper size drill to use for drilling pilot holes used with metal and wood screws. (Various kinds of screws are described in detail later in this module.)

DRILL THIS SIZE HOLE		To Tap For This Size Bolt or Screw	For This Size Wood Screw Pilot in Hard Wood
Drill Size	Dec. Equiv.		
60	.0400		
59	.0410		
58	.0420		
57	.0430		
56	.0465	0 x 80	
3/64	.0469		
55	.0520		
54	.0550	1 x 56	No. 3
53	.0595	1 x 64-72	
1/16	.0625		
52	.0635		No. 4
51	.0670		
50	.0700	2 x 56-64	
49	.0730		No. 5
48	.0760		
5/64	.0781		
47	.0785	3 x 48	No. 6
46	.0810		
45	.0820	3 x 56	
44	.0860	4 x 36	No. 7
43	.0890	4 x 40	
42	.0935	4 x 48	
3/32	.0937		
41	.0960		
40	.0980	5 x 36	No. 8
39	.0995		
38	.1015	5 x 40	
37	.1040	5 x 44	No. 9
36	.1069		
7/64	.1094		
35	.1100	6 x 32	
34	.1110	6 x 36	
33	.1130	6 x 40	No. 10
32	.1160		
31	.1200		No. 11
1/8	.1250	7 x 36	
30	.1285	8 x 30	No. 12
29	.1360	8 x 32-36	
28	.1405	8 x 40	
27	.1440	9 x 30	
26	.1470	3/16 x 24	
25	.1495	10 x 24	No. 14
24	.1520		
23	.1540	10 x 28	
5/32	.1562		
22	.1570	10 x 30	
21	.1590	10 x 32	
20	.1610	3/16 x 32	
19	.1660		
18	.1695		No. 16
11/64	.1719		
17	.1730		
16	.1770	12 x 24	
15	.1800		
14	.1820	12 x 28	
13	.1850	12 x 32	No. 18
3/16	.1875		
12	.1890		

DRILL THIS SIZE HOLE		To Tap For This Size Bolt or Screw	For This Size Wood Screw Pilot in Hard Wood
Drill Size	Dec. Equiv.		
11	.1910		
10	.1935	15 x 20	
9	.1960		
8	.1990		
7	.2010	1/4 x 20	
13/64	.2031		
6	.2040		
5	.2055		
4	.2090	1/4 x 24	No. 20
3	.2130	1/4 x 28	
7/32	.2187	1/4 x 32	
2	.2210		
1	.2280		No. 24
A	.2340		
15/64	.2344		
B	.2380		
C	.2420		
D	.2460		
1/4	.2500		

DRILL THIS SIZE HOLE		To Tap For This Size Bolt or Screw
Drill Size	Dec. Equiv.	
E	.2500	
F	.2570	5/16 x 18
G	.2610	
17/64	.2656	5/16 x 18
H	.2660	
I	.2720	
J	.2770	5/16 x 24-32*
K	.2810	
9/32	.2812	5/16 x 24-32*
L	.2900	
M	.2950	
19/64	.2969	
N	.3020	
5/16	.3125	3/8* x 16-1/8* P
O	.3160	
P	.3230	
21/64	.3281	3/8 x 20-24
Q	.3332	
R	.3390	
11/32	.3437	
S	.3480	
T	.3580	
23/64	.3594	
U	.3680	
3/8	.3750	7/16 x 14
V	.3770	
W	.3860	
25/64	.3906	7/16 x 14
X	.3970	
Y	.4040	
13/32	.4062	
Z	.4130	
27/64	.4219	1/2 x 12-13
7/16	.4375	1/4* Pipe
29/64	.4531	1/2 x 20-24
15/32	.4687	1/2 x 27
31/64	.4844	9/16 x 12
1/2	.5000	

* All tap drill sizes are for 75% full thread except asterisked sizes which are 60% full thread.

Figure 16. Fastener Hole Guide Chart

2.5.1 Torque Tightening

To properly tighten a threaded fastener, two primary factors must be considered:

- The strength of the fastener material
- The degree to which the fastener is tightened

A torque wrench is used to control the degree of tightness. The torque wrench measures how much a fastener is being tightened. *Torque* is the turning force applied to the fastener. Torque is normally expressed in **inch-pounds (in. lbs.)** or **foot-pounds (ft. lbs.)**. A one-pound force applied to a wrench that is one-foot long exerts one foot-pound, or twelve inch-pounds, of torque. The torque reading is shown on the indicator on the torque wrench as the fastener is being tightened. *Figure 17* shows two types of torque wrenches.

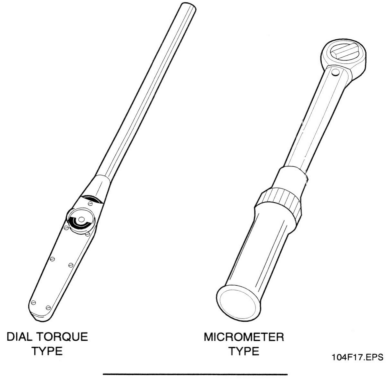

Figure 17. Torque Wrenches

Different types of bolts, nuts, and screws are torqued to different values depending on the application. Always check the project specifications and the manufacturer's manual to determine the proper torque for a particular type of fastener. *Figure 18* shows selected torque values for various graded steel bolts.

TORQUE IN FOOT-POUNDS

FASTENER DIAMETER	THREADS PER INCH	MILD STEEL	STAINLESS STEEL 18-8	ALLOY STEEL
1/4	20	4	6	8
5/16	18	8	11	16
3/8	16	12	18	24
7/16	14	20	32	40
1/2	13	30	43	60
5/8	11	60	92	120
3/4	10	100	128	200
7/8	9	160	180	320
1	8	245	285	490

SUGGESTED TORQUE VALUES FOR GRADED STEEL BOLTS

GRADE	SAE 1 OR 2	SAE 5	SAE 6	SAE 8
TENSILE STRENGTH	64000 PSI	105000 PSI	130000 PSI	150000 PSI

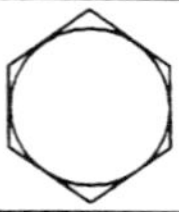

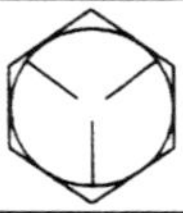

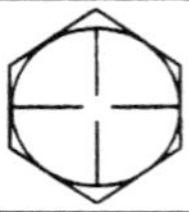

BOLT DIAMETER	THREADS PER INCH	FOOT-POUNDS TORQUE			
1/4	20	5	7	10	10
5/16	18	9	14	19	22
3/8	16	15	25	34	37
7/16	14	24	40	55	60
1/2	13	37	60	85	92
9/16	12	53	88	120	132
5/8	11	74	120	169	180
3/4	10	120	200	280	296
7/8	9	190	302	440	473
1	8	282	466	660	714

Figure 18. Torque Value Chart

2.5.2 Installing Threaded Fasteners

The following general procedure can be used to install threaded fasteners in a variety of applications.

Note: When installing threaded fasteners for a specific job, make sure to check all installation requirements.

WARNING! Follow all safety precautions.

Step 1 Select the proper bolts or screws for the job.

Step 2 Check for damaged or dirty internal and external threads.

Step 3 Clean the bolt or screw threads. Do not lubricate the threads if a torque wrench is to be used to tighten the nuts.

Step 4 Insert the bolts through the pre-drilled holes and tighten the nuts by hand. Or, insert the screws through the holes and start the threads by hand.

Note: Turn the nuts or screws several turns by hand and check for cross threading.

Step 5 Following the proper tightening sequence, tighten the bolts or screws snugly.

Step 6 Check the torque specification. Following the proper tightening sequence, tighten each bolt, nut, or screw several times approaching the specified torque. Tighten to the final torque specification.

Step 7 If required to keep the bolts or nuts from working loose, install jam nuts, cotter pins, or safety wire. *Figure 19* shows fasteners with a safety wire installed.

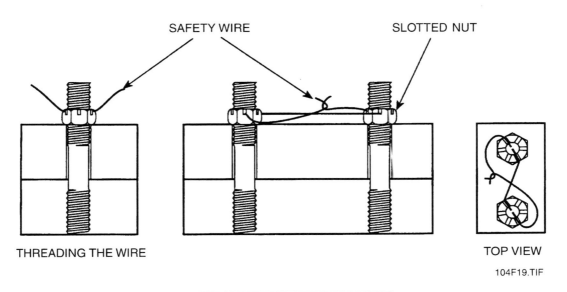

Figure 19. Safety-Wired Fasteners

3.0.0 NON-THREADED FASTENERS

Non-threaded fasteners have many uses. Different types of non-threaded fasteners include retainers, keys, pins, clamps, washers, rivets, and tie wraps.

3.1.0 RETAINER FASTENERS

Retainer fasteners, also called *retaining rings*, are used for both internal and external applications. Some retaining rings are seated in grooves in the fastener. Other types of retainer fasteners are self-locking and do not require a groove. To easily remove internal and external retainer rings without damaging the ring or the fastener, special pliers are used. *Figure 20* shows several types of retainer fasteners.

Note: External retainer fasteners are sometimes called *clips*.

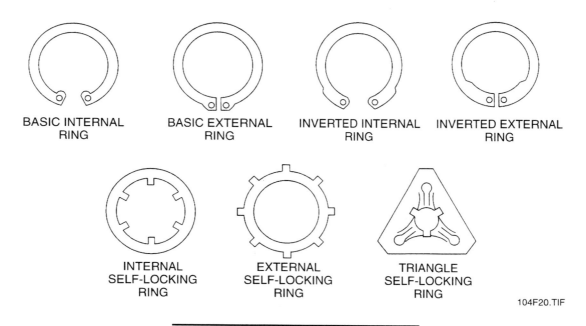

Figure 20. Retainer Fasteners (Rings)

3.2.0 KEYS

To prevent a gear or pulley from rotating on a shaft, keys are inserted. Half of the key fits into a keyseat on the shaft. The other half fits into a **keyway** in the hub of the gear or pulley. The key fastens the two parts together, stopping the gear or pulley from turning on the shaft. *Figure 21* shows several types of keys and keyways and their uses.

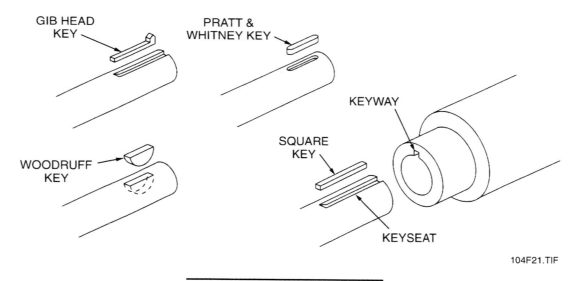

Figure 21. Keys And Keyways

Some different types of keys include:

- *Square key* – Usually one-quarter of the shaft diameter. It may be slightly tapered on the top for easier fitting.
- *Pratt and Whitney key* – Similar to the square key but rounded at both ends. It fits into a keyseat of the same shape.
- *Gib head key* – Interchangeable with the square key. The head design allows easy removal from the assembly.
- *Woodruff key* – Semicircular shape that fits into a keyseat of the same shape. The top of the key fits into the keyway of the mating part.

3.3.0 PIN FASTENERS

Pin fasteners come in several types and sizes. They have a variety of applications. Common uses of pin fasteners include holding moving parts together, aligning mating parts, fastening hinges, holding gears and pulleys on shafts, and securing slotted nuts. *Figure 22* shows several pin fasteners.

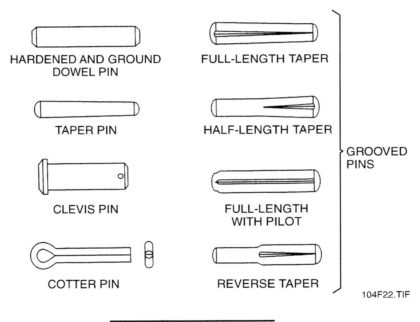

Figure 22. Pin Fasteners

3.3.1 Dowel Pins

Dowel pins are fit into holes to position mating parts. They may also support a portion of the load placed on the parts. *Figure 23* shows an application of dowel pins used to position mating parts.

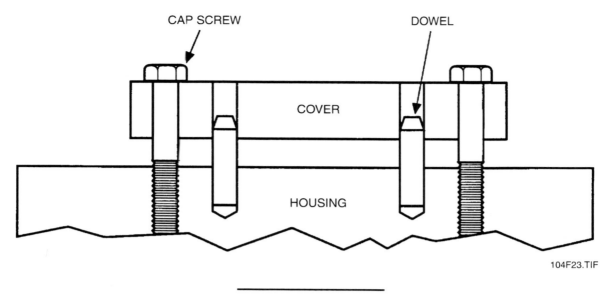

Figure 23. Dowel Pins

3.3.2 Taper And Spring Pins

Taper and spring pins are used to fasten gears, pulleys, and collars to a shaft. *Figure 24* shows how taper and spring pins are used to attach a component to a shaft. The groove in a spring pin allows it to compress against the walls in a spring-like fashion.

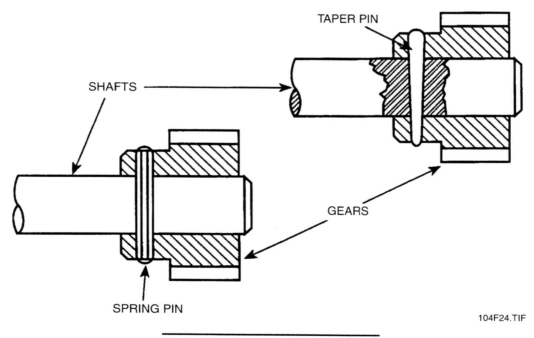

Figure 24. Taper And Spring Pins

3.3.3 Cotter Pins

There are several different types of cotter pins used as a locking device for a variety of applications. Cotter pins are often inserted through a hole drilled crosswise through a shaft to prevent parts from slipping on or off the shaft. They are also used to keep slotted nuts from working loose. Standard cotter pins are general-use pins. When installed, the extended prong is normally bent back over the nut to provide the locking action. If it is ever removed, throw it away and replace it with a new one. The humped, cinch, and hitch-type cotter pins are self-locking pins. The humped and cinch type should also be thrown away and replaced with a new one if removed. The hitch pin, also called a *hair pin*, is a reusable pin made to be installed and removed quickly. *Figure 25* shows several common types of cotter pins.

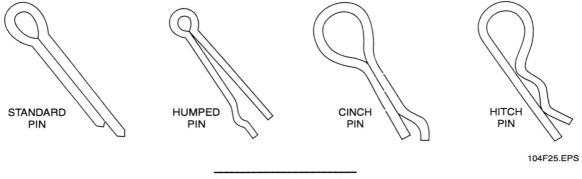

Figure 25. Cotter Pins

3.4.0 BLIND RIVETS

When only one side of a joint can be reached, blind rivets can be used to fasten the parts together. Some applications of blind rivets include fastening light- to heavy-gauge sheet metal, fiberglass, plastics, and belting. Blind rivets are made of a variety of materials and come in several sizes and lengths. They are installed using special riveting tools. *Figure 26* shows a typical blind rivet installation.

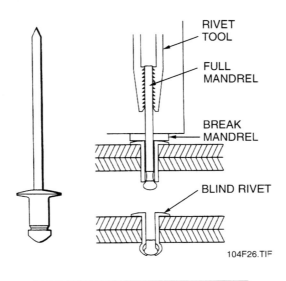

Figure 26. Blind Rivet Installation

FASTENERS AND ANCHORS — TRAINEE TASK MODULE 33104

Blind rivets are installed through drilled or punched holes using a special blind rivet gun. *Figure 27* shows a typical rivet gun.

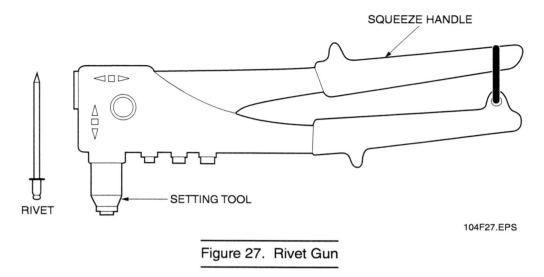

Figure 27. Rivet Gun

Use the following general procedure to install blind rivets.

WARNING!	Follow all safety precautions. Make sure to wear proper eye and face protection when riveting.

Step 1   Select the correct length and diameter of blind rivet to be used.

Step 2   Select the appropriate drill bit for the size of rivet being used.

Step 3   Drill a hole through both parts being connected.

Step 4   Inspect the rivet gun for any defects that might make it unsafe for use.

Step 5   Place the rivet mandrel into the proper size setting tool.

Step 6   Insert the rivet end into the pre-drilled hole.

Step 7   Install the rivet by squeezing the handle of the rivet gun, causing the jaws in the setting tool to grip the mandrel. The mandrel is pulled up, expanding the rivet until it breaks at the shear point. *Figure 28* shows the rivet and tool positioned for joining parts together.

Step 8   Inspect the rivet to make sure the pieces are firmly riveted together and that the rivet is properly installed. *Figure 29* shows a properly installed blind rivet.

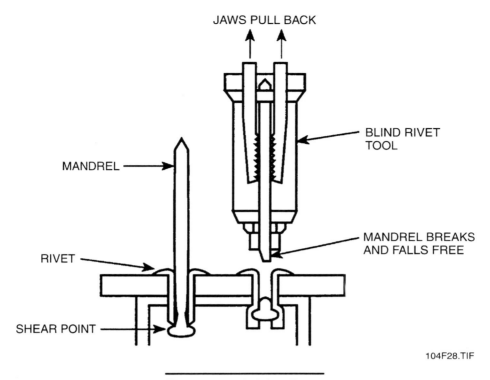

Figure 28. Joining Parts

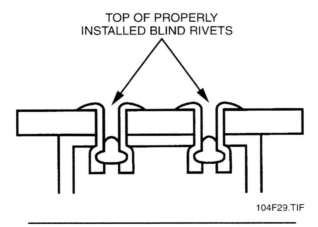

Figure 29. Properly Installed Blind Rivets

3.5.0 TIE WRAPS

A tie wrap is a one-piece, self-locking cable tie, usually made of nylon, that is used to fasten a bundle of wires and cables together. Tie wraps can be quickly installed either manually or using a special installation tool. Black tie wraps resist ultraviolet light and are recommended for outdoor use.

Tie wraps are made in standard, cable strap and clamp, and identification configurations (*Figure 30*). All types function to clamp bundled wires or cables together. In addition, the cable strap and clamp has a molded mounting hole in the head used to secure the tie with a rivet, screw, or bolt after the tie wrap has been installed around the wires or cable. Identification tie wraps have a large flat area provided for imprinting or writing cable identification information. There is also a releasable version available. It is a nonpermanent tie used for bundling wires or cables that may require frequent additions or deletions. Cable ties are made in various lengths ranging from about 3" to 30", allowing them to be used for fastening wires and cables into bundles with diameters ranging from about ½" to 9", respectively. Tie wraps can also be attached to a variety of adhesive mounting bases made for that purpose.

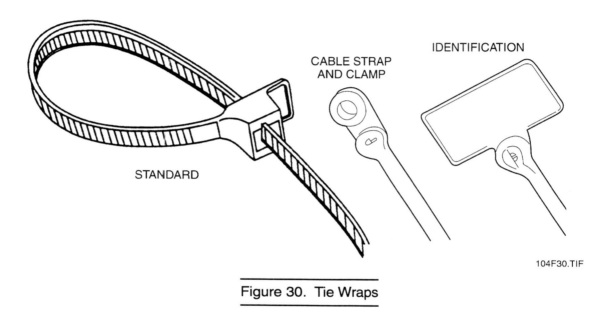

Figure 30. Tie Wraps

4.0.0 SPECIAL THREADED FASTENERS

Special threaded fasteners consist of hardware manufactured in several shapes and sizes and designed to perform specific jobs. Certain types of nuts may be considered special threaded fasteners if they are designed especially for a particular application.

The three types of special threaded fasteners described below are eye bolts, toggle bolts, and J-bolts.

4.1.0 EYE BOLTS

Eye bolts get their name from the eye or loop at one end. The other end of an eye bolt is threaded. There are many types of eye bolts. The eye on some eye bolts is formed and welded while the eye on other types is forged. Shoulder forged eye bolts are commonly used as lifting devices and guides for wires, cables, and cords. *Figure 31* shows some typical eye bolts.

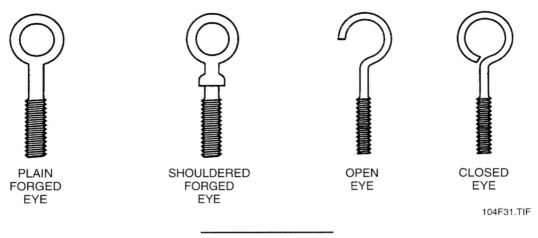

Figure 31. Eye Bolts

4.2.0 ANCHOR BOLTS

An anchor bolt is used to fasten parts, machines, and equipment to concrete or masonry foundations, floors, and walls. There are several types of anchor bolts designed for different applications. *Figure 32* shows a type of anchor bolt for use in wet concrete. If the concrete has already hardened, expansion anchor bolts are used. These are covered later in this module.

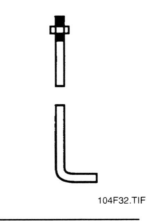

Figure 32. Anchor Bolt

One common method used to install anchor bolts in wet concrete involves making a wooden template to locate the anchor bolts. The template positions the anchor bolts so that they correspond to those in the equipment to be fastened.

4.3.0 J-BOLTS

J-bolts get their name from the curve on one end that gives them a J shape. The other end of a J-bolt is threaded. There are many types of J-bolts. Some J-bolts are used to hold tubing bundles and include a plastic jacket to protect the tubing. Others are used to attach equipment to existing grating. Most J-bolts used in tubing racks are attached using two nuts. The upper nut allows for adjustment. The tubing bundle is clamped firmly, but not flattened. Both nuts are tightened against the tube track for positive holding. *Figure 33* shows a typical J-bolt.

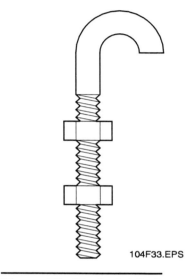

Figure 33. Typical J-Bolt

5.0.0 SCREWS

Screws are made in a variety of shapes and sizes for different fastening jobs. The finish or coating used on a screw determines whether it is for interior or exterior use, corrosion resistant, etc. Screws of all types have heads with different shapes and slots similar to those previously described for machine screws. Some have machine threads and are self-drilling. The size or diameter of a screw body or shank is given in gauge numbers ranging from No. 0 to No. 24, and in fractions of an inch for screws with diameters larger than ¼". The higher the gauge number, the larger the diameter of the shank. Screw lengths range from ¼" to 6", measured from the tip to the part of the head that is flush to the surface when driven in. When choosing a screw for an application, you must consider the type and thickness of the materials to be fastened, the size of the screw, the material it is made of, the shape of its head, and the type of driver. Because of the wide diversity in the types of screws and their application, always follow the manufacturer's recommendation to select the right screw for the job. To prevent damage to the screw head or the material being fastened, always use a screwdriver or power driver bit with the proper size and shape tip to fit the screw.

Some of the more common types of screws are:

- Wood screws
- Lag screws
- Masonry/concrete screws
- Thread-forming and thread-cutting screws
- Deck screws
- Drywall screws
- Drive screws

5.1.0 WOOD SCREWS

Wood screws (*Figure 34*) are typically used to fasten boxes, panel enclosures, etc., to wood framing or structures where greater holding power is needed than can be provided by nails. They are also used to fasten equipment to wood in applications where it may occasionally need to be unfastened and removed. Wood screws are commonly made in lengths from ¼" to 4", with shank gauge sizes ranging from 0 to 24. The shank size used is normally determined by the size hole provided in the box, panel, etc., to be fastened. When determining the length of a wood screw to use, a good rule of thumb is to select screws long enough to allow about ⅔ of the screw length to enter the piece of wood that is being gripped.

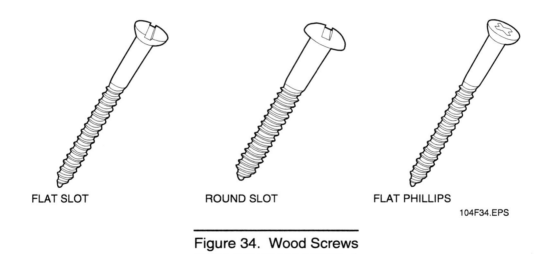

Figure 34. Wood Screws

5.2.0 LAG SCREWS AND SHIELDS

Lag screws (*Figure 35*) or lag bolts are heavy-duty wood screws with square- or hex-shaped heads that provide greater holding power. Lag screws with diameters ranging between ¼" and ½" and lengths ranging from 1" to 6" are common. They are typically used to fasten heavy equipment to wood, but they can also be used to fasten equipment to concrete when a lag shield is used.

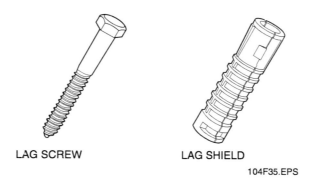

Figure 35. Lag Screw And Shield

A lag shield is a lead tube that is split lengthwise but remains joined at one end. It is placed in a pre-drilled hole in the concrete. When a lag screw is screwed into the lag shield, the shield expands in the hole, firmly securing the lag screw. In hard masonry, short lag shields (typically 1" to 2" long) may be used to minimize drilling time. In soft or weak masonry, long lag shields (typically up to 3" long) should be used to achieve maximum holding strength.

Make sure to use the proper length lag screw to achieve proper expansion. The length of the lag screw used should be equal to the thickness of the component being fastened plus the length of the lag shield. Also, drill the hole in the masonry to a depth approximately ½" longer than the shield being used. If the head of a lag screw rests directly on wood when installed, a flat washer should be placed under the head to prevent the head from digging into the wood as the lag screw is tightened down. Be sure to take the thickness of any washers used into account when selecting the length of the screw.

5.3.0 CONCRETE/MASONRY SCREWS

Concrete/masonry screws (*Figure 36*), commonly called *self-threading anchors*, are used to fasten a device or fixture to concrete, block, or brick. No anchor is needed. To provide a matched tolerance anchoring system, the screws are installed using specially designed carbide drill bits and installation tools made for use with the screws. These tools are typically used with a standard rotary drill hammer. The installation tool, along with an appropriate drive socket or bit, is used to drive the screws directly into pre-drilled holes that have a diameter and depth specified by the screw manufacturer. When being driven into the concrete, the widely spaced threads on the screws cut into the walls of the hole to provide a tight friction fit. Most types of concrete/masonry screws can be removed and reinstalled to allow for shimming and leveling of the fastened device.

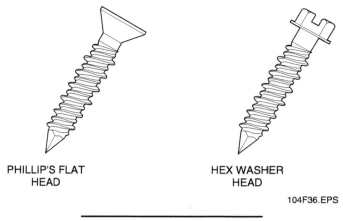

Figure 36. Concrete Screws

5.4.0 THREAD-FORMING AND THREAD-CUTTING SCREWS

Thread-forming screws (*Figure 37*), commonly called *sheet metal screws*, are made of hard metal. They form a thread as they are driven into the work. This thread-forming action eliminates the need to tap a hole before installing the screw. To achieve proper holding, it is important to make sure to use the proper size bit when drilling pilot holes for thread-forming screws. The correct drill bit size used for a specific size screw is usually marked on the box containing the screws. Some types of thread-forming screws also drill their own holes, eliminating drilling, punching, and aligning parts. Thread-forming screws are primarily used to fasten light-gauge metal parts together. They are made in the same diameters and lengths as wood screws.

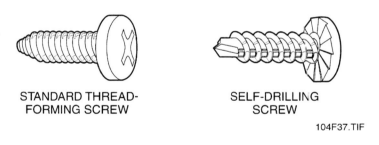

Figure 37. Thread-Forming Screws

Hardened steel thread-cutting metal screws with blunt points and fine threads (*Figure 38*) are used to join heavy-gauge metals, metals of different gauges, and nonferrous metals. They are also used to fasten sheet metal to building structural members. These screws are made of hardened steel that is harder than the metal being tapped. They cut threads by removing and cutting a portion of the metal as they are driven into a pilot hole and through the material.

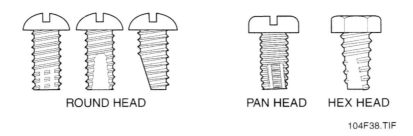

Figure 38. Thread-Cutting Screws

5.5.0 DECK SCREWS

Deck screws (*Figure 39*) are made in a wide variety of shapes and sizes for different indoor and outdoor applications. Some are made to fasten pressure-treated and other types of wood decking to wood framing. Self-drilling types are made to fasten wood decking to different gauges of metal support structures. Similarly, other self-drilling kinds are made to fasten metal decking and sheeting to different gauges and types of metal structural support members. Because of their wide diversity, it is important to follow the manufacturer's recommendations for selection of the proper screw for a particular application. Many manufacturers make a stand-up installation tool used for driving their deck screws. Use of this tool eliminates angle driving, underdriven or overdriven screws, screw wobble, etc. It also reduces operator fatigue.

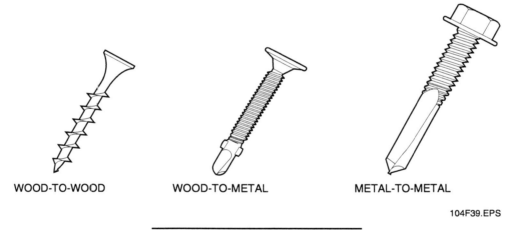

Figure 39. Typical Deck Screws

5.6.0 DRYWALL SCREWS

Drywall screws (*Figure 40*) are thin, self-drilling screws with bugle-shaped heads. Depending on the type of screw, it cuts through the wallboard and anchors itself into wood and/or metal studs, holding the wallboard tight to the stud. Coarse thread screws are normally used to fasten wallboard to wood studs. Fine thread and high-and-low thread types are generally used for fastening to metal studs. Some screws are made for use in either wood or metal. A Phillips or Robertson drive head allows the drywall screw to be countersunk without tearing the surface of the wallboard.

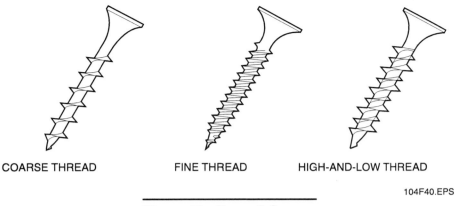

Figure 40. Drywall Screws

5.7.0 DRIVE SCREWS

Drive screws do not require that the hole be tapped. They are installed by hammering the screw into a drilled or punched hole of the proper size. Drive screws are mostly used to fasten parts that will not be exposed to much pressure. A typical use of drive screws is to attach permanent name plates on electric motors and other types of equipment. *Figure 41* shows typical drive screws.

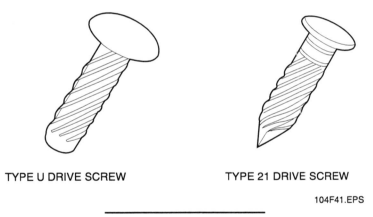

Figure 41. Drive Screws

6.0.0 HAMMER-DRIVEN PINS AND STUDS

Hammer-driven pins or threaded studs (*Figure 42*) can be used to fasten wood or steel to concrete or block without the need to pre-drill holes. The pin or threaded stud is inserted into a hammer-driven tool designed for its use. The pin or stud is inserted in the tool point end out with the washer seated in the recess. The pin or stud is then positioned against the base material where it is to be fastened and the drive rod of the tool tapped lightly until the striker pin contacts the pin or stud. Following this, the tool's drive rod is struck using heavy blows with about a two-pound engineer's hammer. The force of the hammer blows is transmitted through the tool directly to the head of the fastener, causing it to be driven into the concrete or block. For best results, the drive pin or stud should be embedded a minimum of ½" in hard concrete to 1¼" in softer concrete block.

FASTENERS AND ANCHORS — TRAINEE TASK MODULE 33104

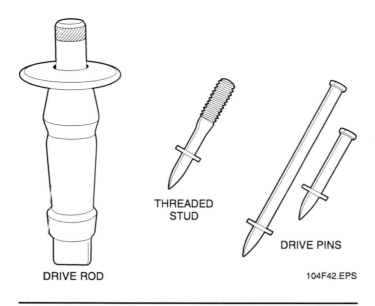

Figure 42. Hammer-Driven Pins And Installation Tool

7.0.0 POWDER-ACTUATED TOOLS AND FASTENERS

Powder-actuated tools (*Figure 43*) can be used to drive a wide variety of specially designed pin and threaded stud-type fasteners into masonry and steel. These tools look and fire like a gun and use the force of a detonated gunpowder load (typically .22, .25, or .27 caliber) to drive the fastener into the material. The depth to which the pin or stud is driven is controlled by the density of the base material in which the pin or stud is being installed and by the power level or strength of the cased powder load.

Powder loads and their cases are designed for use with specific types and/or models of powder-actuated tools and are not interchangeable. Typically, powder loads are made in 12 increasing power or load levels used to achieve the proper penetration. The different power levels are identified by a color-code system and load case types. Note that different manufacturers may use different color codes to identify load strength. Power level 1 is the lowest power level while 12 is the highest. Higher number power levels are used when driving into hard materials or when a deeper penetration is needed. Powder loads are available as single-shot units for use with single-shot tools. They are also made in multi-shot strips or disks for semi-automatic tools.

WARNING! Powder-actuated fastening tools are to be used only by trained and licensed operators and in accordance with the tool operator's manual. You must carry your license with you whenever you are using a powder-actuated tool.

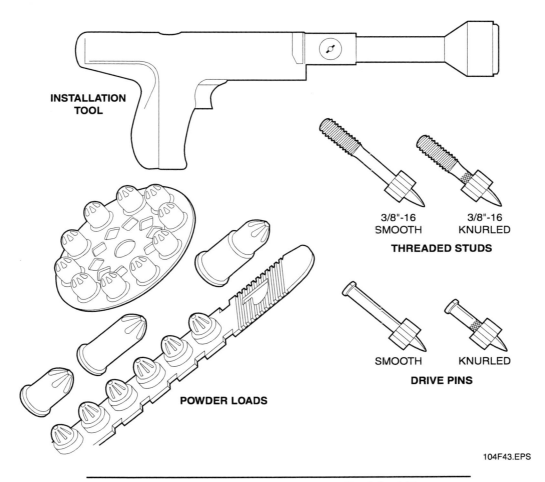

Figure 43. Powder-Actuated Installation Tool And Fasteners

OSHA Standard 29 CFR 1926.302(e) governs the use of powder-actuated tools and requires only operators who have been trained in the operation of the particular tool in use be allowed to operate a powder-actuated tool. Authorized instructors available from the various powder-actuated tool manufacturers generally provide such training and licensing. Trained operators must take precautions to protect both themselves and others in the area when using a powder-actuated driver tool:

- Always use the tool in accordance with the published tool operation instructions. The instructions should be kept with the tool. Never attempt to override the safety features of the tool.
- Never place your hand or other body parts over the front muzzle end of the tool.
- Use only fasteners, powder loads, and tool parts specifically made for use with the tool. Use of other materials can cause improper and unsafe functioning of the tool.
- Operators and bystanders must wear eye and hearing protection along with hard hats. Other personal safety gear, as required, must also be used.
- Always post warning signs which state *Powder-Actuated Tool in Use* within 50' of the area where tools are used.
- Do not guess before fastening into any base material; always perform a center punch test.

- Prior to using a tool, make sure it is unloaded and perform a proper function test. Check the functioning of the unloaded tool as described in the published tool operation instructions.
- Always make a test firing into a suitable base material with the lowest power load recommended for the tool being used. If this does not set the fastener, try the next higher power level. Continue this procedure until the proper fastener penetration is obtained.
- Always point the tool away from operators or bystanders.
- Never use the tool in an explosive or flammable area.
- Never leave a loaded tool unattended. Do not load the tool until you are prepared to complete the fastening. Should you decide not to make a fastening after the tool has been loaded, always remove the powder load first, then the fastener. Always unload the tool before cleaning, servicing, or when changing parts, prior to work breaks, and when storing the tool.
- Always hold the tool perpendicular to the work surface and use the spall (chip or fragment) guard or stop spall whenever possible.
- Always follow the required spacing, edge distance, and base material thickness requirements.
- Never fire through an existing hole or into a weld area.
- In the event of a misfire, always hold the tool depressed against the work surface for at least 30 seconds. If the tool still does not fire, follow the published tool instructions. Never carelessly discard or throw unfired powder loads into a trash receptacle.
- Always store the powder loads and unloaded tool under lock and key.

8.0.0 MECHANICAL ANCHORS

Mechanical anchors are devices used to give fasteners a firm grip in a variety of materials, where the fasteners by themselves would otherwise have a tendency to pull out. Anchors can be classified in many ways by different manufacturers. In this module, anchors have been divided into five broad categories:

- One-step anchors
- Bolt anchors
- Screw anchors
- Self-drilling anchors
- Hollow-wall anchors

8.1.0 ONE-STEP ANCHORS

One-step anchors are designed so that they can be installed through the mounting holes in the component to be fastened. This is because the anchor and the drilled hole into which it is installed have the same size diameter. They come in various diameters ranging from ¼" to 1¼" with lengths ranging from 1¾" to 12". Wedge, stud, sleeve, one-piece, screw, and nail anchors (*Figure 44*) are common types of one-step anchors.

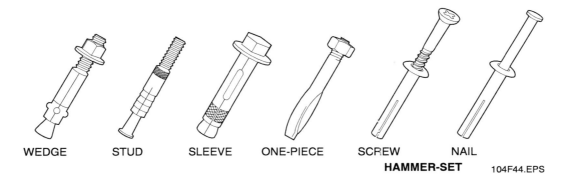

Figure 44. One-Step Anchors

8.1.1 Wedge Anchors

Wedge anchors are heavy-duty anchors supplied with nuts and washers. The drill bit size used to drill the hole is the same diameter as the anchor. The depth of the hole is not critical as long as the minimum length recommended by the manufacturer is drilled. After blowing the hole clean of dust and other material, the anchor is inserted into the hole and driven with a hammer far enough so that at least six threads are below the top surface of the component. Then, the component is fastened by tightening the anchor nut to expand the anchor and tighten it in the hole.

8.1.2 Stud Bolt Anchors

Stud bolt anchors are heavy-duty threaded anchors. Because this type of anchor is made to bottom in its mounting hole, it is a good choice to use when jacking or leveling of the fastened component is needed. The depth of the hole drilled in the masonry must be as specified by the manufacturer in order to achieve proper expansion. After blowing the hole clean of dust and other material, the anchor is inserted in the hole with the expander plug end down. Following this, the anchor is driven into the hole with a hammer (or setting tool) to expand the anchor and tighten it in the hole. The anchor is fully set when it can no longer be driven into the hole. The component is fastened using the correct size and thread bolt for use with the anchor stud.

8.1.3 Sleeve Anchors

Sleeve anchors are multi-purpose anchors. The depth of the anchor hole is not critical as long as the minimum length recommended by the manufacturer is drilled. After blowing the hole clean of dust and other material, the anchor is inserted into the hole and tapped until flush with the component. Then, the anchor nut or screw is tightened to expand the anchor and tighten it in the hole.

FASTENERS AND ANCHORS — TRAINEE TASK MODULE 33104

8.1.4 One-Piece Anchors

One-piece anchors are multi-purpose anchors. They work on the principle that as the anchor is driven into the hole, the spring force of the expansion mechanism is compressed and flexes to fit the size of the hole. Once set, it tries to regain its original shape. The depth of the hole drilled in the masonry must be at least ½" deeper than the required embedment. The proper depth is crucial. Overdrilling is as bad as underdrilling. After blowing the hole clean of dust and other material, the anchor is inserted through the component and driven with a hammer into the hole until the head is firmly seated against the component. It is important to make sure that the anchor is driven to the proper embedment depth. Note that manufacturers also make specially designed drivers and manual tools that are used instead of a hammer to drive one-piece anchors. These tools allow the anchors to be installed in confined spaces and help prevent damage to the component from stray hammer blows.

8.1.5 Hammer-Set Anchors

Hammer-set anchors are made for use in concrete and masonry. There are two types: nail and screw. An advantage of the screw-type anchors is that they are removable. Both types have a diameter the same size as the anchoring hole. For both types, the anchor hole must be drilled to the diameter of the anchor and to a depth of at least ¼" deeper than that required for embedment. After blowing the hole clean of dust and other material, the anchor is inserted into the hole through the mounting holes in the component to be fastened, then the screw or nail is driven into the anchor body to expand it. It is important to make sure that the head is seated firmly against the component and is at the proper embedment.

8.2.0 BOLT ANCHORS

Bolt anchors are designed to be installed flush with the surface of the base material. They are used in conjunction with threaded machine bolts or screws. In some types, they can be used with threaded rod. Drop-in, single and double expansion, and caulk-in anchors (*Figure 45*) are commonly used types of bolt anchors.

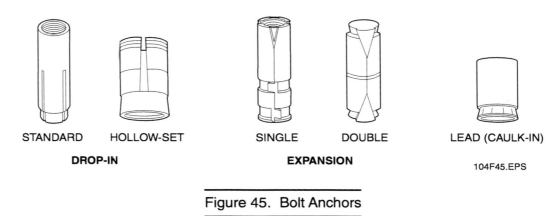

Figure 45. Bolt Anchors

8.2.1 Drop-In Anchors

Drop-in anchors are typically used as heavy-duty anchors. There are two types of drop-in anchors. The first type, made for use in solid concrete and masonry, has an internally threaded expansion anchor with a pre-assembled internal expander plug. The anchor hole must be drilled to the specific diameter and depth specified by the manufacturer. After blowing the hole clean of dust and other material, the anchor is inserted into the hole and tapped until it is flush with the surface. Following this, a setting tool supplied with the anchor is driven into the anchor to expand it. The component to be fastened is positioned in place and fastened by threading and tightening the correct size machine bolt or screw into the anchor.

The second type, called a *hollow set drop-in anchor*, is made for use in hollow concrete and masonry base materials. Hollow set drop-in anchors have a slotted, tapered expansion sleeve and a serrated expansion cone. They come in various lengths compatible with the outer wall thickness of most hollow base materials. They can also be used in solid concrete and masonry. The anchor hole must be drilled to the diameter specified by the manufacturer. When installed in hollow base materials, the hole is drilled into the cell or void. After blowing the hole clean of dust and other material, the anchor is inserted into the hole and tapped until it is flush with the surface. Following this, the component to be fastened is positioned in place, then the proper size machine bolt or screw is threaded into the anchor and tightened to expand the anchor in the hole.

8.2.2 Single- And Double-Expansion Anchors

Single- and double-expansion anchors are both made for use in concrete and other masonry. The double-expansion anchor is used mainly when fastening into concrete or masonry of questionable strength. For both types, the anchor hole must be drilled to the specific diameter and depth specified by the manufacturer. After blowing the hole clean of dust and other material, the anchor is inserted into the hole, threaded cone end first. It is then tapped until it is flush with the surface. Following this, the component to be fastened is positioned in place, then the proper size machine bolt or screw is threaded into the anchor and tightened to expand the anchor in the hole.

8.2.3 Lead (Caulk-In) Anchors

Lead (caulk-in) anchors are a cast-type anchor used in concrete and masonry. They consist of an internally threaded expander cone with a series of vertical internal ribs and a lead sleeve. The vertical internal ribs prevent the cone from turning in the sleeve as the anchor is tightened. The anchor hole must be drilled to the specific diameter and depth specified by the manufacturer. However, in weak or soft masonry, a slightly deeper hole can be drilled to countersink the anchor below the surface. After blowing the hole clean of dust and other material, the anchor is inserted into the hole, threaded cone end first. Following this, a setting tool supplied with the anchor is driven into the anchor to expand it. The component to be fastened is positioned in place and fastened by threading and tightening the correct size machine bolt or screw into the anchor.

8.3.0 SCREW ANCHORS

Screw anchors are lighter-duty anchors made to be installed flush with the surface of the base material. They are used in conjunction with sheet metal, wood, or lag screws depending on the anchor type. Fiber, lead, and plastic anchors are common types of screw anchors (*Figure 46*). The lag shield anchor used with lag screws was described earlier in this module.

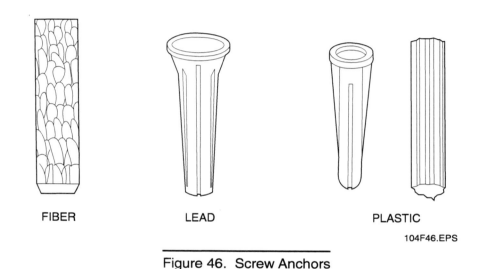

Figure 46. Screw Anchors

Fiber, lead, and plastic anchors are typically used in concrete and masonry. Plastic anchors are also commonly used in wallboard and similar base materials. The installation of all types is simple. The anchor hole must be drilled to the diameter specified by the manufacturer. The minimum depth of the hole must equal the anchor length. After blowing the hole clean of dust and other material, the anchor is inserted into the hole and tapped until it is flush with the surface. Following this, the component to be fastened is positioned in place, then the proper type and size screw is driven through the component mounting hole and into the anchor to expand the anchor in the hole.

8.4.0 SELF-DRILLING ANCHORS

Some anchors made for use in masonry are self-drilling anchors. *Figure 47* is typical of those in common use. This fastener has a cutting sleeve that is first used as a drill bit and later becomes the expandable fastener itself. A rotary hammer is used to drill the hole in the concrete using the anchor sleeve as the drill bit. After the hole is drilled, the anchor is pulled out and the hole cleaned. This is followed by inserting the anchor's expander plug into the cutting end of the sleeve. The anchor sleeve and expander plug are driven back into the hole with the rotary hammer until they are flush with the surface of the concrete. As the fastener is hammered down, it hits the bottom where the tapered expander causes the fastener to expand and lock into the hole. The anchor is then snapped off at the shear point with a quick lateral movement of the hammer. The component to be fastened can then be attached to the anchor using the proper size bolt.

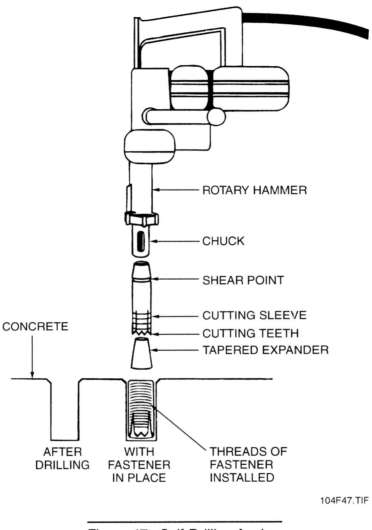

Figure 47. Self-Drilling Anchor

8.5.0 GUIDELINES FOR DRILLING ANCHOR HOLES IN HARDENED CONCRETE OR MASONRY

When selecting masonry anchors, regardless of the type, always take into consideration and follow the manufacturer's recommendations pertaining to hole diameter and depth, minimum embedment in concrete, maximum thickness of material to be fastened, and the pullout and shear load capacities.

When installing anchors and/or anchor bolts in hardened concrete, make sure the area where the equipment or component is to be fastened is smooth so that it will have solid footing. Uneven footing might cause the equipment to twist, warp, not tighten properly, or vibrate when in operation. Before starting, carefully inspect the rotary hammer or hammer drill and the drill bit(s) to ensure they are in good operating condition. Be sure to use the type of carbide-tipped masonry or percussion drill bits recommended by the drill/hammer or anchor manufacturer because these bits are made to take the higher impact of the masonry materials. Also, it is recommended that the drill or hammer tool depth gauge be set to the depth of the hole needed. The trick to using masonry drill bits is not to force them into the material by pushing down hard on the drill. Use a little pressure and let the drill do the work. For large holes, start with a smaller bit, then change to a larger bit.

The methods for installing the different types of anchors in hardened concrete or masonry were briefly described in the sections above. Always install the selected anchors according to the manufacturer's directions. Here is an example of a typical procedure used to install many types of expansion anchors in hardened concrete or masonry. Refer to *Figure 48* as you study the procedure.

WARNING!   Drilling in concrete generates noise, dust, and flying particles. Always wear safety goggles, ear protectors, and gloves. Make sure other workers in the area also wear protective equipment.

Step 1   Drill the anchor bolt hole the same size as the anchor bolt. The hole must be deep enough for six threads of the bolt to be below the surface of the concrete (see *Figure 48*, Step 1). Clean out the hole using a squeeze bulb.

Step 2   Drive the anchor bolt into the hole using a hammer (*Figure 48*, Step 2). Protect the threads of the bolt with a nut that does not allow any threads to be exposed.

Step 3   Put a washer and nut on the bolt, and tighten the nut with a wrench until the anchor is secure in the concrete (*Figure 48*, Step 3).

8.6.0 HOLLOW-WALL ANCHORS

Hollow-wall anchors are used in hollow materials such as concrete plank, block, structural steel, wallboard, and plaster. Some types can also be used in solid materials. Toggle bolts, sleeve-type wall anchors, wallboard anchors, and metal drive-in anchors are common anchors used when fastening to hollow materials.

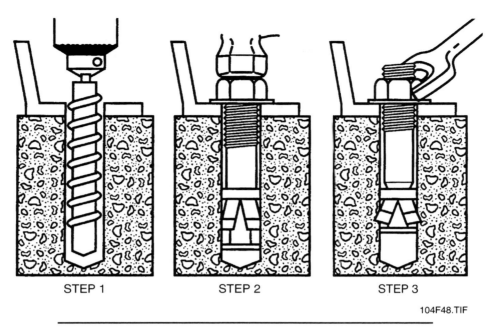

Figure 48. Installing An Anchor Bolt In Hardened Concrete

When installing anchors in hollow walls or ceilings, regardless of the type, always follow the manufacturer's recommendations pertaining to use, hole diameter, wall thickness, grip range (thickness of the anchoring material), and the pullout and shear load capacities.

8.6.1 Toggle Bolts

Toggle bolts (*Figure 49*) are used to fasten equipment, hangers, supports, and similar items into hollow surfaces such as walls and ceilings. They consist of a slotted bolt or screw and spring-loaded wings. When inserted through the item to be fastened, then through a pre-drilled hole in the wall or ceiling, the wings spring apart and provide a firm hold on the inside of the hollow wall or ceiling as the bolt is tightened. Note that the hole drilled in the wall or ceiling should be just large enough for the compressed wing-head to pass through. Once the toggle bolt is installed, be careful not to completely unscrew the bolt because the wings will fall off, making the fastener useless. Screw-actuated plastic toggle bolts are also made. These are similar to metal toggle bolts, but they come with a pointed screw and do not require as large a hole. Unlike the metal version, the plastic wings remain in place if the screw is removed.

Toggle bolts are used to fasten a part to hollow block, wallboard, plaster, panel, or tile. The following general procedure can be used to install toggle bolts.

WARNING!   Follow all safety precautions.

Step 1   Select the proper size drill bit or punch and toggle bolt for the job.

Step 2   Check the toggle bolt for damaged or dirty threads or a malfunctioning wing mechanism.

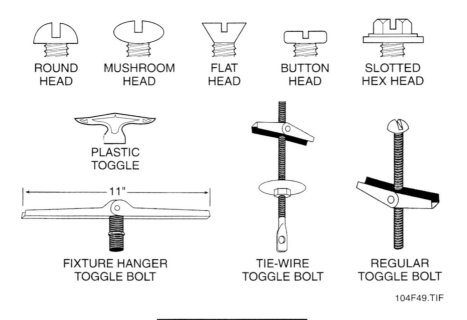

Figure 49. Toggle Bolts

Step 3 Drill a hole completely through the surface to which the part is to be fastened.

Step 4 Insert the toggle bolt through the opening in the item to be fastened.

Step 5 Screw the toggle wing onto the end of the toggle bolt, ensuring that the flat side of the toggle wing is facing the bolt head.

Step 6 Fold the wings completely back and push them through the drilled hole until the wings spring open.

Step 7 Pull back on the item to be fastened in order to hold the wings firmly against the inside surface to which the item is being attached.

Step 8 Tighten the toggle bolt with a screwdriver until it is snug.

8.6.2 Sleeve-Type Wall Anchors

Sleeve-type wall anchors (*Figure 50*) are suitable for use in concrete, block, plywood, wallboard, hollow tile, and similar materials. The two types made are standard and drive. The standard type is commonly used in walls and ceilings and is installed by drilling a mounting hole to the required diameter. The anchor is inserted into the hole and tapped until the gripper prongs embed in the base material. Following this, the anchor's screw is tightened to draw the anchor tight against the inside of the wall or ceiling. Note that the drive-type anchor is hammered into the material without the need for drilling a mounting hole. After the anchor is installed, the anchor screw is removed, the component being fastened is positioned in place, then the screw is reinstalled through the mounting hole in the component and into the anchor. The screw is tightened into the anchor to secure the component.

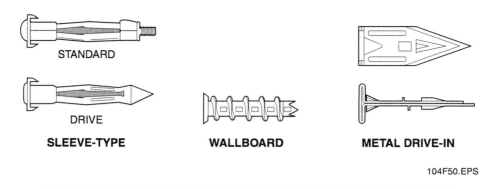

Figure 50. Sleeve-Type, Wallboard, And Metal Drive-In Anchors

8.6.3 Wallboard Anchors

Wallboard anchors (*Figure 50*) are self-drilling medium- and light-duty anchors used for fastening in wallboard. The anchor is driven into the wall with a Phillips head manual or cordless screwdriver until the head of the anchor is flush with the wall or ceiling surface. Following this, the component being fastened is positioned over the anchor, then secured with the proper size sheet metal screw driven into the anchor.

8.6.4 Metal Drive-In Anchors

Metal drive-in anchors (*Figure 50*) are used to fasten light to medium loads to wallboard. They have two pointed legs that stay together when the anchor is hammered into a wall and spread out against the inside of the wall when a No. 6 or 8 sheet metal screw is driven in.

9.0.0 EPOXY ANCHORING SYSTEMS

Epoxy resin compounds can be used to anchor threaded rods, dowels, and similar fasteners in solid concrete, hollow wall, and brick. For one manufacturer's product, a two-part epoxy is packaged in a two-chamber cartridge that keeps the resin and hardener ingredients separated until use. This cartridge is placed into a special tool similar to a caulking gun. When the gun handle is pumped, the epoxy resin and hardener components are mixed within the gun, then the epoxy is ejected from the gun nozzle.

To use the epoxy to install an anchor in solid concrete (*Figure 51*), a hole of the proper size is drilled in the concrete and cleaned using a nylon (not metal) brush. Following this, a small amount of epoxy is dispensed from the gun to make sure that the resin and hardener have mixed properly. This is indicated by the epoxy being of a uniform color. The gun nozzle is then placed into the hole, and the epoxy is injected into the hole until half the depth of the hole is filled. Following this, the selected fastener is pushed into the hole with a slow twisting motion to make sure that the epoxy fills all voids and crevices, then it is set to the required plumb (or level) position. After the recommended cure time for the epoxy has elapsed, the fastener nut can be tightened to secure the component or fixture in place.

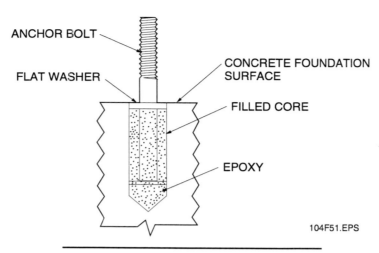

Figure 51. Fastener Anchored In Epoxy

The procedure for installing a fastener in a hollow wall or brick using epoxy is basically the same as described above. The difference is that the epoxy is first injected into an anchor screen to fill the screen, then the anchor screen is installed into the drilled hole. Use of the anchor screen is necessary to hold the epoxy intact in the hole until the anchor is inserted into the epoxy.

SUMMARY

Fasteners and anchors are used for a variety of tasks in low voltage work. In this module, you learned about various types of fasteners and anchors and their uses. Basic installation procedures were also included. Selecting the correct fastener or anchor for a particular job is required to perform high-quality work. It is important to be familiar with the correct terms used to describe fasteners and anchors. Using the proper technical terms helps avoid confusion and improper selection. Installation techniques for fasteners and anchors may vary depending on the job. Make sure to check the project specifications and manufacturer's information when installing any fastener or anchor.

New fasteners and anchors are being developed every day. Your local distributor/manufacturer is an excellent source of information.

References

For advanced study of topics covered in this task module, the following books are suggested:

American Electrician's Handbook, Latest Edition, McGraw-Hill, New York, NY.
Carpentry, Latest Edition, Delmar Publishers, New York, NY.
The Sheet Metal Toolbox Manual, Latest Edition, Prentice Hall, New York, NY.

REVIEW/PRACTICE QUESTIONS

1. The quality of some fasteners can be determined by _____ on the head of a bolt or screw.
   a. the number of grooves cut
   b. the number of sides
   c. the length of the lines
   d. the grade markings

2. The purpose of a jam nut is to _____.
   a. hold a piece of material stationary
   b. lock a standard nut in place
   c. stop rotation of a machine quickly for safety reasons
   d. compress a lock washer in place

3. Washers are used to _____.
   a. distribute the load over a larger area
   b. attach an item to a hollow surface
   c. anchor materials that expand due to temperature changes
   d. allow the bolts to expand with temperature changes

4. Torque is normally expressed in _____.
   a. pounds per square inch (psi)
   b. gallons per minute (gpm)
   c. foot-pounds (ft. lbs.)
   d. cubic feet per minute (cpm)

5. When tightening bolts, nuts, or screws, always use the proper _____.
   a. tightening sequence
   b. bolt pattern
   c. ratchet wrench
   d. lubrication

6. What type of fasteners are commonly used as lifting devices?
   a. Toggle bolts
   b. Anchor bolts
   c. J-bolts
   d. Eye bolts

7. When installing a panel in concrete using a lag screw, you would also use a _____.
   a. drop-in anchor
   b. lag shield
   c. caulk-in anchor
   d. double-expansion anchor

8. When fastening a device or fixture to concrete using concrete/masonry screws, the screws _____.
   a. need not be installed in pre-drilled holes because they are self-tapping
   b. are installed using specially designed carbide drill bits and installation tools made for use with the screws
   c. are installed by hammering them into a drilled or punched hole of the proper size
   d. are installed in a hole drilled with a standard masonry drill bit

9. When using powder-actuated tools and fasteners, _____.
   a. higher-numbered power level loads are used to drive into soft materials
   b. use only powder loads that are specifically designed for use with the installation tool
   c. use only fasteners that are specifically designed for use with the installation tool
   d. Both b and c.

10. Wedge and sleeve anchors are classified as _____ anchors.
    a. bolt
    b. hollow-wall
    c. one-step
    d. screw

ANSWERS TO REVIEW/PRACTICE QUESTIONS

Answer	Section Reference
1. d	2.1.4
2. b	2.3.1
3. a	2.4.0
4. c	2.5.1
5. a	2.5.2
6. d	4.1.0
7. b	5.2.0
8. b	5.3.0
9. d	7.0.0
10. c	8.1.0

APPENDIX A

MECHANICAL ANCHORS AND THEIR USES

Anchor Type	Typically Used In	Use With Fastener	Typical Working Load Range*
One-Step Anchors			
Wedge	Concrete **Stone	None	Light, medium, and heavy duty
Stud	Concrete **Stone, solid brick and block	None	Light, medium, and heavy duty
Sleeve	Concrete, solid brick and block **Stone, hollow brick and block	None	Light and medium duty
One-piece	Concrete, solid block **Stone, solid and hollow brick, hollow block	None	Light and medium duty
Hammer-set	Concrete, solid block **Stone, solid and hollow brick, hollow block	None	Light duty
Bolt Anchors			
Drop-in	Concrete **Stone, solid brick	Machine screw or bolt	Light, medium, and heavy duty
Hollow-set drop-in	Concrete, solid brick and block **Stone, hollow brick and block	Machine screw or bolt	Light and medium duty
Single-expansion	Concrete, solid brick and block **Stone, hollow brick and block	Machine screw or bolt	Light and medium duty
Double-expansion	Concrete, solid brick and block **Stone, hollow brick and block	Machine screw or bolt	Light and medium duty
Lead (caulk-in)	Concrete, solid brick and block **Stone, hollow brick and block	Machine screw or bolt	Light and medium duty

FASTENERS AND ANCHORS — TRAINEE TASK MODULE 33104

MECHANICAL ANCHORS AND THEIR USES (Continued)

Anchor Type	Typically Used In	Use With Fastener	Typical Working Load Range*
Screw Anchors			
Lag shield	Concrete **Stone, solid and hollow brick and block	Lag screw	Light and medium duty
Fiber	Concrete, stone, solid brick and block **Hollow brick and block, wallboard	Wood, sheet metal, or lag screw	Light and medium duty
Lead	Concrete, solid brick and block **Hollow brick and block, wallboard	Wood or sheet metal screw	Light duty
Plastic	Concrete, stone, solid brick and block **Hollow brick and block, wallboard	Wood or sheet metal screw	Light duty
Hollow-Wall Anchors			
Toggle bolts	Concrete, plank, hollow block, wallboard, plywood/paneling	Included	Light and medium duty
Plastic toggle bolts	Wallboard, plywood/paneling **Hollow block, structural tile	Wood or sheet metal screw	Light duty
Sleeve-type wall	Wallboard, plywood/paneling **Hollow block, structural tile	Included	Light duty
Wallboard	Wallboard	Sheet metal screw	Light duty
Metal drive-in	Wallboard	Sheet metal screw	Light duty

* Anchor working loads given in the table are defined below. These are approximate loads only. Actual allowable loads depend on such factors as the anchor style and size, base material strength, spacing and edge distance, and the type of service load applied. Always consult the anchor manufacturer's product literature to determine the correct type of anchor and size to use for a specific application.

- Light duty – Less than 400 lbs.
- Medium duty – 400 to 4,000 lbs.
- Heavy duty – Above 4,000 lbs.

** Indicates use may be suitable depending on the application.

NCCER CRAFT TRAINING USER UPDATES

The NCCER makes every effort to keep these manuals up-to-date and free of technical errors. We appreciate your help in this process. If you have an idea for improving this manual, or if you find an error, a typographical mistake, or an inaccuracy in the NCCER's Craft Training Manuals, please write us, using this form or a photocopy. Be sure to include the exact module number, page number, a description of the problem, and the correction, if possible. Your input will be brought to the attention of the Technical Review Committee. Thank you for your assistance.

Instructors – If you found that additional materials were necessary in order to teach this module effectively, please let us know so that we may include them in the Equipment/Materials list in the Instructor's Guide.

Write: Curriculum Revision and Development Department
National Center for Construction Education and Research
P.O. Box 141104
Gainesville, FL 32614-1104
Fax: 352-334-0932

Craft _____ Module Name _____

Copyright Date _____ Module Number _____ Page Number(s) _____

Description of Problem
_____
_____
_____
_____
_____

(Optional) Correction of Problem
_____
_____
_____

(Optional) Your Name and Address
_____
_____
_____

FASTENERS AND ANCHORS — TRAINEE TASK MODULE 33104

notes

Hand Bending of Conduit
Module 33105

Electronic Systems Technician Trainee Task Module 33105

HAND BENDING OF CONDUIT

NATIONAL
CENTER FOR
CONSTRUCTION
EDUCATION AND
RESEARCH

OBJECTIVES

Upon completion of this module, the trainee will be able to:

1. Identify the methods of hand bending conduit.
2. Identify the various methods used to install conduit.
3. Use math formulas to determine conduit bends.
4. Make 90° bends, back-to-back bends, offsets, kicks, and saddle bends using a hand bender.
5. Cut, ream, and thread conduit.

Prerequisites

Successful completion of the following Task Modules is recommended before beginning study of this Task Module: Core Curricula; Electronic Systems Technician Level One, Modules 33101 through 33104

Required Trainee Materials

1. Trainee Task Module
2. Copy of the latest edition of the *National Electrical Code*
3. *OSHA Electrical Safety Guidelines* (pocket guide)
4. Appropriate Personal Protective Equipment

Note: The designations "National Electrical Code," "NE Code," and "NEC," where used in this document, refer to the *National Electrical Code®*, which is a registered trademark of the National Fire Protection Association, Quincy, MA. *All National Electrical Code (NEC) references in this module refer to the 1999 edition of the NEC.*

COURSE MAP

This course map shows all of the modules in the first level of the Electronic Systems Technician curricula. The suggested training order begins at the bottom and proceeds up. Skill levels increase as a trainee advances on the course map. The training order may be adjusted by the local Training Program Sponsor.

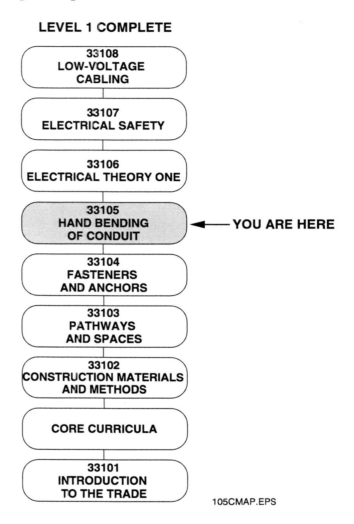

TABLE OF CONTENTS

Section	Topic	Page
	Course Map	2
1.0.0	Introduction	4
2.0.0	Hand Bending Equipment	4
2.1.0	Geometry Required To Make A Bend	7
2.2.0	Making A 90° Bend	9
2.3.0	Gain	11
2.4.0	Back-To-Back 90° Bends	11
2.5.0	Making An Offset	12
2.6.0	Parallel Offsets	14
2.7.0	Saddle Bends	16
2.8.0	Four-Bend Saddles	17
3.0.0	Cutting, Reaming, And Threading Conduit	19
3.1.0	Hacksaw Method Of Cutting Conduit	19
3.2.0	Pipe Cutter Method	20
3.3.0	Reaming Conduit	21
3.4.0	Threading Conduit	22
3.5.0	Cutting And Joining PVC Conduit	23
	Summary	25
	Review/Practice Questions	26
	Answers To Review/Practice Questions	30
	Appendix A	31
	Appendix B	34

Trade Terms Introduced In This Module

90° bend: A bend that changes the direction of the conduit by 90°.

Back-to-back bend: Any bend formed by two 90° bends with a straight section of conduit between the bends.

Concentric bends: Making 90° bends in two or more parallel runs of conduit and increasing the radius of each conduit from inside of the run toward the outside.

Developed length: The actual length of the conduit that will be bent.

Gain: Because a conduit bends in a radius and not at right angles, the length of conduit needed for a bend will not equal the total determined length. Gain is the distance saved by the arc of a 90° bend.

Offsets: An offset (kick) is two bends placed in a piece of conduit to change elevation to go over or under obstructions or for proper entry into boxes, cabinets, etc.

Rise: The length of the bent section of conduit measured from the bottom, centerline, or top of the straight section to the end of the bent section.

Segment bend: A large bend formed by multiple short bends or *shots*.

Stub-up: Another name for the rise in a section of conduit. Also, a term used for conduit penetrating a slab or the ground.

1.0.0 INTRODUCTION

The art of conduit bending depends on the skills of the technician and requires a working knowledge of basic terms and proven procedures. Practice, knowledge, and training will help you gain the skills necessary for proper conduit bending and installation. You will be able to practice conduit bending in the lab and in the field under the supervision of experienced co-workers. In this module, the techniques for using hand-operated and step conduit benders such as the hand bender and the hickey will be covered. The process of hand bending conduit, as well as cutting, reaming, and threading conduit will also be explained.

2.0.0 HAND BENDING EQUIPMENT

Figure 1 shows a hand bender. Hand benders are convenient to use on the job because they are portable and no electrical power is required. Hand benders have a shape that supports the walls of the conduit being bent.

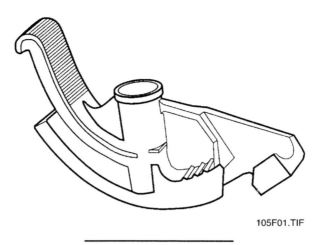

Figure 1. Hand Bender

These benders are used to make various bends in smaller-size conduit (½" to 1¼"). Most hand benders are sized to bend rigid conduit and electrical metallic tubing (EMT) of corresponding sizes. For example, a single hand bender can bend either ¾" EMT or ½" rigid conduit. The next larger size of hand bender will bend either 1" EMT or ¾" rigid conduit. This is because the corresponding sizes of conduit have nearly equal outside diameters.

The first step in making a good bend is familiarizing yourself with the bender. The manufacturer of the bender will typically provide documentation indicating starting points, distance between **offsets**, **gains**, and other important values associated with that particular bender. There is no substitute for taking the time to review this information. It will make the job go faster and result in better bends.

CAUTION: When making bends, be sure you have a firm grip on the handle to avoid slippage and possible injury.

When performing a bend, it is important to keep the conduit on a stable, firm, flat surface for the entire duration of the bend. Hand benders are designed to have force applied using one foot and the hands. See *Figure 2*. It is important to use constant foot pressure as well as force on the handle to achieve uniform bends. Allowing the conduit to rise up or performing the bend on soft ground can result in distorting the conduit outside the bender.

Note: Bends should be made in accordance with the guidelines of **NEC Article 345** (intermediate metal conduit or IMC), **Article 346** (rigid metal conduit), **Article 347** (rigid nonmetallic conduit such as polyvinyl chloride or PVC), or **Article 348** (electrical metallic tubing).

A hickey should not be confused with a hand bender. The hickey, which is used for rigid conduit only, functions quite differently. See *Figure 3*.

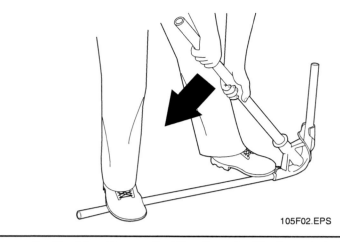

Figure 2. Pushing Down On The Bender To Complete The Bend

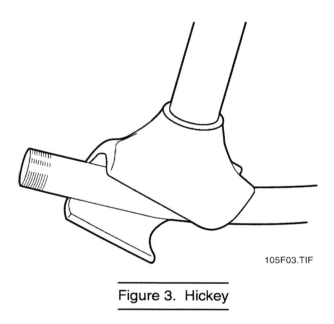

Figure 3. Hickey

When you use a hickey to bend conduit, you are forming the bend as well as the radius. When using a hickey, be careful not to flatten or kink the conduit. Hickeys should only be used with rigid conduit because very little support is given to the walls of the conduit being bent. A hickey is a segment bending device. First, a small bend of about 10° is made. Then, the hickey is moved to a new position and another small bend is made. This process is continued until the bend is completed. A hickey can be used for conduit **stub-ups** in slabs and decks.

PVC conduit is bent using a heating unit (*Figure 4*). The PVC must be rotated regularly while it is in the heater so that it heats evenly. Once heated, the PVC is removed, and the bending is performed by hand. Some units use an electric heating element, while others use liquid propane (LP). After bending, a damp sponge or cloth is often used so that the PVC sets up faster.

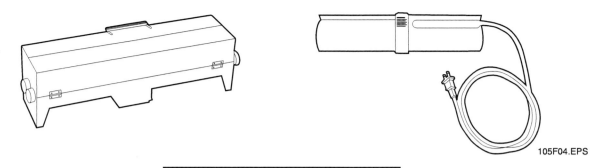

Figure 4. Typical PVC Heating Units

CAUTION: Avoid contact with the case of the heating unit; it can become very hot and cause burns. Also, to avoid a fire hazard, ensure that the unit is cool before storage. If using an LP unit, keep a fire extinguisher nearby.

When bending PVC that is 2" or larger in diameter, there is a risk of wrinkling or flattening the bend. A plug set eliminates this problem (*Figure 5*). A plug is inserted into each end of the piece of PVC being bent. Then, a hand pump is used to pressurize the conduit before bending it. The pressure is about 3 to 5 psi.

Note: The plugs must remain in place until the pipe is cooled and set.

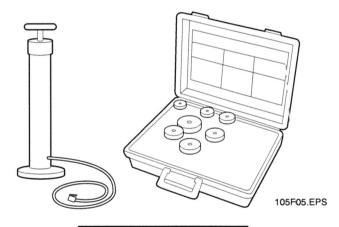

Figure 5. Typical Plug Set

2.1.0 GEOMETRY REQUIRED TO MAKE A BEND

Bending conduit requires that you use some basic geometry. You may already be familiar with most of the concepts needed; however, here is a review of the concepts directly related to this task. A right triangle is defined as any triangle with a 90° angle. The side directly opposite the 90° angle is called the *hypotenuse,* and the side on which the triangle sits is the *base*. The vertical side is called the *height*. On the job, you will apply the relationships in a right triangle when making an offset bend. The offset forms the hypotenuse of a right triangle (*Figure 6*).

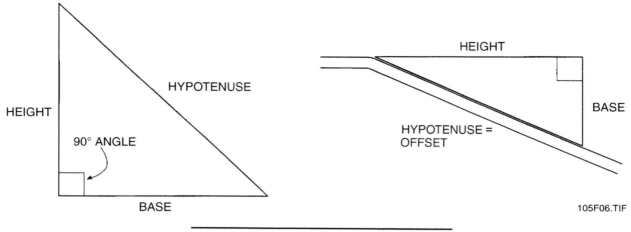

Figure 6. Right Triangle And Offset Bend

Note: There are reference tables for sizing offset bends based on these relationships (see *Appendix A*).

A circle is defined as a closed curved line whose points are all the same distance from its center. The distance from the center point to the edge of the circle is called the *radius*. The length from one edge of the circle to the other edge is the *diameter*. The distance around the circle is called the *circumference*. A circle can be divided into four equal quadrants. Each quadrant accounts for 90°, making a total of 360°. When you make a **90° bend**, you will use ¼ of a circle, or one quadrant. Concentric circles are circles that have a common center but different radii. The concept of concentric circles can be applied to **concentric bends** in conduit. The angle of each bend is 90°. Such bends have the same center point, but the radius of each is different. See *Figure 7*.

To calculate the circumference of a circle, use the following formula:

$C = \pi \times D$ or $C = \pi D$

In this formula, C = circumference, D = diameter, and π = 3.14. Another way of stating the formula for circumference is $C = 2\pi R$, where R equals the radius or ½ the diameter.

To figure the arc of a quadrant use:

Length of arc = $(.25) 2\pi R = 1.57R$

For this formula, the arc of a quadrant equals ¼ the circumference of the circle or 1.57 times the radius.

A bending radius table is included in *Appendix B*.

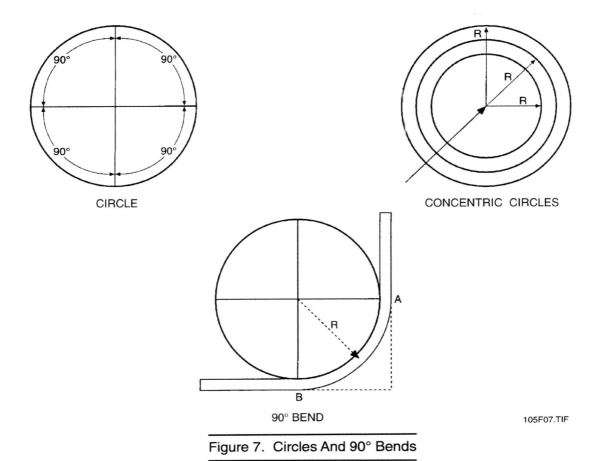

Figure 7. Circles And 90° Bends

2.2.0 MAKING A 90° BEND

The 90° stub bend is probably the most basic bend of all. The stub bend is used much of the time, regardless of the type of conduit being installed. Before beginning to make the bend, you need to know two measurements:

- Desired **rise** or stub-up
- Take-up distance of the bender

The desired rise is the height of the stub-up. The *take-up* is the amount of conduit the bender will use to form the bend. Take-up distances are usually listed in the manufacturer's instruction manual. Typical bender take-up distances are shown in *Table 1*.

EMT	Rigid/IMC	Take-Up
½"	—	5"
¾"	½"	6"
1"	¾"	8"
1¼"	1"	11"

Table 1. Typical Bender Take-Up Distances

Once you have determined the take-up, subtract it from the stub-up height. Mark that distance on the conduit (all the way around) at that distance from the end. The mark will indicate the point at which you will begin to bend the conduit. Line up the starting point on the conduit with the starting point on the bender. Most benders have a mark, like an arrow, to indicate this point. *Figure 8* shows the take-up required to achieve an 18" stub-up on a piece of ½" EMT.

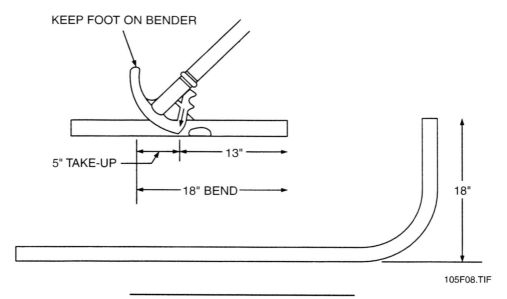

Figure 8. Bending An 18" Stub-Up

Once you have lined up the bender, use one foot to hold the conduit steady. Keep your heel on the floor for balance. Apply pressure on the bender foot pedal with your other foot. Make sure you hold the bender handle level, as far up as possible, to get maximum leverage. Then, bend the conduit in one smooth motion, pulling as steadily as possible. Avoid overstretching.

Note: When bending conduit using the take-up method, always place the bender on the conduit and make the bend facing the hook of the conduit from which the measurements were taken.

After finishing the bend, check to make sure you have the correct angle and measurement. Use the following steps to check a 90° bend:

Step 1 With the back of the bend on the floor, measure to the end of the conduit stub-up to make sure it is the right length.

Step 2 Check the 90° angle of the bend with a square or at the angle formed by the floor and a wall. A torpedo level may also be used.

Note: If you overbend a conduit slightly past the desired angle, you can use the bender to bend the conduit back to the correct angle.

The above procedure will produce a 90° *one-shot bend*. That means that it took a single bend to form the conduit bend. A **segment bend** is any bend that is formed by a series of bends of a few degrees each, rather than a single one-shot bend. A shot is actually one bend in a segment bend. Segment or sweep bends must conform to the provisions of the NEC.

2.3.0 GAIN

The gain is the distance saved by the arc of a 90° bend. Knowing the gain can help you to pre-cut, ream, and pre-thread both ends of the conduit before you bend it. This will make your work go more quickly because it is easier to work with conduit while it is straight. *Figure 9* shows that the overall **developed length** of a piece of conduit with a 90° bend is less than the sum of the horizontal and vertical distances when measured square to the corner. This is shown by the following equation:

Developed length = (A + B) − gain

An example of a manufacturer's gain table is also shown. These tables are used to determine the gain for a certain size conduit.

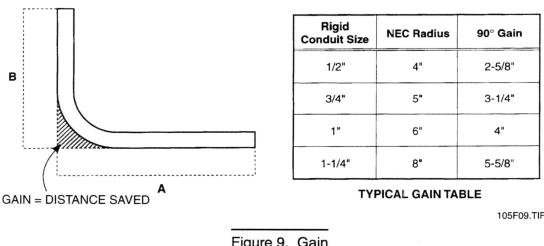

TYPICAL GAIN TABLE

Rigid Conduit Size	NEC Radius	90° Gain
1/2"	4"	2-5/8"
3/4"	5"	3-1/4"
1"	6"	4"
1-1/4"	8"	5-5/8"

Figure 9. Gain

2.4.0 BACK-TO-BACK 90° BENDS

A **back-to-back bend** consists of two 90° bends made on the same piece of conduit and placed back-to-back (*Figure 10*).

To make a back-to-back bend, make the first bend (labeled *X* in *Figure 10*) in the usual manner. To make the second bend, measure the required distance between the bends from the back of the first bend. This distance is labeled *L* in the figure. Reverse the bender on the conduit, as shown in *Figure 10*. Place the bender's back-to-back indicating mark at point Y on the conduit. Note that outside measurements from point X to point Y are used. Holding the bender in the reverse position and properly aligned, apply foot pressure and complete the second bend.

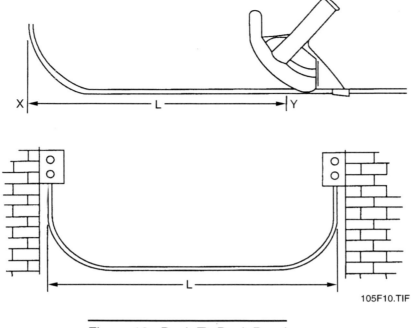

Figure 10. Back-To-Back Bend

2.5.0 MAKING AN OFFSET

Many situations require that the conduit be bent so that it can pass over objects such as beams and other conduits, or enter meter cabinets and junction boxes. Bends used for this purpose are called *offsets (kicks)*. To produce an offset, two equal bends of less than 90° are required, a specified distance apart, as shown in *Figure 11*.

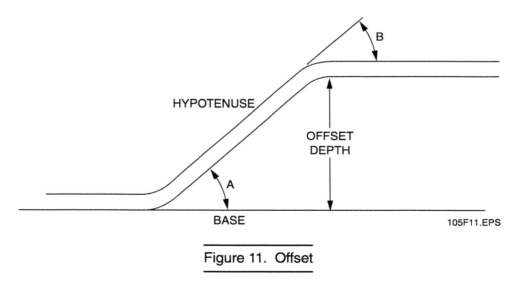

Figure 11. Offset

Offsets are a tradeoff between space and the effort it will take to pull the wire. The larger the degree of bend, the harder it will be to pull the wire. The smaller the degree of bend, the easier it will be to pull the wire. Use the shallowest degree of bend that will still allow the conduit to bypass the obstruction and fit in the given space.

When conduit is offset, some of the conduit length is used. If the offset is made into the area, an allowance must be made for this shrinkage. If the offset angle is away from the obstruction, the shrinkage can be ignored. *Table 2* shows the amount of shrinkage per inch of rise for common offset angles.

Offset Angle	Multiplier	Shrinkage (per inch of rise)
10° × 10°	6	1/16"
22½° × 22½°	2.6	3/16"
30° × 30°	2	¼"
45° × 45°	1.4	⅜"
60° × 60°	1.2	½"

Table 2. Shrinkage Calculation

The formula for figuring the distance between bends is as follows:

Distance between bends = depth of offset × multiplier

The distance between the offset bends can generally be found in the manufacturer's documentation for the bender. *Table 3* shows the distance between bends for the most common offset angles.

	22½°		30°		45°		60°	
Offset Depth	Between Bends	Shrinkage	Between Bends	Shrinkage	Between Bends	Shrinkage	Between Bends	Shrinkage
2	5¼	⅜	—	—	—	—	—	—
3	7¾	9/16	6	¾	—	—	—	—
4	10½	¾	8	1	—	—	—	—
5	13	1 5/16	10	1¼	7	1⅞	—	—
6	15½	1⅛	12	1½	8½	2¼	7¼	3
7	18¼	1 5/16	14	1¾	9¾	2⅝	8⅜	3½
8	20¾	1½	16	2	11¼	3	9⅝	4
9	23½	1¾	18	2¼	12½	3⅜	10⅞	4½
10	26	1⅞	20	2½	14	3¾	12	5

Table 3. Common Offset Factors (In Inches)

Calculations related to offsets are derived from the branch of mathematics known as *trigonometry*, which deals with triangles. The multipliers shown in *Table 2* represent the *cosecant* (COS) of the related offset angle. The multiplier is determined by dividing the depth of the offset by the hypotenuse of the triangle created by the offset (*Figure 11*).

Basic trigonometry (trig) functions are briefly covered in *Appendix A*. As you will see in the next section, the *tangent* (TAN) of the offset angle is also used in calculating parallel offsets. Understanding trig functions will help you understand how offsets are determined. If you have a scientific calculator and understand these functions, you can calculate offset angles when you know the dimensions of the triangle created by the offset and the obstacle.

2.6.0 PARALLEL OFFSETS

Often, multiple pieces of conduit must be bent around a common obstruction. In this case, parallel offsets are made. Since the bends are laid out along a common radius, an adjustment must be made to ensure that the ends do not come out uneven, as shown in *Figure 12*.

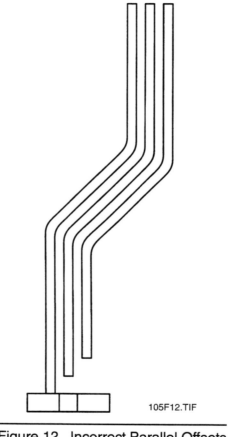

Figure 12. Incorrect Parallel Offsets

The center of the first bend of the innermost conduit is found first, as shown in *Figure 13*. Each successive conduit must have its centerline moved farther away from the end of the pipe, as shown in *Figure 14*. The amount to add is calculated as follows:

Amount added = center-to-center spacing × tangent (TAN) of ½ offset angle

Tangents can be found using the trig tables provided in *Appendix A*.

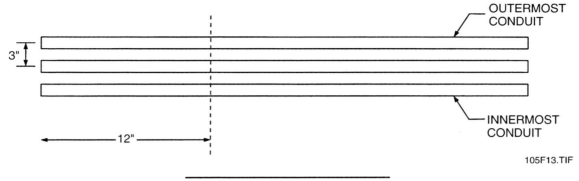

Figure 13. Center Of First Bend

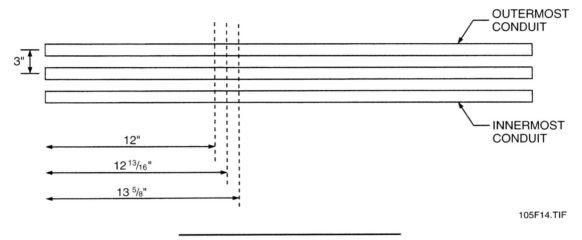

Figure 14. Successive Centerlines

For example, *Figure 15* shows three pipes laid out as parallel and offset. The angle of the offset is 30°. The center-to-center spacing is 3". The start of the innermost pipe's first bend is 12".

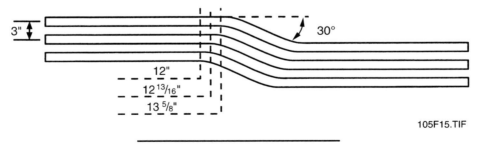

Figure 15. Parallel Offset Pipes

The starting point of the second pipe will be:

12" + [center-to-center spacing × TAN (½ offset angle)]

12" + (3" × TAN 15°) = 12" + (3" × .2679) = 12" + .8037"

This is approximately 12¹³⁄₁₆".

HAND BENDING OF CONDUIT — TRAINEE TASK MODULE 33105

The starting point for the outermost pipe is:

12¹³⁄₁₆" + ¹³⁄₁₆" = 13⅝"

2.7.0 SADDLE BENDS

A saddle bend is used to go around obstructions. *Figure 16* illustrates an example of a saddle bend that is required to clear a pipe obstruction. Making a saddle bend will cause the center of the saddle to shorten ³⁄₁₆" for every inch of saddle depth (see *Table 4*). For example, if the pipe diameter is 2", this would cause a ⅜" shortening of the conduit on each side of the bend.

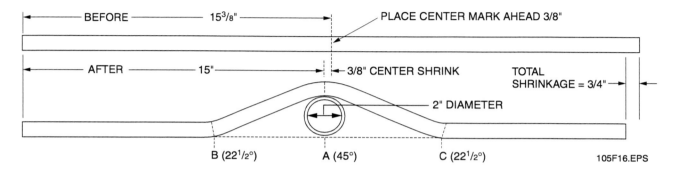

Figure 16. Saddle Measurement

Obstruction Depth	Shrinkage Amount (Move Center Mark Forward)	Make Outside Marks from *New* Center Mark
1	³⁄₁₆"	2½"
2	⅜"	5"
3	⁹⁄₁₆"	7½"
4	¾"	10"
5	¹⁵⁄₁₆"	12½"
6	1⅛"	15"
For each additional inch, add	³⁄₁₆"	2½"

Table 4. Shrinkage Chart For Saddle Bends With A 45° Center Bend And Two 22½° Bends

When making saddle bends, the following steps should apply:

Step 1 Locate the center mark A on the conduit by using the size of the obstruction (i.e., pipe diameter) and calculate the shrink rate of the obstruction (for example, if the pipe diameter is 2", ⅜" of conduit will be lost on each side of the bend for a total shrinkage of ¾"). This figure will be added to the measurement from the end of the conduit to the centerline of the obstruction (for example, if the distance measured from the conduit end and the obstruction centerline was 15", the distance to A would be 15⅜").

Step 2 Locate marks B and C on the conduit by measuring 2½" for every 1" of saddle depth *from* the A mark (i.e., for the saddle depth of 2", the B mark would be 5" before the A mark and the C mark would be 5" after the A mark). See *Figure 17*.

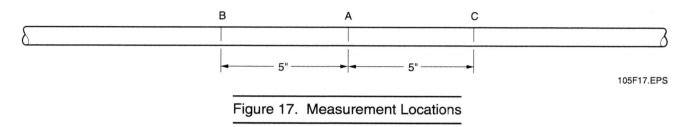

Figure 17. Measurement Locations

Step 3 Refer to *Figure 18* and make a 45° bend at point A, make a 22½° bend at point B, and make a 22½° bend at point C. (Be sure to check the manufacturer's specifications.)

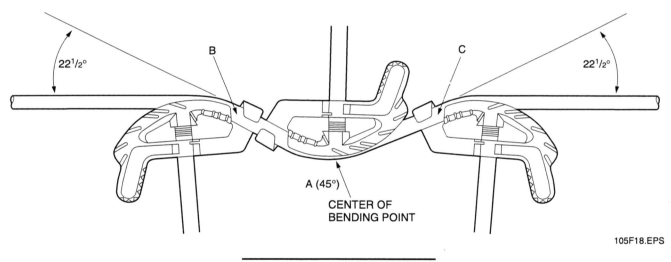

Figure 18. Location Of Bends

2.8.0 FOUR-BEND SADDLES

Four-bend saddles can be difficult. The reason is that four bends must be aligned exactly on the same plane. Extra time spent laying it out and performing the bends will pay off in not having to scrap the whole piece and start over.

Figure 19 illustrates that the four-bend saddle is really two offsets formed back-to-back. Working left to right, the procedure for forming this saddle is as follows:

Step 1 Determine the height of the offset.

Step 2 Determine the correct spacing for the first offset and mark the conduit.

Step 3 Bend the first offset.

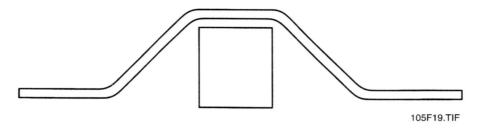

Figure 19. Typical Four-Bend Saddle

Step 4 Mark the start point for the second offset at the trailing edge of the obstruction.

Step 5 Mark the spacing for the second offset.

Step 6 Bend the second offset.

Using *Figure 20* as an example, a four-bend saddle using ½" EMT is laid out as follows:

- Height of the box = 6"
- Width of the box = 8"
- Distance to the obstruction = 36"

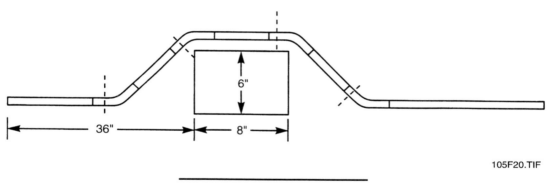

Figure 20. Four-Bend Saddle

Two 30° offsets will be used to form the saddle. It is created as follows:

Step 1 See *Figure 21*. Working from left to right, calculate the start point for the first bend. The distance to the obstruction is 36", the offset is 6", and the 30° multiplier from *Table 2* is 2.00:

Distance to the obstruction − (offset × constant for the angle) + shrinkage = distance to the first bend

36" − (6" × 2.00) + 1½" = 25½"

Step 2 Determine where the second bend will end to ensure the conduit clears the obstruction. See *Figure 22*.

Distance to the first bend + distance to second bend + shrinkage = total length of the first offset

25½" + 12" + 1½" = 39"

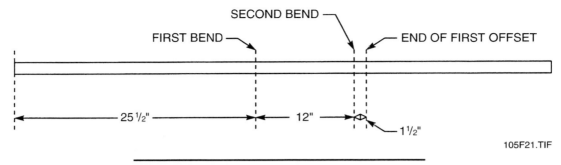

Figure 21. Four-Bend Saddle Measurements

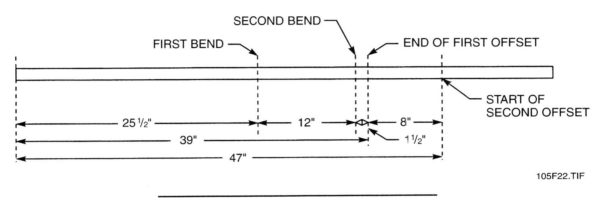

Figure 22. Bend And Offset Measurements

Step 3 Determine the start point of the second offset. The width of the box is 8"; therefore, the start point of the second offset should be 8" beyond the end of the first offset.

8" + 39" = 47"

Step 4 Determine the spacing for the second offset. Since the first and second offsets have the same rise and angle, the distance between bends will be the same, or 12".

3.0.0 CUTTING, REAMING, AND THREADING CONDUIT

Rigid conduit, IMC, and EMT are available in standard 10' lengths. When installing conduit, it is cut to fit the job requirements.

3.1.0 HACKSAW METHOD OF CUTTING CONDUIT

Conduit is normally cut using a hacksaw. To cut conduit with a hacksaw:

Step 1 Inspect the blade of the hacksaw and replace it, if needed. A blade with 18, 24, or 32 cutting teeth per inch is recommended for conduit. Use a higher tooth count for EMT and a lower tooth count for rigid conduit and IMC. If the blade needs to be replaced, point the teeth toward the front of the saw when installing the new blade.

Step 2 Secure the conduit in a pipe vise.

Step 3 Rest the middle of the hacksaw blade on the conduit where the cut is to be made. Position the saw so the end of the blade is pointing slightly down and the handle is pointing slightly up. Push forward gently until the cut is started. Make even strokes until the cut is finished.

CAUTION: To avoid bruising your knuckles on the newly cut pipe, use gentle strokes for the final cut.

Step 4 Check the cut. The end of the conduit should be straight and smooth. *Figure 23* shows correct and incorrect cuts. Ream the conduit.

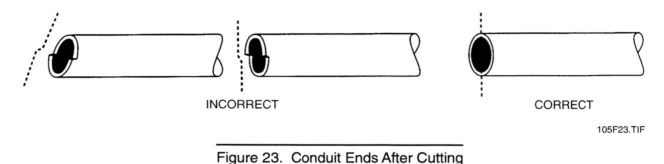

Figure 23. Conduit Ends After Cutting

3.2.0 PIPE CUTTER METHOD

A pipe cutter can also be used to cut rigid IMC conduit. To use a pipe cutter:

Step 1 Secure the conduit in a pipe vise and mark a place for the cut.

Step 2 Open the cutter and place it over the conduit with the cutter wheel on the mark.

Step 3 Tighten the cutter by rotating the screw handle.

CAUTION: Do not overtighten the cutter. Overtightening can break the cutter wheel and distort the wall of the conduit.

Step 4 Rotate the cutter counterclockwise to start the cut. *Figure 24* shows the proper way to rotate the cutter.

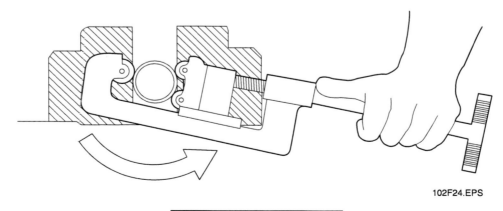

Figure 24. Cutter Rotation

Step 5 Tighten the cutter handle ¼ turn for each full turn around the conduit. Again, make sure that you do not overtighten it.

Step 6 Add a few drops of cutting oil to the groove and continue cutting. Avoid skin contact with the oil.

Step 7 When the cut is almost finished, stop cutting and snap the conduit to finish the cut. This reduces the ridge that can be formed on the inside of the conduit.

Step 8 Clean the conduit and cutter with a shop towel rag.

Step 9 Ream the conduit.

3.3.0 REAMING CONDUIT

When the conduit is cut, the inside edge is sharp. This edge will damage the insulation of the wire when it is pulled through. To avoid this damage, the inside edge must be smoothed or *reamed* using a reamer (*Figure 25*).

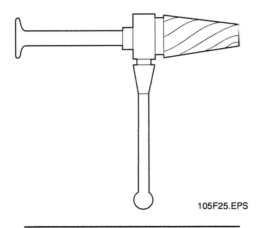

Figure 25. Rigid Conduit Reamer

To ream the inside edge of a piece of conduit using a hand reamer, proceed as follows:

Step 1 Place the conduit in a pipe vise.

Step 2 Insert the reamer tip in the conduit.

Step 3 Apply light forward pressure and start rotating the reamer. *Figure 26* shows the proper way to rotate the reamer. It should be rotated using a downward motion. The reamer can be damaged if you rotate it in the wrong direction. The reamer should bite as soon as you apply the proper pressure.

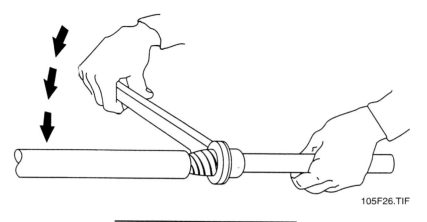

Figure 26. Reamer Rotation

Step 4 Remove the reamer by pulling back on it while continuing to rotate it. Check the progress and then reinsert the reamer. Rotate the reamer until the inside edge is smooth. You should stop when all burrs have been removed.

Note: If a conduit reamer is not available, use a half-round file (tang of file must have a handle attached). EMT may be reamed using the nose of diagonal cutters or small hand reamers.

3.4.0 THREADING CONDUIT

After conduit is cut and reamed, it is usually threaded so it can be properly joined. Only rigid conduit and IMC have walls thick enough for threading.

The tool used to cut threads in conduit is called a *die*. Conduit dies are made to cut a taper of ¾" per foot. The number of threads per inch varies from 8 to 18, depending upon the diameter of the conduit. A thread gauge is used to measure how many threads per inch are cut.

The threading dies are contained in a die head. The die head can be used with a hand-operated ratchet threader (*Figure 27*) or with a portable power drive.

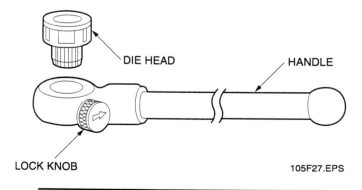

Figure 27. Hand-Operated Ratchet Threader

To thread conduit using a hand-operated threader, proceed as follows:

Step 1   Insert the conduit in a pipe vise. Make sure the vise is fastened to a strong surface. Place supports, if necessary, to help secure the conduit.

Step 2   Determine the correct die and head. Inspect the die for damage such as broken teeth. Never use a damaged die.

Step 3   Insert the die securely in the head. Make sure the proper die is in the appropriately numbered slot on the head.

Step 4   Determine the correct thread length to cut for the conduit size used (match the manufacturer's thread length).

Step 5   Lubricate the die with cutting oil at the beginning and throughout the threading operation. Avoid skin contact with the oil.

Step 6   Cut threads to the proper length. Make sure that the conduit enters the tapered side of the die. Apply pressure and start turning the head. You should back off the head each ¼ turn to clear away chips.

Step 7   Remove the die when the proper cut is made. Threads should be cut only to the length of the die. Overcutting will leave the threads exposed to corrosion.

Step 8   Inspect the threads to make sure they are clean, sharp, and properly made. Use a thread gauge to measure the threads. The finished end should allow for a wrench-tight fit with one or two threads exposed.

Note: The conduit should be reamed again after threading to remove any burrs and edges. Cutting oil must be swabbed from the inside and outside of the conduit. Use a sandbox or drip pan under the threader to collect drips and shavings.

Die heads can also be used with portable power drives. You will follow the same steps when using a portable power drive. Threading machines are often used on larger conduit and where frequent threading is required. Threading machines hold and rotate the conduit while the die is fed onto the conduit for cutting. When using a threading machine, make sure you secure the legs properly and follow the manufacturer's instructions.

3.5.0 CUTTING AND JOINING PVC CONDUIT

PVC conduit may be easily cut with a fine-tooth handsaw. To ensure square cuts, a miter box or similar device is recommended for cutting 2" and larger PVC. You can deburr the cut ends using a pocket knife. Smaller diameter PVC conduit, up to 1½", may be cut using a PVC cutter.

Use the following steps to join PVC conduit sections or attachments to plastic boxes:

Step 1   Wipe all the contacting surfaces clean and dry.

Step 2   Apply a coat of cement (a brush or aerosol can is recommended) to the male end to be attached.

Step 3   Press the conduit and fitting together and rotate about a half-turn to evenly distribute the cement.

Note: Cementing the PVC must be done quickly. The aerosol spray cans of cement or the cement/brush combination are usually provided by the PVC manufacturer. Make sure you use the recommended cement.

Forming PVC in the field requires a special tool called a *hot box* or other specialized methods. PVC may not be threaded when it is used for electrical applications.

CAUTION: Solvents and cements used with PVC are hazardous. Wear gloves and follow the product instructions.

SUMMARY

You must choose a conduit bender to suit the kind of conduit being installed and the type of bend to be made. Some knowledge of the geometry of right triangles and circles needs to be mastered to make the necessary calculations. You must be able to calculate, lay out, and perform bending operations on a single run of conduit and also on two or more parallel runs of conduit. At times, data tables for the figures may be consulted for the calculations. All work must conform to the requirements of the NEC.

References

For advanced study of topics covered in this task module, the following books are suggested:

Benfield Conduit Bending Manual, Latest Edition, McGraw-Hill Publishing Company, New York, NY.

National Electrical Code Handbook, Latest Edition, National Fire Protection Association, Quincy, MA.

Tom Henry's Conduit Bending Package (includes video, book, and bending chart), Code Electrical Classes, Inc., Winter Park, FL.

REVIEW/PRACTICE QUESTIONS

1. The field bending of PVC requires a _____.
   a. hickey
   b. heating unit
   c. segmented bender
   d. one-shot bender

2. After bending PVC, the bend can be set by using _____.
   a. a damp sponge or cloth
   b. dry ice
   c. ice cold water
   d. a blow dryer

3. A hickey is used for bending _____.
   a. rigid conduit
   b. EMT and IMC
   c. PVC conduit
   d. rigid conduit, EMT, IMC, and PVC conduit

4. A plug set is typically used to prevent _____ PVC when bending it.
   a. overpressurizing
   b. flattening
   c. corroding
   d. cutting

5. What is the key to accurate bending with a hand bender?
   a. Correct size and length of handle
   b. Constant foot pressure on the back piece
   c. Using only the correct brand of bender
   d. Correct inverting of the conduit bender

6. In a right triangle, the side directly opposite the 90° angle is called the _____.
   a. right side
   b. hypotenuse
   c. altitude
   d. base

7. The formula for calculating the circumference of a circle is _____.
   a. $\pi \times R^2$
   b. $2\pi \times R^2$
   c. $\pi \times D$
   d. $2\pi \times D$

8. Prior to making a 90° bend, what two measurements must be known?
   a. Length of conduit and size of conduit
   b. Desired rise and length of conduit
   c. Size of bender and size of conduit
   d. Stub-up distance and take-up distance

9. A back-to-back bend is _____.
   a. a two-shot 90° bend
   b. two 90° bends made back-to-back
   c. an offset with four bends
   d. a segmented bend

10. To ensure the conduit enters straight into the junction box, a(n) _____ may be required.
    a. back-to-back bend
    b. saddle bend
    c. offset
    d. take-up

11. To prevent the ends of the conduit from being staggered, what additional information must be used when making parallel offset bends?
    a. Center-to-center spacing and tangent of ½ the offset angle
    b. Length of conduit and size of conduit
    c. Stub-up distance and take-up distance
    d. Offset angle and length of conduit

12. When making a saddle bend, the center of the saddle will cause the conduit to shrink _____ for every inch of saddle depth.
    a. 3/8"
    b. 3/16"
    c. 3/4"
    d. 3/32"

13. When using a pipe cutter, always rotate the cutter _____ to start the cut.
    a. in a clockwise direction
    b. with the grain
    c. in a counterclockwise direction
    d. against the grain

HAND BENDING OF CONDUIT — TRAINEE TASK MODULE 33105

14. You would use _____ to smooth the sharp inside edge of metal conduit after it has been cut.
    a. a flat file
    b. rough sandpaper
    c. a reamer
    d. a pocket knife

15. What tool is used to cut threads in rigid conduit or IMC?
    a. A thread gauge
    b. A cutter
    c. A tap
    d. A die

notes

ANSWERS TO REVIEW/PRACTICE QUESTIONS

Answer	Section Reference
1. b	2.0.0
2. a	2.0.0
3. a	2.0.0
4. b	2.0.0
5. b	2.0.0
6. b	2.1.0
7. c	2.1.0
8. d	2.2.0
9. b	2.4.0
10. c	2.5.0
11. a	2.6.0
12. b	2.7.0
13. c	3.2.0
14. c	3.3.0
15. d	3.4.0

APPENDIX A

USING TRIGONOMETRY TO DETERMINE OFFSET ANGLES AND MULTIPLIERS

YOU DO NOT HAVE TO BE A MATHEMATICIAN TO USE TRIGONOMETRY. UNDERSTANDING THE BASIC TRIG FUNCTIONS AND HOW TO USE THEM CAN HELP YOU CALCULATE UNKNOWN DISTANCES OR ANGLES. ASSUME THAT THE RIGHT TRIANGLE BELOW REPRESENTS A CONDUIT OFFSET. IF YOU KNOW THE LENGTH OF ONE SIDE AND THE ANGLE, YOU CAN CALCULATE THE LENGTH OF THE OTHER SIDES, OR IF YOU KNOW THE LENGTH OF ANY TWO OF THE SIDES OF THE TRIANGLE, YOU CAN THEN FIND THE OFFSET ANGLE USING ONE OR MORE OF THESE TRIG FUNCTIONS. YOU CAN USE A TRIG TABLE SUCH AS THAT SHOWN ON THE FOLLOWING PAGES OR A SCIENTIFIC CALCULATOR TO DETERMINE THE OFFSET ANGLE. FOR EXAMPLE, IF THE COSECANT OF ANGLE A IS 2.6, THE TRIG TABLE TELLS YOU THAT THE OFFSET ANGLE IS 22½°.

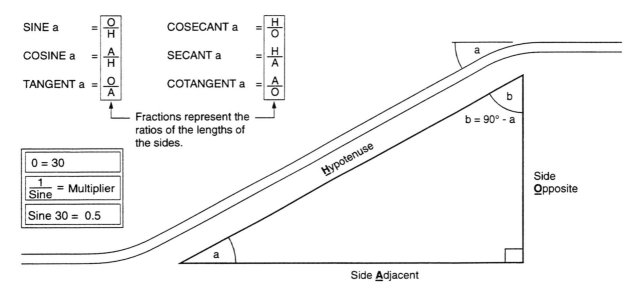

To determine the multiplier for the distance between bends in an offset:

1. Determine the angle of the offset: 30°
2. Find the sine of the angle: 0.5
3. Find the inverse (reciprocal) of the sine: $\frac{1}{0.5}$ = 2. This is also listed in trig tables as the cosecant of the angle.
4. This number multiplied by the height of the offset gives the hypotenuse of the triangle, which is equal to the distance between bends.

TRIGONOMETRY TABLE

Angle	Sine	Cosine	Tangent	Cotangent	Cosecant
1°	0.0175	0.9998	0.0175	57.2900	57.2987
2°	0.0349	0.9994	0.0349	28.6363	28.6537
3°	0.0523	0.9986	0.0524	19.0811	19.1073
4°	0.0698	0.9976	0.0699	14.3007	14.3356
5°	0.0872	0.9962	0.0875	11.4301	11.4737
6°	0.1045	0.9945	0.1051	9.5144	9.5668
7°	0.1219	0.9925	0.1228	8.1443	8.2055
8°	0.1392	0.9903	0.1405	7.1154	7.1853
9°	0.1564	0.9877	0.1584	6.3138	6.3925
10°	0.1736	0.9848	0.1763	5.6713	5.7588
11°	0.1908	0.9816	0.1944	5.1446	5.2408
12°	0.2079	0.9781	0.2126	4.7046	4.8097
13°	0.2250	0.9744	0.2309	4.3315	4.4454
14°	0.2419	0.9703	0.2493	4.0108	4.1336
15°	0.2588	0.9659	0.2679	3.7321	3.8637
16°	0.2756	0.9613	0.2867	3.4874	3.6280
17°	0.2924	0.9563	0.3057	3.2709	3.4203
18°	0.3090	0.9511	0.3249	3.0777	3.2361
19°	0.3256	0.9455	0.3443	2.9042	3.0716
20°	0.3420	0.9397	0.3640	2.7475	2.9238
21°	0.3584	0.9336	0.3839	2.6051	2.7904
22°	0.3746	0.9272	0.4040	2.4751	2.6695
23°	0.3907	0.9205	0.4245	2.3559	2.5593
24°	0.4067	0.9135	0.4452	2.2460	2.4586
25°	0.4226	0.9063	0.4663	2.1445	2.3662
26°	0.4384	0.8988	0.4877	2.0503	2.2812
27°	0.4540	0.8910	0.5095	1.9626	2.2027
28°	0.4695	0.8829	0.5317	1.8807	2.1301
29°	0.4848	0.8746	0.5543	1.8040	2.0627
30°	0.5000	0.8660	0.5774	1.7321	2.0000
31°	0.5150	0.8572	0.6009	1.6643	1.9416
32°	0.5299	0.8480	0.6249	1.6003	1.8871
33°	0.5446	0.8387	0.6494	1.5399	1.8361
34°	0.5592	0.8290	0.6745	1.4826	1.7883
35°	0.5736	0.8192	0.7002	1.4281	1.7434
36°	0.5878	0.8090	0.7265	1.3764	1.7013
37°	0.6018	0.7986	0.7536	1.3270	1.6616
38°	0.6157	0.7880	0.7813	1.2799	1.6243
39°	0.6293	0.7771	0.8098	1.2349	1.5890
40°	0.6428	0.7660	0.8391	1.1918	1.5557
41°	0.6561	0.7547	0.8693	1.1504	1.5243
42°	0.6691	0.7431	0.9004	1.1106	1.4945
43°	0.6820	0.7314	0.9325	1.0724	1.4663
44°	0.6947	0.7193	0.9657	1.0355	1.4396
45°	0.7071	0.7071	1.0000	1.0000	1.4142

TRIGONOMETRY TABLE (Continued)

Angle	Sine	Cosine	Tangent	Cotangent	Cosecant
46°	0.7193	0.6947	1.0355	0.9657	1.3902
47°	0.7314	0.6820	1.0724	0.9325	1.3673
48°	0.7431	0.6691	1.1106	0.9004	1.3456
49°	0.7547	0.6561	1.1504	0.8693	1.3250
50°	0.7660	0.6428	1.1918	0.8391	1.3054
51°	0.7771	0.6293	1.2349	0.8098	1.2868
52°	0.7880	0.6157	1.2799	0.7813	1.2690
53°	0.7986	0.6018	1.3270	0.7536	1.2521
54°	0.8090	0.5878	1.3764	0.7265	1.2361
55°	0.8192	0.5736	1.4281	0.7002	1.2208
56°	0.8290	0.5592	1.4826	0.6745	1.2062
57°	0.8387	0.5446	1.5399	0.6494	1.1924
58°	0.8480	0.5299	1.6003	0.6249	1.1792
59°	0.8572	0.5150	1.6643	0.6009	1.1666
60°	0.8660	0.5000	1.7321	0.5774	1.1547
61°	0.8746	0.4848	1.8040	0.5543	1.1434
62°	0.8829	0.4695	1.8807	0.5317	1.1326
63°	0.8910	0.4540	1.9626	0.5095	1.1223
64°	0.8988	0.4384	2.0503	0.4877	1.1126
65°	0.9063	0.4226	2.1445	0.4663	1.1034
66°	0.9135	0.4067	2.2460	0.4452	1.0946
67°	0.9205	0.3907	2.3559	0.4245	1.0864
68°	0.9272	0.3746	2.4751	0.4040	1.0785
69°	0.9336	0.3584	2.6051	0.3839	1.0711
70°	0.9397	0.3420	2.7475	0.3640	1.0642
71°	0.9455	0.3256	2.9042	0.3443	1.0576
72°	0.9511	0.3090	3.0777	0.3249	1.0515
73°	0.9563	0.2924	3.2709	0.3057	1.0457
74°	0.9613	0.2756	3.4874	0.2867	1.0403
75°	0.9659	0.2588	3.7321	0.2679	1.0353
76°	0.9703	0.2419	4.0108	0.2493	1.0306
77°	0.9744	0.2250	4.3315	0.2309	1.0263
78°	0.9781	0.2079	4.7046	0.2126	1.0223
79°	0.9816	0.1908	5.1446	0.1944	1.0187
80°	0.9848	0.1736	5.6713	0.1763	1.0154
81°	0.9877	0.1564	6.3138	0.1584	1.0125
82°	0.9903	0.1392	7.1154	0.1405	1.0098
83°	0.9925	0.1219	8.1443	0.1228	1.0075
84°	0.9945	0.1045	9.5144	0.1051	1.0055
85°	0.9962	0.0872	11.4301	0.0875	1.0038
86°	0.9976	0.0698	14.3007	0.0699	1.0024
87°	0.9986	0.0523	19.0811	0.0524	1.0014
88°	0.9994	0.0349	28.6363	0.0349	1.0006
89°	0.9998	0.0175	57.2900	0.0175	1.0002
90°	1.0000	0.0000	∞	0.0000	1.0000

APPENDIX B

BENDING RADIUS TABLE

RADIUS (INCHES)	RADIUS INCREMENTS (INCHES)									
	0	1	2	3	4	5	6	7	8	9
0	0.00	1.57	3.14	4.71	6.28	7.85	9.42	11.00	12.57	14.14
10	15.71	17.28	18.85	20.42	21.99	23.56	25.13	26.70	28.27	29.85
20	31.42	32.99	34.56	36.13	37.70	39.27	40.84	42.41	43.98	45.55
30	47.12	48.69	50.27	51.84	53.41	54.98	56.55	58.12	59.69	61.26
40	62.83	64.40	65.97	67.54	69.11	70.69	72.26	73.83	75.40	76.97
50	78.54	80.11	81.68	83.25	84.82	86.39	87.96	89.54	91.11	92.68
60	94.25	95.82	97.39	98.96	100.53	102.10	103.67	105.24	106.81	108.38
70	109.96	111.53	113.10	114.67	116.24	117.81	119.38	120.95	122.52	124.09
80	125.66	127.23	128.81	130.38	131.95	133.52	135.09	136.66	138.23	139.80
90	141.37	142.94	144.51	146.08	147.65	149.23	150.80			

Developed length for following angles ues fraction of 90° chart.

For	15°	22.5°	30°	45°	60°	67.5°	75°	90°
Take	1/6	1/4	1/3	1/2	2/3	3/4	5/6	See Chart

For any other degrees: Developed length = 0.1745 x radius x degrees.

NCCER CRAFT TRAINING USER UPDATES

The NCCER makes every effort to keep these manuals up-to-date and free of technical errors. We appreciate your help in this process. If you have an idea for improving this manual, or if you find an error, a typographical mistake, or an inaccuracy in the NCCER's Craft Training Manuals, please write us, using this form or a photocopy. Be sure to include the exact module number, page number, a description of the problem, and the correction, if possible. Your input will be brought to the attention of the Technical Review Committee. Thank you for your assistance.

Instructors – If you found that additional materials were necessary in order to teach this module effectively, please let us know so that we may include them in the Equipment/Materials list in the Instructor's Guide.

Write: Curriculum Revision and Development Department
National Center for Construction Education and Research
P.O. Box 141104
Gainesville, FL 32614-1104
Fax: 352-334-0932

Craft _____ Module Name _____

Copyright Date _____ Module Number _____ Page Number(s) _____

Description of Problem
_____
_____
_____
_____

(Optional) Correction of Problem
_____
_____
_____

(Optional) Your Name and Address
_____
_____
_____

notes

Electrical Theory One
Module 33106

Electronic Systems Technician Trainee Task Module 33106

ELECTRICAL THEORY ONE

NATIONAL
CENTER FOR
CONSTRUCTION
EDUCATION AND
RESEARCH

OBJECTIVES

Upon completion of this module, the trainee will be able to:

1. Recognize what atoms are and how they are constructed.
2. Define voltage and identify the ways in which it can be produced.
3. Explain the difference between conductors and insulators.
4. Define the units of measurement that are used to measure the properties of electricity.
5. Explain how voltage, current, and resistance are related to each other.
6. Using the formula for Ohm's Law, calculate an unknown value.
7. Explain the different types of meters used to measure voltage, current, and resistance.
8. Using the power formula, calculate the amount of power used by a circuit.

Prerequisites

Successful completion of the following Task Modules is recommended before beginning study of this Task Module: Core Curricula; Electronic Systems Technician Level One, Modules 33101 through 33105.

Required Trainee Materials

1. Trainee Task Module
2. Copy of the latest edition of the *National Electrical Code*
3. Appropriate Personal Protective Equipment

Note: The designations "National Electrical Code," "NE Code," and "NEC," where used in this document, refer to the *National Electrical Code®*, which is a registered trademark of the National Fire Protection Association, Quincy, MA. *All National Electrical Code (NEC) references in this module refer to the 1999 edition of the NEC.*

COURSE MAP

This course map shows all of the modules in the first level of the Electronic Systems Technician curricula. The suggested training order begins at the bottom and proceeds up. Skill levels increase as a trainee advances on the course map. The training order may be adjusted by the local Training Program Sponsor.

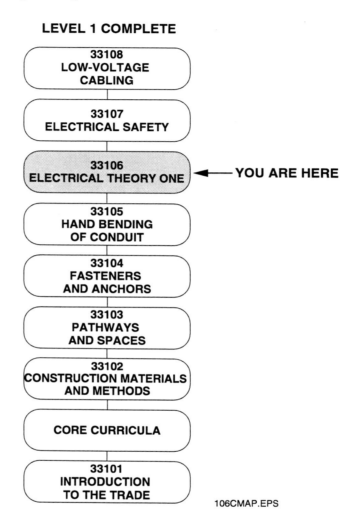

TABLE OF CONTENTS

Section	Topic	Page
	Course Map	2
1.0.0	Introduction	6
2.0.0	Conductors And Insulators	6
2.1.0	The Atom	6
2.1.1	The Nucleus	7
2.1.2	Electrical Charges	8
2.2.0	Conductors And Insulators	8
3.0.0	Electric Charge And Current	9
3.1.0	Current Flow	10
3.2.0	Voltage	11
4.0.0	Resistance	11
4.1.0	Characteristics Of Resistance	11
4.2.0	Ohm's Law	13
5.0.0	Schematic Representation Of Circuit Elements	14
6.0.0	Resistors	16
6.1.0	Resistor Color Codes	17
7.0.0	Measuring Voltage, Current, And Resistance	18
7.1.0	Basic Meter Operation	18
7.2.0	Voltmeter	19
7.3.0	Ammeter	20
7.4.0	Ohmmeter	21
8.0.0	Electrical Power	21
8.1.0	Power Equation	23
8.2.0	Power Rating Of Resistors	26
	Summary	27
	Review/Practice Questions	28
	Answers To Review/Practice Questions	30

Trade Terms Introduced In This Module

Ammeter: An instrument for measuring electrical current.

Ampere (A): A unit of electrical current. For example, one volt across one ohm of resistance causes a current flow of one ampere.

Atom: The smallest particle into which an element may be divided and still retain the properties of the element.

Battery: A DC voltage source consisting of two or more cells that convert chemical energy into electrical energy.

Circuit: A complete path for current flow.

Conductor: A material that offers very little resistance to current flow.

Coulomb: 6.25×10^{18} electrons or 6,250,000,000,000,000,000 electrons. A coulomb is the common unit of quantity used for specifying the size of a given charge.

Current: The movement, or flow, of electrons in a circuit. Current (I) is measured in amperes.

Electron: A negatively charged particle that orbits the nucleus of an atom.

Insulator: A material that offers resistance to current flow.

Joule (J): A unit of measurement that represents one newton-meter (Nm), which is a unit of measure for doing work.

Kilo: A prefix used to indicate one thousand; for example, one kilowatt is equal to one thousand watts.

Matter: Any substance that has mass and occupies space.

Mega: A prefix used to indicate one million; for example, one megawatt is equal to one million watts.

Micro: A prefix used to indicate one-millionth; for example, one microwatt is equal to one-millionth of a watt.

Neutrons: Electrically neutral particles (neither positive nor negative) that have the same mass as a proton and are found in the nucleus of an atom.

Nucleus: The center of an atom. It contains the protons and neutrons of the atom.

Ohm (Ω): The basic unit of measurement for resistance.

Ohmmeter: An instrument used for measuring resistance.

Ohm's Law: A statement of the relationships among current, voltage, and resistance in an electrical circuit: current (I) equals voltage (E) divided by resistance (R). Generally expressed as a mathematical formula: I = E/R.

Power: The rate of doing work or the rate at which energy is used or dissipated. Electrical power is the rate of doing electrical work. Electrical power is measured in watts.

Protons: The smallest positively charged particles of an atom. Protons are contained in the nucleus of an atom.

Resistance: An electrical property that opposes the flow of current through a circuit. Resistance (R) is measured in ohms (Ω).

Resistor: Any device in a circuit that resists the flow of electrons.

Schematic: A type of drawing in which symbols are used to represent the components in a system.

Series circuit: A circuit with only one path for current flow.

Valence shell: The outermost ring of electrons that orbit about the nucleus of an atom.

Volt (V): The unit of measurement for voltage (electromotive force). One volt is equivalent to the force required to produce a current of one ampere through a resistance of one ohm.

Voltage: The driving force that makes current flow in a circuit. Voltage (E) is also referred to as *electromotive force* or *potential*.

Voltage drop: The change in voltage across a component that is caused by the current flowing through it and the amount of resistance opposing it.

Voltmeter: An instrument for measuring voltage. The resistance of the voltmeter is fixed. When the voltmeter is connected to a circuit, the current passing through the meter will be directly proportional to the voltage at the connection points.

Watt (W): The basic unit of measurement for electrical power.

1.0.0 INTRODUCTION

As an Electronic Systems Technician, you must work with a force that cannot be seen. However, electricity is there on the job, every day of the year. It is necessary that you understand the forces of electricity so that you will be safe on the job. The first step is a basic understanding of the principles of electricity.

The relationships among **current**, **voltage**, **resistance**, and **power** in a basic direct current (DC) **series circuit** are common to all types of electrical **circuits**. This module provides a general introduction to the electrical concepts used in **Ohm's Law**. It also presents the opportunity to practice applying these basic concepts to DC series circuits. In this way, you can prepare for further study in electrical and electronics theory and maintenance techniques. By practicing these techniques for all combinations of DC circuits, you will be prepared to work on any DC circuits you might encounter.

2.0.0 CONDUCTORS AND INSULATORS

2.1.0 THE ATOM

The **atom** is the smallest part of an element that enters into a chemical change, but it does so in the form of a charged particle. These charged particles are called *ions*, and are of two types—positive and negative. A positive ion may be defined as an atom that has become positively charged. A negative ion may be defined as an atom that has become negatively charged. One of the properties of charged ions is that ions of the same charge tend to repel one another, whereas ions of unlike charge will attract one another. The term *charge* can be taken to mean a quantity of electricity that is either positive or negative.

The structure of an atom is best explained by a detailed analysis of the simplest of all atoms, that of the element hydrogen. The hydrogen atom in *Figure 1* is composed of a **nucleus** containing one **proton** and a single orbiting **electron**. As the electron revolves around the nucleus, it is held in this orbit by two counteracting forces. One of these forces is called *centrifugal force*, which is the force that tends to cause the electron to fly outward as it travels around its circular orbit. The second force acting on the electron is *electrostatic force*. This force tends to pull the electron in toward the nucleus and is provided by the mutual attraction between the positive nucleus and the negative electron. At some given radius, the two forces will balance each other, providing a stable path for the electron.

- A proton (+) repels another proton (+).
- An electron (–) repels another electron (–).
- A proton (+) attracts an electron (–).

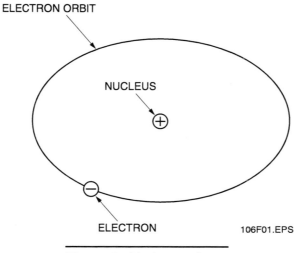

Figure 1. Hydrogen Atom

Basically, an atom contains three types of subatomic particles that are of interest in electricity: electrons, protons, and **neutrons**.

The protons and neutrons are located in the center, or nucleus, of the atom, and the electrons travel about the nucleus in orbits.

Because protons are relatively heavy, the repulsive force they exert on one another in the nucleus of an atom has little effect.

The attracting and repelling forces on charged materials occur because of the electrostatic lines of force that exist around the charged materials. In a negatively charged object, the lines of force of the excess electrons add to produce an electrostatic field that has lines of force coming into the object from all directions. In a positively charged object, the lines of force of the excess protons add to produce an electrostatic field that has lines of force going out of the object in all directions. The electrostatic fields either aid or oppose each other to attract or repel.

2.1.1 The Nucleus

The nucleus is the central part of the atom. It is made up of heavy particles called *protons* and *neutrons*. The proton is a charged particle containing the smallest known unit of positive electricity. The neutron has no electrical charge. The number of protons in the nucleus determines how the atom of one element differs from the atom of another element.

Although a neutron is actually a particle by itself, it is generally thought of as an electron and proton combined and is electrically neutral. Since neutrons are electrically neutral, they are not considered important to the electrical nature of atoms.

2.1.2 Electrical Charges

The negative charge of an electron is equal but opposite to the positive charge of a proton. The charges of an electron and a proton are called *electrostatic charges*. The lines of force associated with each particle produce electrostatic fields. Because of the way these fields act together, charged particles can attract or repel one another. The Law of Electrical Charges states that particles with like charges repel each other and those with unlike charges attract each other. This is shown in *Figure 2*.

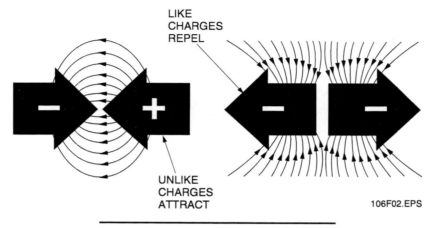

Figure 2. Law Of Electrical Charges

2.2.0 CONDUCTORS AND INSULATORS

The difference between atoms, with respect to chemical activity and stability, depends on the number and position of the electrons included within the atom. In general, the electrons reside in groups of orbits called *shells*. The shells are arranged in steps that correspond to fixed energy levels.

The number of electrons in the outermost shell determines the valence of an atom. For this reason, the outer shell of an atom is called the **valence shell**, and the electrons contained in this shell are called *valence electrons* (*Figure 3*). The valence of an atom determines its ability to gain or lose an electron, which in turn determines the chemical and electrical properties of the atom. An atom that is lacking only one or two electrons from its outer shell will easily gain electrons to complete its shell, but a large amount of energy is required to free any of its electrons. An atom having a relatively small number of electrons in its outer shell in comparison to the number of electrons required to fill the shell will easily lose these valence electrons.

It is the valence electrons that we are most concerned with in electricity. These are the electrons that are easiest to break loose from their parent atom. Normally, a **conductor** has three or less valence electrons, an **insulator** has five or more valence electrons, and semiconductors usually have four valence electrons.

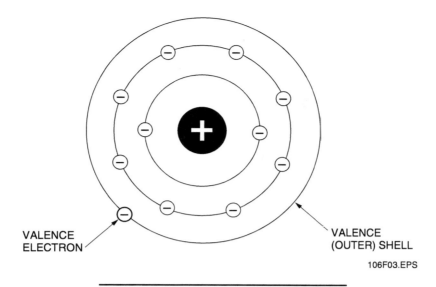

Figure 3. Valence Shell And Electrons

All the elements of which **matter** is made may be placed into one of three categories: conductors, insulators, and semiconductors.

Conductors, for example, are elements such as copper and silver that will conduct a flow of electricity very readily. Because of their good conducting abilities, they are formed into wire and used whenever it is desired to transfer electrical energy from one point to another.

Insulators, on the other hand, do not conduct electricity to any great degree and are used when it is desirable to prevent the flow of electricity. Compounds such as porcelain and plastic are good insulators.

Materials such as germanium and silicon are not good conductors but cannot be used as insulators either, since their electrical characteristics fall between those of conductors and those of insulators. These in-between materials are classified as *semiconductors*. As you will learn later in your training, semiconductors play a crucial role in electronic circuits.

3.0.0 ELECTRIC CHARGE AND CURRENT

An electric charge has the ability to do the work of moving another charge by attraction or repulsion. The ability of a charge to do work is called its *potential*. When one charge is different from another, there must be a difference in potential between them. The sum of the difference of potential of all the charges in the electrostatic field is referred to as *electromotive force (emf)* or *voltage*. Voltage is frequently represented by the letter E.

Electric charge is measured in **coulombs**. An electron has 1.6×10^{-19} coulombs of charge. Therefore, it takes 6.25×10^{18} electrons to make up one coulomb of charge, as shown below.

$$\frac{1}{1.6 \times 10^{-19}} = 6.25 \times 10^{18} \text{ electrons}$$

If two particles, one having charge Q_1 and the other charge Q_2, are a distance (d) apart, then the force between them is given by Coulomb's Law, which states that the force is directly proportional to the product of the two charges and inversely proportional to the square of the distance between them:

$$\text{Force} = \frac{Q_1 \times Q_2}{d^2}$$

If Q_1 and Q_2 are both positive or both negative, then the force is positive; it is repulsive. If Q_1 and Q_2 are of opposite charges, then the force is negative; it is attractive.

3.1.0 CURRENT FLOW

The movement of the flow of electrons is called *current*. To produce current, the electrons are moved by a potential difference. Current is represented by the letter *I*. The basic unit in which current is measured is the **ampere (A)**, also called the *amp*. The symbol for the ampere is *A*. One ampere of current is defined as the movement of one coulomb past any point of a conductor during one second of time. One coulomb is equal to 6.25×10^{18} electrons; therefore, one ampere is equal to 6.25×10^{18} electrons moving past any point of a conductor during one second of time.

The definition of current can be expressed as an equation:

$$I = \frac{Q}{T}$$

Where:

I = current (amperes)
Q = charge (coulombs)
T = time (seconds)

Charge differs from current in that Q is an accumulation of charge, while I measures the intensity of moving charges.

In a conductor, such as copper wire, the free electrons are charges that can be forced to move with relative ease by a potential difference. If a potential difference is connected across two ends of a copper wire, as shown in *Figure 4*, the applied voltage forces the free electrons to move. This current is a flow of electrons from the point of negative charge (−) at one end of the wire, moving through the wire to the positive charge (+), at the other end. The direction of the electron flow is from the negative side of the **battery**, through the wire, and back to the positive side of the battery. The direction of current flow is therefore from a point of negative potential to a point of positive potential.

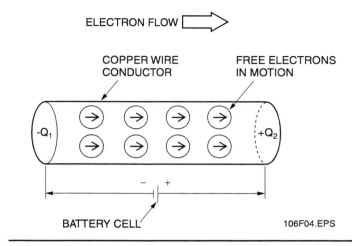

Figure 4. Potential Difference Causing Electric Current

3.2.0 VOLTAGE

The force that causes electrons to move is called *voltage, potential difference,* or *electromotive force (emf)*. One **volt (V)** is the potential difference between two points for which one coulomb of electricity will do one **joule (J)** of work. A battery is one of several means of creating voltage. It chemically creates a large reserve of free electrons at the negative (−) terminal. The positive (+) terminal has electrons chemically removed and will therefore accept them if an external path is provided from the negative (−) terminal. When a battery is no longer able to chemically deposit electrons at the negative (−) terminal, it is said to be dead, or in need of recharging. Batteries are normally rated in volts. Large batteries are also rated in ampere-hours, where one ampere-hour is a current of one amp supplied for one hour.

4.0.0 RESISTANCE

4.1.0 CHARACTERISTICS OF RESISTANCE

Resistance is directly related to the ability of a material to conduct electricity. Conductors have very low resistance; insulators have very high resistance.

Resistance can be defined as the opposition to current flow. To add resistance to a circuit, electrical components called **resistors** are used. A resistor is a device whose resistance to current flow is a known, specified value. Resistance is measured in **ohms (Ω)** and is represented by the symbol R in equations. One ohm is defined as that amount of resistance that will limit the current in a conductor to one ampere when the voltage applied to the conductor is one volt.

The resistance of a wire is proportional to the length of the wire, inversely proportional to the cross-sectional area of the wire, and dependent upon the kind of material of which the wire is made. The relationship for finding the resistance of a wire is:

$$R = \rho \frac{L}{A}$$

Where:

R = resistance (ohms)
L = length of wire (feet)
A = area of wire (circular mils, CM, or cm^2)
ρ = specific resistance (ohm-CM/ft. or microhm-cm)

A *mil* equals 0.001 inch; a circular mil is the cross-sectional area of a wire one mil in diameter.

The specific resistance is a constant that depends on the material of which the wire is made. *Table 1* shows the properties of various wire conductors.

Metal	Specific Resistance (Resistance of 1 CM/ft. in ohms)	
	32°F or 0°C or	75°F or 23.8°C
Silver, pure annealed	8.831	9.674
Copper, pure annealed	9.39	10.351
Copper, annealed	9.59	10.505
Copper, hard-drawn	9.81	10.745
Gold	13.216	14.404
Aluminum	15.219	16.758
Zinc	34.595	37.957
Iron	54.529	62.643

Table 1. Conductor Properties

Table 1 shows that at 75°F, a one-mil diameter, pure annealed copper wire that is one foot long has a resistance of 10.351Ω; while a one-mil diameter, one-foot-long aluminum wire has a resistance of 16.758Ω. Temperature is important in determining the resistance of a wire. The hotter a wire, the greater its resistance.

4.2.0 OHM'S LAW

Ohm's Law defines the relationship between current, voltage, and resistance. There are three ways to express Ohm's Law mathematically.

- The current in a circuit is equal to the voltage applied to the circuit divided by the resistance of the circuit:

$$I = \frac{E}{R}$$

- The resistance of a circuit is equal to the voltage applied to the circuit divided by the current in the circuit:

$$R = \frac{E}{I}$$

- The applied voltage to a circuit is equal to the product of the current and the resistance of the circuit:

$$E = I \times R = IR$$

Where:

- I = current (amperes)
- R = resistance (ohms)
- E = voltage or emf (volts)

If any two of the quantities E, I, or R are known, the third can be calculated.

The Ohm's Law equations can be memorized and practiced effectively by using an Ohm's Law circle, as shown in *Figure 5*. To find the equation for E, I, or R when two quantities are known, cover the unknown third quantity. The other two quantities in the circle will indicate how the covered quantity may be found.

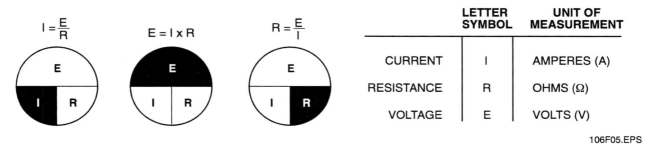

Figure 5. Ohm's Law Circle

Example 1:

Find I when E = 120V and R = 30Ω.

$$I = \frac{E}{R}$$
$$I = \frac{120V}{30\Omega}$$
$$I = 4A$$

This formula shows that in a DC circuit, current (I) is directly proportional to voltage (E) and inversely proportional to resistance (R).

Example 2:

Find R when E = 240V and I = 20A.

$$R = \frac{E}{I}$$
$$R = \frac{240V}{20A}$$
$$R = 12\Omega$$

Example 3:

Find E when I = 15A and R = 8Ω.

$$E = I \times R$$
$$E = 15A \times 8\Omega$$
$$E = 120V$$

5.0.0 SCHEMATIC REPRESENTATION OF CIRCUIT ELEMENTS

A simple electric circuit is shown in both pictorial and **schematic** forms in *Figure 6*. The schematic diagram is a shorthand way to draw an electric circuit, and circuits are usually represented in this way. In addition to the connecting wire, three components are shown symbolically: the battery, the switch, and the lamp. Note the positive (+) and negative (−) markings in both the pictorial and schematic representations of the battery. The schematic components represent the pictorial components in a simplified manner. A schematic diagram is one that shows, by means of graphic symbols, the electrical connections and functions of the different parts of a circuit.

The standard graphic symbols for commonly used electrical and electronic components are shown in *Figure 7*.

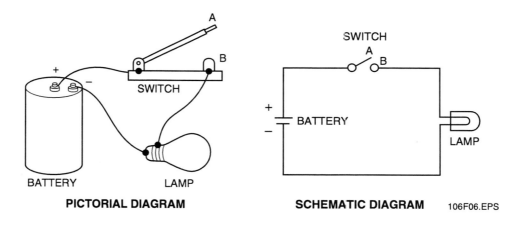

Figure 6. Simple Electrical Symbols

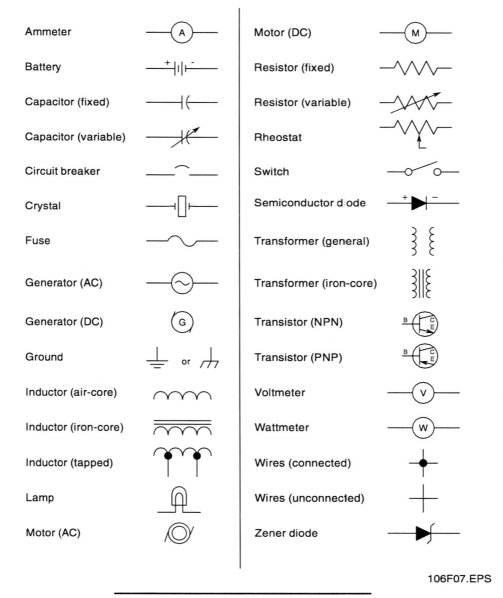

Figure 7. Standard Schematic Symbols

6.0.0 RESISTORS

The function of a resistor is to offer a particular resistance to current flow. For a given current and known resistance, the change in voltage across the component, or **voltage drop**, can be predicted using Ohm's Law. Voltage drop refers to a specific amount of voltage used, or developed, by that component. An example is a very basic circuit of a 10V battery and a single resistor in a series circuit. The voltage drop across that resistor is 10V because it is the only component in the circuit and all voltage must be dropped across that resistor. Similarly, for a given applied voltage, the current that flows may be predetermined by selection of the resistor value. The required power dissipation largely dictates the construction and physical size of a resistor.

The two most common types of electronic resistors are *wire-wound* and *carbon composition construction*. A typical wire-wound resistor consists of a length of nickel wire wound on a ceramic tube and covered with porcelain. Low-resistance connecting wires are provided, and the resistance value is usually printed on the side of the component. *Figure 8* illustrates the construction of typical resistors. Carbon composition resistors are constructed by molding mixtures of powdered carbon and insulating materials into a cylindrical shape. An outer sheath of insulating material affords mechanical and electrical protection, and copper connecting wires are provided at each end. Carbon composition resistors are smaller and less expensive than the wire-wound type. However, the wire-wound type is the more rugged of the two and is able to survive much larger power dissipations than the carbon composition type.

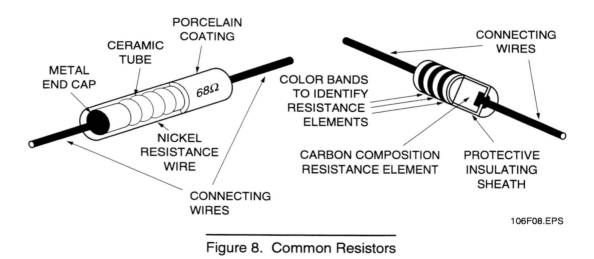

Figure 8. Common Resistors

Most resistors have standard fixed values, so they can be termed *fixed resistors*. Variable resistors, also known as *adjustable resistors*, are used a great deal in electronics. Two common symbols for a variable resistor are shown in *Figure 9*.

A variable resistor consists of a coil of closely-wound insulated resistance wire formed into a partial circle. The coil has a low-resistance terminal at each end, and a third terminal is connected to a movable contact with a shaft adjustment facility. The movable contact can be set to any point on a connecting track that extends over one (uninsulated) edge of the coil.

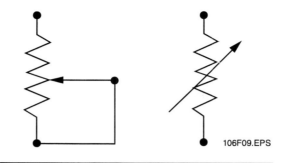

Figure 9. Symbols Used For Variable Resistors

Using the adjustable contact, the resistance from either end terminal to the center terminal may be adjusted from zero to the maximum coil resistance.

Another type of variable resistor is known as a *decade resistance box*. This is a laboratory component that contains precise values of switched series-connected resistors.

6.1.0 RESISTOR COLOR CODES

Because carbon composition resistors are physically small (some are less than 1 cm in length), it is not convenient to print the resistance value on the side. Instead, a color code in the form of colored bands is employed to identify the resistance value and tolerance. The color code is illustrated in *Figure 10*. Starting from one end of the resistor, the first two bands identify the first and second digits of the resistance value, and the third band indicates the number of zeros. An exception to this is when the third band is either silver or gold, which indicates a 0.01 or 0.1 multiplier, respectively. The fourth band is always either silver or gold, and in this position, silver indicates a ±10% tolerance and gold indicates a ±5% tolerance. Where no fourth band is present, the resistor tolerance is ±20%.

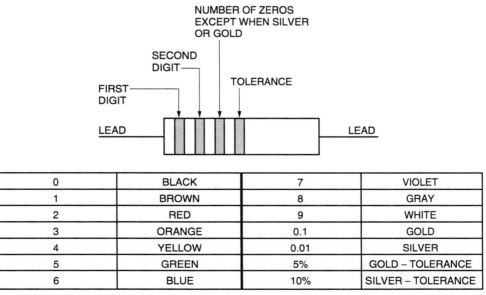

0	BLACK	7	VIOLET
1	BROWN	8	GRAY
2	RED	9	WHITE
3	ORANGE	0.1	GOLD
4	YELLOW	0.01	SILVER
5	GREEN	5%	GOLD – TOLERANCE
6	BLUE	10%	SILVER – TOLERANCE

Figure 10. Resistor Color Codes

ELECTRICAL THEORY ONE — TRAINEE TASK MODULE 33106

We can put this information to practical use by determining the range of values for the carbon resistor in *Figure 11*.

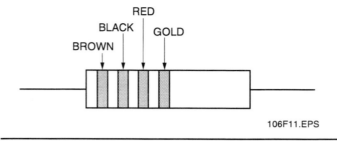

Figure 11. Sample Color Codes On A Fixed Resistor

The color code for this resistor is as follows:

- Brown = 1, black = 0, red = 2, gold = a tolerance of ±5%
- First digit = 1, second digit = 0, number of zeros (2) = 1,000Ω

Since this resistor has a value of 1,000Ω ±5%, the resistor can range in value from 950Ω to 1,050Ω.

7.0.0 MEASURING VOLTAGE, CURRENT, AND RESISTANCE

Working with electricity requires making accurate measurements. This section will discuss the basic meters used to measure voltage, current, and resistance: the **voltmeter**, **ammeter**, and **ohmmeter**.

WARNING! Only qualified individuals may use these meters. Consult your company's safety policy for applicable rules.

7.1.0 BASIC METER OPERATION

When troubleshooting or testing equipment, you will need various meters to check for proper circuit voltages, currents, and resistances and to determine if the wiring is defective. Meters are used in repairing, maintaining, and troubleshooting electrical circuits and equipment. The best and most expensive measuring instrument is of no use to you unless you know what you are measuring and what each reading indicates. Remember that the purpose of a meter is to measure quantities existing within a circuit. For this reason, when the meter is connected to a circuit, it must not change the condition of the circuit.

The three basic electrical quantities discussed in this section are current, voltage, and resistance. Actually, it is really current that causes the meter to respond even when voltage or resistance is being measured. In a basic meter, the measurement of current can be calibrated to indicate almost any electrical quantity based on the principle of Ohm's Law. The amount of current that flows through a meter is determined by the voltage applied to the meter and the resistance of the meter, as stated by $I = E/R$.

For a given meter resistance, different values of applied voltage will cause specific values of current to flow. Although the meter actually measures current, the meter scale can be calibrated in units of voltage. Similarly, for a given applied voltage, different values of resistance will cause specific values of current to flow; therefore, the meter scale can also be calibrated in units of resistance rather than current. The same holds true for power, since power is proportional to current, as stated by $P = EI$. It is on this principle that the meter was developed, and its construction allows for the measurement of various parameters by actually measuring current.

You must understand the purpose and function of each individual piece of test equipment and any limitations associated with it. It is also extremely important that you understand how to safely use each piece of equipment. If you understand the capabilities of the test equipment, you can better use the equipment, better understand the indications on the equipment, and know what substitute or backup meters can be used.

7.2.0 VOLTMETER

A simple voltmeter consists of the meter movement in series with the internal resistance of the voltmeter itself. For example, a meter with a 50-microamp (µA) meter movement and a 1,000Ω internal resistance can be used to directly measure voltages up to 0.05V, as shown in *Figure 12*. (The prefix **micro** means one-millionth.) When the meter is placed across the voltage source, a current determined by the internal resistance of the meter flows through the meter movement. A voltmeter's internal resistance is typically high to minimize meter loading effects on the source.

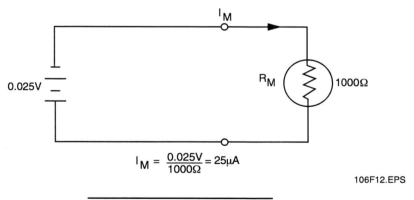

Figure 12. Simple Voltmeter

To measure larger voltages, a multiplier resistor is used. This increased series resistance limits the current that can flow through the meter movement, thus extending the range of the meter.

To avoid damage to the meter movement, the following precautions should be observed when using a voltmeter:

- Always set the full-scale voltage of the meter to be larger than the expected voltage to be measured.
- Always ensure that the internal resistance of the voltmeter is much greater than the resistance of the component to be measured. This means that the current it takes to drive the voltmeter (about 50µA) should be a negligible fraction of the current flowing through the circuit element being measured.
- If you are unsure of the level of the voltage to be measured, take a reading at the highest range of the voltmeter and progressively (step-by-step) lower the range until the reading is obtained.

In most commercial voltmeters, the internal resistance is expressed by the ohms-per-volt rating of the meter. A typical meter has a rating of 20,000 ohms-per-volt with a 50µA movement. This quantity tells what the internal resistance of the meter is on any particular full-scale setting. In general, the meter's internal resistance is the ohms-per-volt rating multiplied by the full-scale voltage. The higher the ohms-per-volt rating, the higher the internal resistance of the meter and the smaller the effect of the meter on the circuit.

7.3.0 AMMETER

There are two kinds of meters used to measure current. The in-line meter, which is connected in series with the circuit being tested, is covered in this module. In a later module, we will cover the clamp-on ammeter. The in-line ammeter is used by placing the meter in series with the wire through which the current is flowing. This method of connection is shown in *Figure 13*. Notice how the magnitude of load current will flow through the ammeter. Because of this, an ammeter's internal resistance must be low to minimize the circuit-loading effects as seen by the source. Also, high current magnitudes flowing through an ammeter can damage it. For this reason, ammeter shunts are employed to reduce the ammeter circuit current to a fraction of the current flowing through the load.

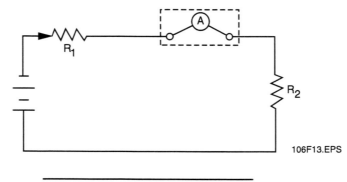

Figure 13. Ammeter Connection

To avoid damage to the meter movement, the following precautions should be observed when taking current measurements with an ammeter:

- Always check the polarity of the ammeter. Make certain that the meter is connected to the circuit so that electrons flow into the negative lead and out of the positive lead. It is easy to tell which is the positive lead because it is normally red. The negative lead is usually black.
- Always set the full-scale deflection of the meter to be larger than the expected current. To be safe, set the full-scale current several times larger than the expected current, and then slowly increase the meter sensitivity to the appropriate scale.
- Always connect the ammeter in series with the circuit element through which the current to be measured is flowing. Never connect the ammeter in parallel. When an ammeter is connected across a constant-potential source of appreciable voltage, the low internal resistance of the meter bypasses the circuit resistance. This results in the application of the source voltage directly to the meter terminals. The resulting excess current will burn out the meter coil.

7.4.0 OHMMETER

An ohmmeter is used to measure resistance and check continuity. The deflection of the pointer of an ohmmeter is controlled by the amount of battery current passing through the coil. Current flow depends on the applied voltage and the circuit resistance. By applying a constant source voltage to the circuit under test, the resultant current flow depends only on circuit resistance. This magnitude of current will create meter movement. By knowing the relationship between current and resistance, an ohmmeter's scale can be calibrated to indicate circuit resistance based on the magnitude of current for a constant source voltage. Refer to *Figure 14*, a simple ohmmeter circuit.

8.0.0 ELECTRICAL POWER

Power is defined as the rate of doing work. This is equivalent to the rate at which energy is used or dissipated. Electrons passing through a resistance dissipate energy in the form of heat. In electrical circuits, power is measured in units called **watts (W)**. The power in watts equals the rate of energy conversion. One watt of power equals the work done in one second by one volt of potential difference in moving one coulomb of charge. One coulomb per second is an ampere; therefore, power in watts equals the product of amperes times volts.

The work done in an electrical circuit can be useful work or it can be wasted work. In both cases, the rate at which the work is done is still measured in power. The turning of an electric motor is useful work. On the other hand, the heating of wires or resistors in a circuit is wasted work, since no useful function is performed by the heat.

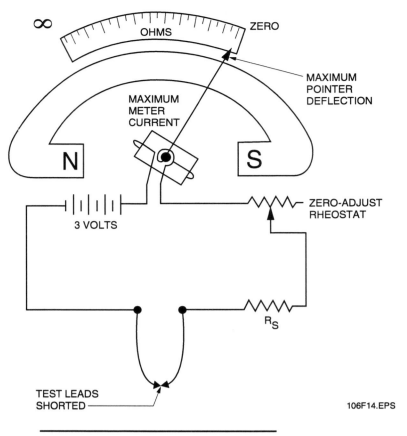

Figure 14. Simple Ohmmeter Circuit

The unit of electrical work is the joule. This is the amount of work done by one coulomb flowing through a potential difference of one volt. Thus, if five coulombs flow through a potential difference of one volt, five joules of work are done. The time it takes these coulombs to flow through the potential difference has no bearing on the amount of work done.

It is more convenient when working with circuits to think of amperes of current rather than coulombs. As previously discussed, one ampere equals one coulomb passing a point in one second. Using amperes, one joule of work is done in one second when one ampere moves through a potential difference of one volt. This rate of one joule of work in one second is the basic unit of power and is called a *watt*. Therefore, a watt is the power used when one ampere of current flows through a potential difference of one volt, as shown in *Figure 15*.

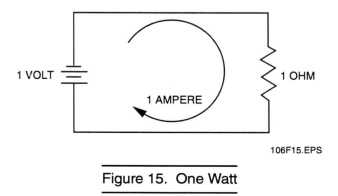

Figure 15. One Watt

Mechanical power is usually measured in units of horsepower (hp). To convert from horsepower to watts, multiply the number of horsepower by 746. To convert from watts to horsepower, divide the number of watts by 746. Conversions for common units of power are given in *Table 2*.

1,000 Watts (W)	= 1 Kilowatt (kW)
1,000,000 Watts (W)	= 1 Megawatt (MW)
1,000 Kilowatts (kW)	= 1 Megawatt (MW)
1 Watt (W)	= 0.00134 Horsepower (hp)
1 Horsepower (hp)	= 746 Watts (W)

Table 2. Conversion Table

The kilowatt-hour (kWh) is commonly used for large amounts of electrical work or energy. (The prefix **kilo** means one thousand.) The amount is calculated simply as the product of the power in kilowatts multiplied by the time in hours during which the power is used. If a light bulb uses 300W or 0.3kW for 4 hours, the amount of energy is 0.3×4, which equals 1.2kWh.

Very large amounts of electrical work or energy are measured in megawatts (MW). (The prefix **mega** means one million.)

Electricity usage is figured in kilowatt-hours of energy. The power line voltage is fairly constant at 120V. Suppose the total load current in the main line equals 20A. Then, the power in watts from the 120V line is:

P = 120V × 20A
P = 2,400W or 2.4kW

If this power is used for five hours, then the energy of work supplied equals:

2.4 × 5 = 12kWh

8.1.0 POWER EQUATION

When one ampere flows through a difference of two volts, two watts must be used. In other words, the number of watts used is equal to the number of amperes of current times the potential difference. This is expressed in equation form as:

P = I × E or P = IE

ELECTRICAL THEORY ONE — TRAINEE TASK MODULE 33106

Where:

    P = power used in watts

    I = current in amperes

    E = potential difference in volts

The equation is sometimes called *Ohm's Law for Power* because it is similar to Ohm's Law. This equation is used to find the power consumed in a circuit or load when the values of current and voltage are known. The second form of the equation is used to find the voltage when the power and current are known:

$$E = \frac{P}{I}$$

The third form of the equation is used to find the current when the power and voltage are known:

$$I = \frac{P}{E}$$

Using these three equations, the power, voltage, or current in a circuit can be calculated whenever any two of the values are already known.

Example 1:

Calculate the power in a circuit where the source of 100V produces 2A in a 50Ω resistance.

    P = IE
    P = 2 × 100
    P = 200W

This means the source generates 200W of power while the resistance dissipates 200W in the form of heat.

Example 2:

Calculate the source voltage in a circuit that consumes 600W at a current of 5A.

$$E = \frac{P}{I}$$

$$E = \frac{600}{5}$$

    E = 120V

Example 3:

Calculate the current in a circuit that consumes 600W with a source voltage of 120V.

$$I = \frac{P}{E}$$

$$I = \frac{600}{120}$$

$$I = 5A$$

Components that use the power dissipated in their resistance are generally rated in terms of power. The power is rated at normal operating voltage, which is usually 120V. For instance, an appliance that draws 5A at 120V would dissipate 600W. The rating for the appliance would then be 600W/120V.

To calculate I or R for components rated in terms of power at a specified voltage, it may be convenient to use the power formula in different forms. There are actually three basic power formulas, but each can be rearranged into three other forms for a total of nine combinations:

$$P = IE \qquad P = I^2R \qquad P = \frac{E^2}{R}$$

$$I = \frac{P}{E} \qquad R = \frac{P}{I^2} \qquad R = \frac{E^2}{P}$$

$$E = \frac{P}{I} \qquad I = \sqrt{\frac{P}{R}} \qquad E = \sqrt{PR}$$

Note that all of these formulas are based on Ohm's Law (E = IR) and the power formula (P = I × E). *Figure 16* shows all of the applicable power, voltage, resistance, and current equations.

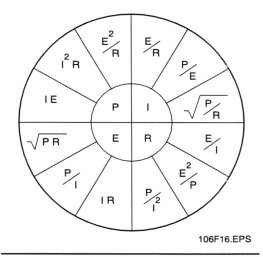

Figure 16. Expanded Ohm's Law Circle

8.2.0 POWER RATING OF RESISTORS

If too much current flows through a resistor, the heat caused by the current will damage or destroy the resistor. This heat is caused by I^2R heating, which is power loss expressed in watts. Therefore, every resistor is given a wattage, or power rating, to show how much I^2R heating it can take before it burns out. This means that a resistor with a power rating of one watt will burn out if it is used in a circuit where the current causes it to dissipate heat at a rate greater than one watt.

If the power rating of a resistor is known, the maximum current it can carry is found by using an equation derived from $P = I^2R$:

$P = I^2R$ becomes $I^2 = P/R$, which becomes $I = \sqrt{P/R}$

Using this equation, find the maximum current that can be carried by a 1Ω resistor with a power rating of 4W:

$I = \sqrt{P/R} = \sqrt{4/1} = 2A$

If such a resistor conducts more than 2A, it will dissipate more than its rated power and burn out.

Power ratings assigned by resistor manufacturers are usually based on the resistors being mounted in an open location where there is free air circulation and where the temperature is not higher than 104°F (40°C). Therefore, if a resistor is mounted in a small, crowded, enclosed space, or where the temperature is higher than 104°F, there is a good chance it will burn out even before its power rating is exceeded. Also, some resistors are designed to be attached to a chassis or frame that will carry away the heat.

SUMMARY

The relationships among current, voltage, resistance, and power are consistent for all types of DC circuits and can be calculated using Ohm's Law and Ohm's Law for Power. Understanding and being able to apply these concepts is necessary for effective circuit analysis and troubleshooting.

References

For advanced study of topics covered in this task module, the following books are suggested:

Electronics Fundamentals, Latest Edition, Prentice Hall, New York, NY.
Principles of Electric Circuits, Latest Edition, Prentice Hall, New York, NY.

REVIEW/PRACTICE QUESTIONS

1. A negative charge that is a type of subatomic particle is a(n) _____.
   a. proton
   b. neutron
   c. electron
   d. atom

2. A proton has a _____ charge.
   a. negative
   b. positive
   c. neutral
   d. Both b and c.

3. Like charges _____ each other.
   a. attract
   b. repel
   c. have no effect on
   d. complement

4. An electron has _____ coulombs of charge.
   a. 1.6×10^{-9}
   b. 1.6×10^{9}
   c. 1.6×10^{19}
   d. 1.6×10^{-19}

5. The quantity that Ohm's Law does not express a relationship for in an electrical circuit is _____.
   a. charge
   b. resistance
   c. voltage
   d. current

6. The color band that represents tolerance on a resistor is the _____.
   a. 4th band
   b. 3rd band
   c. 2nd band
   d. 1st band

7. A resistor with a color code of red/red/orange indicates a value of _____.
    a. 22,000Ω
    b. 66Ω
    c. 223Ω
    d. 220Ω

8. An ammeter is used to measure _____.
    a. voltage
    b. current
    c. resistance
    d. power

9. The basic unit of power is the _____ .
    a. volt
    b. ampere
    c. coulomb
    d. watt

10. The power in a circuit with 120V and 5A is _____.
    a. 24W
    b. 600W
    c. 6,000W
    d. $\frac{1}{24}$W

ANSWERS TO REVIEW/PRACTICE QUESTIONS

Answer	Section Reference
1. c	Terms/2.1.0
2. b	Terms/2.1.0
3. b	2.1.0
4. d	3.0.0
5. a	4.2.0
6. a	6.1.0
7. a	6.1.0
8. b	7.3.0
9. d	8.0.0
10. b	8.1.0

NCCER CRAFT TRAINING USER UPDATES

The NCCER makes every effort to keep these manuals up-to-date and free of technical errors. We appreciate your help in this process. If you have an idea for improving this manual, or if you find an error, a typographical mistake, or an inaccuracy in the NCCER's Craft Training Manuals, please write us, using this form or a photocopy. Be sure to include the exact module number, page number, a description of the problem, and the correction, if possible. Your input will be brought to the attention of the Technical Review Committee. Thank you for your assistance.

Instructors – If you found that additional materials were necessary in order to teach this module effectively, please let us know so that we may include them in the Equipment/Materials list in the Instructor's Guide.

Write: Curriculum Revision and Development Department
National Center for Construction Education and Research
P.O. Box 141104
Gainesville, FL 32614-1104
Fax: 352-334-0932

Craft _____ Module Name _____

Copyright Date _____ Module Number _____ Page Number(s) _____

Description of Problem
_____
_____
_____
_____
_____

(Optional) Correction of Problem
_____
_____
_____

(Optional) Your Name and Address
_____
_____
_____

notes

Electrical Safety
Module 33107

Electronic Systems Technician Trainee Task Module 33107

ELECTRICAL SAFETY

NATIONAL
CENTER FOR
CONSTRUCTION
EDUCATION AND
RESEARCH

OBJECTIVES

Upon completion of this module, the trainee will be able to:

1. Demonstrate safe working procedures in a construction environment.
2. Explain the purpose of OSHA and how it promotes safety on the job.
3. Identify electrical hazards and how to avoid or minimize them in the workplace.
4. Explain safety issues concerning lockout/tagout procedures, personal protection using assured grounding and isolation programs, confined space entry, respiratory protection, and fall protection systems.

Prerequisites

Successful completion of the following Task Modules is recommended before beginning study of this Task Module: Core Curricula; Electronic Systems Technician Level One, Modules 33101 through 33106.

Required Trainee Materials

1. Trainee Task Module
2. Copy of the latest edition of the *National Electrical Code*
3. *OSHA Electrical Safety Guidelines* (pocket guide)
4. Appropriate Personal Protective Equipment

Note: The designations "National Electrical Code," "NE Code," and "NEC," where used in this document, refer to the *National Electrical Code®*, which is a registered trademark of the National Fire Protection Association, Quincy, MA. *All National Electrical Code (NEC) references in this module refer to the 1999 edition of the NEC.*

COURSE MAP

This course map shows all of the modules in the first level of the Electronic Systems Technician curricula. The suggested training order begins at the bottom and proceeds up. Skill levels increase as a trainee advances on the course map. The training order may be adjusted by the local Training Program Sponsor.

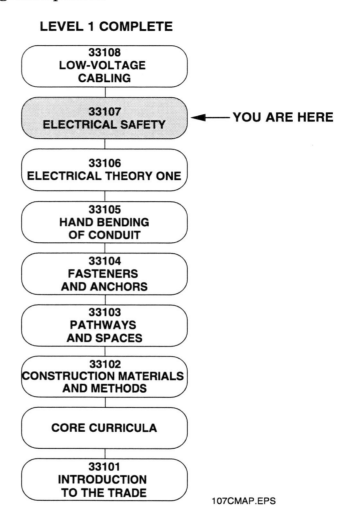

TABLE OF CONTENTS

Section	Topic	Page
	Course Map	2
1.0.0	Introduction	5
2.0.0	Electrical Shock	6
2.1.0	The Effect Of Current	7
2.1.1	Body Resistance	7
2.1.2	Burns	9
3.0.0	Reducing Your Risk	10
3.1.0	Protective Equipment	10
3.1.1	Protective Apparel	12
3.1.2	Personal Clothing	12
3.1.3	Fuse Pullers	13
3.1.4	Eye And Face Protection	13
3.2.0	Verify Circuits Are Deenergized	13
3.3.0	Other Precautions	13
4.0.0	OSHA	14
4.1.0	Safety Standards	14
4.1.1	1910.302-308/1926.402-408 Design Safety Standards For Electrical Systems	15
4.1.2	1910.302/1926.402 Electric Utilization Systems	15
4.1.3	1910.303/1926.403 General Requirements	15
4.1.4	1910.304/1926.404 Wiring Design And Protection	16
4.1.5	1910.305/1926.405 Wiring Methods, Components, And Equipment For General Use	16
4.1.6	1910.306/1926.406 Specific Purpose Equipment And Installations	16
4.1.7	1910.307/1926.407 Hazardous (Classified) Locations	17
4.1.8	1910.308/1926.408 Special Systems	17
4.1.9	1910.331/1926.416 Scope	17
4.1.10	1910.332 Training	18
4.1.11	1910.333/1926.416-417 Selection And Use Of Work Practices	18
4.1.12	1910.334/1926.431 Use Of Equipment	18
4.1.13	1910.335/1926.416 Safeguards For Personnel Protection	18
4.2.0	Safety Philosophy And General Safety Precautions	19
4.3.0	Electrical Regulations	21
4.3.1	OSHA Lockout/Tagout Rule	22
4.4.0	Other OSHA Regulations	27
4.4.1	Testing For Voltage	27
5.0.0	Ladders And Scaffolding	28
5.1.0	Ladders	28
5.1.1	Straight And Extension Ladders	29
5.1.2	Step Ladders	30
5.2.0	Scaffolding	30

TABLE OF CONTENTS (Continued)

Section	Topic	Page
6.0.0	Lifts, Hoists, And Cranes	31
7.0.0	Lifting	31
8.0.0	Basic Tool Safety	32
9.0.0	Confined Space Entry Procedures	32
9.1.0	General Guidelines	32
9.2.0	Confined Space Hazard Review	33
9.3.0	Entry And Work Procedures	34
10.0.0	First Aid	36
11.0.0	Solvents And Toxic Vapors	36
11.1.0	Precautions When Using Solvents	37
11.2.0	Respiratory Protection	38
12.0.0	Asbestos	39
12.1.0	Monitoring	39
12.2.0	Regulated Areas	39
12.3.0	Methods Of Compliance	41
13.0.0	Batteries	41
13.1.0	Acids	42
13.2.0	Wash Stations	42
14.0.0	PCBs	42
15.0.0	Fall Protection	42
15.1.0	Fall Protection Procedures	42
15.2.0	Types Of Fall Protection Systems	43
	Summary	44
	Review/Practice Questions	45
	Answers To Review/Practice Questions	48
	Appendix A	49

Trade Terms Introduced In This Module

Double-insulated/ungrounded tool: An electrical tool that is constructed so that the case is insulated from electrical energy. The case is made of a nonconductive material.

Fibrillation: Very rapid irregular contractions of the muscle fibers of the heart that result in the heartbeat and pulse going out of rhythm with each other.

Ground fault circuit interrupter (GFCI): A protective device that functions to deenergize a circuit or portion thereof within an established period of time when a current to ground exceeds some predetermined value that is less than that required to operate the overcurrent protective device of the supply circuit.

Grounded tool: An electrical tool with a three-prong plug at the end of its power cord or some other means to ensure that stray current travels to ground without passing through the body of the user. The ground plug is bonded to the conductive frame of the tool.

Polychlorinated biphenyls (PCBs): Toxic chemicals that may be contained in liquids used to cool certain types of large transformers and capacitors.

1.0.0 INTRODUCTION

In your work as an electronic systems technician, you will occasionally be exposed to potentially hazardous conditions on the job site. No training manual, set of rules and regulations, or listing of hazards can make working conditions completely safe. However, it is possible to work a full career without serious accident or injury. To reach this goal, you need to be aware of potential hazards and stay constantly alert to these hazards. You must take the proper precautions and practice the basic rules of safety. You must be safety-conscious at all times. Safety should become a habit. Keeping a safe attitude on the job will go a long way in reducing the number and severity of accidents. Remember that your safety is up to you.

As an apprentice, you need to be especially careful. You should only work under the direction of experienced personnel who are familiar with the various job site hazards and the means of avoiding them.

The most life-threatening hazards on a construction site are:
- Falls when you are working in high places
- Electrocution caused by coming into contact with live electrical circuits
- The possibility of being crushed by falling materials or equipment
- The possibility of being struck by flying objects or moving equipment/vehicles such as trucks, forklifts, and construction equipment

Other hazards include cuts, burns, back sprains, and getting chemicals or objects in your eyes. Most injuries, both those that are life-threatening and those that are less severe, are preventable if the proper precautions are taken.

2.0.0 ELECTRICAL SHOCK

Electricity can be described as a potential that results in the movement of electrons in a conductor. This movement of electrons is called *electrical current*. Some substances, such as silver, copper, steel, and aluminum, are excellent conductors. The human body is also a conductor. The conductivity of the human body greatly increases when the skin is wet or moistened with perspiration.

Electrical current flows along the path of least resistance to return to its source. The source return point is called the *neutral* or *ground* of a circuit. If the human body contacts an electrically energized point and is also in contact with the ground or another point in the circuit, the human body becomes a path for the current to return to its source. *Table 1* shows the effects of current passing through the human body. One mA is one milliamp, or one one-thousandth of an ampere.

Current Value	Typical Effects
Less than 1mA	No sensation.
1 to 20mA	Sensation of shock, possibly painful. May lose some muscular control between 10 and 20mA.
20 to 50mA	Painful shock, severe muscular contractions, breathing difficulties.
50 to 200mA	Same symptoms as above, only more severe, up to 100mA. Between 100 and 200mA, ventricular fibrillation may occur. Typically results in almost immediate death unless special medical equipment and treatment are available.
Over 200mA	Severe burns and muscular contractions. The chest muscles contract and stop the heart for the duration of the shock.

Table 1. Current Level Effects On The Human Body

A primary cause of death from electrical shock is when the heart's rhythm is overcome by an electrical current. Normally, the heart's operation uses a very low level electrical signal to cause the heart to contract and pump blood. When an abnormal electrical signal, such as current from an electrical shock, reaches the heart, the low level heartbeat signals are overcome. The heart begins twitching in an irregular manner and goes out of rhythm with the pulse. This twitching is called **fibrillation**. Unless the normal heartbeat rhythm is restored using special defibrillation equipment (paddles), the individual will die. No known case of heart fibrillation has ever been corrected without the use of defibrillation equipment by a qualified medical practitioner. Other effects of electrical shock may include immediate heart stoppage and burns. In addition, the body's reaction to the shock can cause a fall or other accident. Delayed internal problems can also result.

Note: Electric shocks or burns are a major cause of accidents in our industry.

2.1.0 THE EFFECT OF CURRENT

The amount of current measured in amperes that passes through a body determines the outcome of an electrical shock. The higher the voltage, the greater the chance for a fatal shock. In a one-year study in California, the following results were observed by the State Division of Industry Safety:

- Thirty percent of all electrical accidents were caused by contact with conductors. Of these accidents, 66% involved low-voltage conductors [those carrying 600 volts (V) or less].
- Portable, electrically operated hand tools made up the second largest number of injuries (15%). Almost 70% of these injuries happened when the frame or case of the tool became energized. These injuries could have been prevented by following proper safety practices, using grounded or double-insulated tools, and using **ground fault circuit interrupter (GFCI)** protection.

In one 10-year study, investigators found 9,765 electrical injuries that occurred in accidents. Over 18% of these injuries involved contact with voltage levels of over 600V. A little more than 13% of these high-voltage injuries resulted in death. These high-voltage totals included limited-amperage contacts, which are often found on electronic equipment. When tools or equipment touch high-voltage overhead lines, the chance that a resulting injury will be fatal climbs to 28%. Of the low-voltage injuries, 1.4% were fatal.

CAUTION: High voltage, defined as 600V or more, is almost ten times as likely to kill as low voltage. However, on the job you spend most of your time working on or near lower voltages. Due to the frequency of contact, most electrocution deaths actually occur at low voltages. Attitude about the harmlessness of lower voltages undoubtedly contributes to this statistic.

These statistics have been included to help you gain respect for the environment where you work and to stress how important safe working habits really are.

2.1.1 Body Resistance

Electricity travels in closed circuits, and its normal route is through a conductor. Shock occurs when the body becomes part of the electric circuit (*Figure 1*). The current must enter the body at one point and leave at another. Shock normally occurs in one of three ways: the person must come in contact with both wires of the electric circuit, one wire of the electric circuit and the ground, or a metallic part that has become *hot* by being in contact with an energized wire while the person is also in contact with the ground.

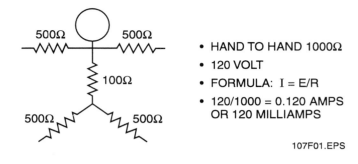

Figure 1. Body Resistance

To fully understand the harm done by electrical shock, we need to understand something about the physiology of certain body parts: the skin, heart, and muscles.

- Skin covers the body and is made up of three layers. The most important layer, as far as electric shock is concerned, is the outer layer of dead cells referred to as the *horny layer*. This layer is composed mostly of a protein called *keratin*, and it is the keratin that provides the largest percentage of the body's electrical resistance. When it is dry, the outer layer of skin may have a resistance of several thousand ohms, but when it is moist there is a radical drop in resistance, as is also the case if there is a cut or abrasion that pierces the horny layer. The amount of resistance provided by the skin will vary widely from individual to individual. A worker with a thick horny layer will have a much higher resistance than a child. The resistance will also vary widely at different parts of the body. For instance, the worker with high-resistance hands may have low-resistance skin on the back of his calf. The skin, like any insulator, has a breakdown voltage at which it ceases to act as a resistor and is simply *punctured*, leaving only the lower-resistance body tissue to impede the flow of current in the body. The breakdown voltage will vary with the individual but is in the area of 600V. Since most industrial power distribution systems operate at 480V or higher, technicians who may come into contact with voltages at these levels need to have special awareness of the shock potential.

- The heart is the pump that sends life-sustaining blood to all parts of the body. The blood flow is caused by the contractions of the heart muscle, which is controlled by electrical impulses. The electrical impulses are delivered by an intricate system of nerve tissue with built-in timing mechanisms which make the chambers of the heart contract at exactly the right time. An outside electric current of as little as 75 milliamperes can upset the rhythmic, coordinated beating of the heart by disturbing the nerve impulses. When this happens, the heart is said to be in *fibrillation*, and the pumping action stops. Death will occur quickly if the normal beat is not restored. Remarkable as it may seem, what is needed to defibrillate the heart is a shock of an even higher intensity.

- The other muscles of the body are also controlled by electrical impulses delivered by nerves. Electric shock can cause loss of muscular control, resulting in the inability to let go of an electrical conductor. Electric shock can also cause injuries of an indirect nature in which involuntary muscle reactions from the electric shock can cause bruises, fractures, and even deaths resulting from collisions or falls.

The severity of shock received when a person becomes a part of an electric circuit is affected by three primary factors: the amount of current flowing through the body (measured in amperes), the path of the current through the body, and the length of time the body is in the circuit. Other factors which may affect the severity of the shock are the frequency of the current, the phase of the heart cycle when shock occurs, and the general health of the person prior to the shock. Effects can range from a barely perceptible tingle to immediate cardiac arrest. Although there are no absolute limits, or even known values that show the exact injury at any given amperage range, *Table 1* lists the general effects of electric current on the body for different current levels. As this table illustrates, a difference of only 100 milliamperes exists between a current that is barely perceptible and one that can kill.

A severe shock can cause considerably more damage to the body than is visible. For example, a person may suffer internal hemorrhages and destruction of tissues, nerves, and muscle. In addition, shock is often only the beginning in a chain of events. The final injury may well be from a fall, cuts, burns, or broken bones.

2.1.2 Burns

The most common shock-related injury is a burn. Burns suffered in electrical accidents may be of three types: electrical burns, arc burns, and thermal contact burns.

- Electrical burns are the result of electric current flowing through the tissues or bones. Tissue damage is caused by the heat generated by the current flow through the body. An electrical burn is one of the most serious injuries you can receive, and it should be given immediate attention. Since the most severe burning is likely to be internal, what may appear at first to be a small surface wound could, in fact, be an indication of severe internal burns.

- Arc burns make up a substantial portion of the injuries from electrical malfunctions. The electric arc between metals can be up to 35,000°F, which is about four times hotter than the surface of the sun. Workers several feet from the source of the arc can receive severe or fatal burns. Since most electrical safety guidelines recommend safe working distances based on shock considerations, workers can be following these guidelines and still be at risk from arc. Electric arcs can occur due to poor electrical contact or failed insulation. Electrical arcing is caused by the passage of substantial amounts of current through the vaporized terminal material (usually metal or carbon).

CAUTION: Since the heat of the arc is dependent on the short circuit current available at the arcing point, arcs generated on 480V systems can be just as dangerous as those generated at 13,000V.

- The third type of burn is a thermal contact burn. It is caused by contact with objects thrown during the blast associated with an electric arc. This blast comes from the pressure developed by the near-instantaneous heating of the air surrounding the arc, and from the expansion of the metal as it is vaporized. (Copper expands by a factor in excess of 65,000 times in boiling.) These pressures can be great enough to hurl people, switchgear, and cabinets considerable distances. Another hazard associated with the blast is the hurling of molten metal droplets, which can also cause thermal contact burns and associated damage. A possible beneficial side effect of the blast is that it could hurl a nearby person away from the arc, thereby reducing the effect of arc burns.

3.0.0 REDUCING YOUR RISK

There are many things that can be done to greatly reduce the chance of receiving an electrical shock. Always comply with your company's safety policy and all applicable rules and regulations, including job site rules. In addition, the Occupational Safety and Health Administration (OSHA) publishes the *Code of Federal Regulations (CFR)*. CFR Part 1910 covers the OSHA standards for general industry and CFR Part 1926 covers the OSHA standards for the construction industry.

Do not approach any electrical conductors closer than indicated in *Table 2* unless you are sure they are deenergized and your company has designated you as a qualified individual. Also, the values given in the table are *minimum* safe clearance distances; if you already have standard distances established, these are provided only as supplemental information. These distances are listed in CFR 1910.333/1926.416.

Voltage Range (Phase-to-Phase)	Minimum Approach Distance
300V and less	Avoid contact
Over 300V, not over 750V	1 ft. 0 in. (30.5 cm)
Over 750V, not over 2kV	1 ft. 6 in. (46 cm)
Over 2kV, not over 15kV	2 ft. 0 in. (61 cm)
Over 15kV, not over 37kV	3 ft. 0 in. (91 cm)
Over 37kV, not over 87.5kV	3 ft. 6 in. (107 cm)
Over 87.5kV, not over 121kV	4 ft. 0 in. (122 cm)
Over 121kV, not over 140kV	4 ft. 6 in. (137 cm)

Table 2. Approach Distances For Qualified Employees – Alternating Current

3.1.0 PROTECTIVE EQUIPMENT

You should also become familiar with appropriate personal protective equipment. In particular, know the voltage rating of each piece of equipment. Rubber gloves are used to prevent the skin from coming into contact with energized circuits. A separate leather cover protects the rubber glove from punctures and other damage (see *Figure 2*). OSHA addresses the use of protective equipment, apparel, and tools in CFR 1910.335(a). This article is divided into two sections: *Personal Protective Equipment* and *General Protective Equipment and Tools*.

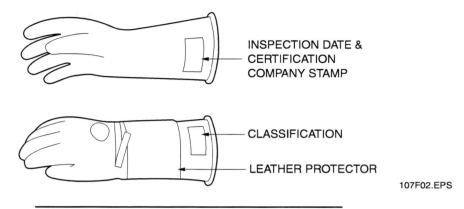

Figure 2. Rubber Gloves And Leather Protector

The first section, *Personal Protective Equipment*, includes the following requirements:

- Employees working in areas where there are potential electrical hazards shall be provided with, and shall use, electrical protective equipment that is appropriate for the specific parts of the body to be protected and for the work to be performed.
- Protective equipment shall be maintained in a safe, reliable condition and shall be periodically inspected or tested, as required by CFR 1910.137/1926.95.
- If the insulating capability of protective equipment may be subject to damage during use, the insulating material shall be protected.
- Employees shall wear nonconductive head protection wherever there is a danger of head injury from electric shock or burns due to contact with exposed energized parts.
- Employees shall wear protective equipment for the eyes and face wherever there is danger of injury to the eyes or face from electric arcs or flashes or from flying objects resulting from electrical explosion.

The second section, *General Protective Equipment and Tools*, includes the following requirements:

- When working near exposed energized conductors or circuit parts, each employee shall use insulated tools or handling equipment if the tools or handling equipment might make contact with such conductors or parts. If the insulating capability of insulated tools or handling equipment is subject to damage, the insulating material shall be protected.
- Fuse handling equipment insulated for the circuit voltage shall be used to remove or install fuses when the fuse terminals are energized.
- Ropes and handlines used near exposed energized parts shall be nonconductive.

- Protective shields, protective barriers, or insulating materials shall be used to protect each employee from shock, burns, or other electrically related injuries while that employee is working near exposed energized parts that might be accidentally contacted or where dangerous electric heating or arcing might occur. When normally enclosed live parts are exposed for maintenance or repair, they shall be guarded to protect unqualified persons from contact with the live parts.

The types of electrical safety equipment, protective apparel, and protective tools available are quite varied. We will discuss the most common types of safety equipment, which include:

- Protective apparel
- Personal clothing
- Fuse pullers
- Eye and face protection

3.1.1 Protective Apparel

Besides rubber gloves, there are other types of special application protective apparel, such as fire suits, face shields, and rubber sleeves.

Manufacturing plants should have other types of special application protective equipment available for use, such as high-voltage sleeves, high-voltage boots, nonconductive protective helmets, nonconductive eyewear and face protection, and switchboard blankets.

All equipment should be inspected before use and during use, as necessary. The equipment used and the extent of the precautions taken depend on each individual situation; however, it is better to be overprotected than underprotected when you are trying to prevent electrocution.

When working with high voltages, flash suits may be required in some applications. Some plants require them to be worn for all switching and rack-in or rack-out operations.

Face shields should also be worn during all switching operations where arcs are a possibility. The thin plastic type of face shield should be avoided because it will melt when exposed to the extremely high temperatures of an electrical arc.

Rubber sleeves are another type of protective apparel that should be worn during switching operations and breaker racking. Sleeves must be inspected yearly.

3.1.2 Personal Clothing

Any individual who will perform work in an electrical environment or in plant substations should dress accordingly. Avoid wearing synthetic fiber clothing; these types of materials will melt when exposed to high temperatures and will actually increase the severity of a burn. Wear cotton clothing, fiberglass-toe boots or shoes, and hard hats. Use hearing protection where needed.

3.1.3 Fuse Pullers

Use the plastic or fiberglass style of fuse puller for removing and installing low-voltage cartridge fuses. All fuse pulling and replacement operations must be done using fuse pullers.

The best type of fuse puller is one that has a spread guard installed. This prevents the puller from opening if resistance is met when installing fuses.

3.1.4 Eye And Face Protection

National Fire Protection Agency code NFPA 70-E requires that protective equipment for the eyes and face shall be used whenever there is danger of injury to the eyes or face from electrical arcs or flashes, or from flying or falling objects resulting from electrical explosion.

3.2.0 VERIFY CIRCUITS ARE DEENERGIZED

You should always assume that all the circuits are energized until you have verified that the circuit is deenergized. Follow these steps to verify that a circuit is deenergized:

Step 1   Ensure the circuit is properly tagged and locked out (OSHA 1910.333/1926.417).

Step 2   Verify the test instrument operation on a known source.

Step 3   Using the test instrument, check the circuit to be deenergized. The voltage should be zero.

Step 4   Verify the test instrument operation once again on a known power source.

3.3.0 OTHER PRECAUTIONS

There are several other precautions you can take to help make your job safer. For example:

- Always remove all jewelry (e.g., rings, watches, bracelets, and necklaces) before working on electrical equipment. Most jewelry is made of conductive material and wearing it can result in a shock, as well as other injuries if the jewelry gets caught in moving components.
- When working on energized equipment, it is safer to work in pairs. In doing so, if one of the workers experiences a harmful electrical shock, the other worker can quickly deenergize the circuit and call for help.
- Plan each job before you begin it. Make sure you understand exactly what it is you are going to do. If you are not sure, ask your supervisor.
- You will need to look over the appropriate prints and drawings to locate isolation devices and potential hazards. Never defeat safety interlocks. Remember to plan your escape route before starting work. Know where the nearest phone is and the emergency number to dial for assistance.

- If you realize that the work will go beyond the scope of what was planned, stop and get instructions from your supervisor before continuing. Do not attempt to plan as you go.
- It is critical that you stay alert. Work places are dynamic, and situations relative to safety are always changing. If you leave the work area to pick up material, take a break, or have lunch, reevaluate your surroundings when you return. Remember, plan ahead.

4.0.0 OSHA

The purpose of the Occupational Safety and Health Administration (OSHA) is to assure safe and healthful working conditions for working men and women. OSHA is authorized to enforce standards and assist and encourage the states in their efforts to ensure safe and healthful working conditions. OSHA assists states by providing for research, information, education, and training in the field of occupational safety and health.

The law that established OSHA specifies the duties of both the employer and employee with respect to safety. Some of the key requirements are outlined below. This list does not include everything, nor does it override the procedures called for by your employer.

- Employers shall provide a place of employment free from recognized hazards likely to cause death or serious injury.
- Employers shall comply with the standards of the act.
- Employers shall be subject to fines and other penalties for violation of those standards.

WARNING! OSHA states that employees have a duty to follow the safety rules laid down by the employer. Additionally, some states can reduce the amount of benefits paid to an injured employee if that employee was not following known, established safety rules. Your company may also terminate you if you violate an established safety rule.

4.1.0 SAFETY STANDARDS

The OSHA standards are split into several sections. As discussed earlier, the two which affect you the most are CFR 1926, construction specific, and CFR 1910, which is the standard for general industry. Either or both may apply depending on where you are working and what you are doing. If a job site condition is covered in the 1926 book, then that standard takes precedent. However, if a more stringent requirement is listed in the 1910 standard, it should also be met. An excellent example is the current difference in the two standards on confined spaces; if someone gets hurt or killed, the decision to use the less stringent 1926 standard could be called into question. OSHA's *General Duty Clause* states that an employer should have known all recognized hazards and removed the hazard or protected the employee.

To protect workers from the occupational injuries and fatalities caused by electrical hazards, OSHA has issued a set of design safety standards for electrical utilization systems. These standards are 1926.400-449 and 1910.302-308. OSHA also currently recognizes the 1984 edition of the *National Electrical Code (NEC)* for installations with exceptions for 1926.404(b)(1) and 1926.405(a)(2)(ii)(E)(F)(G) and (J). The OSHA electrical standards for construction are included in *Appendix A* in the back of this module.

Note: OSHA does *not* recognize the current edition of the NEC; it generally takes several years for updated NEC codes to be incorporated.

The CFR 1910 standard must be followed whenever the construction standard CFR 1926 does not address an issue which is covered by CFR 1910, or for a pre-existing installation. If the CFR 1910 standard is more stringent then CFR 1926, then the more stringent standard should be followed. OSHA does not update their standards in a timely manner, and as such, there are often differences in similar sections of the two standards. Safety should always be a priority, and the more protective work rules should always be chosen.

4.1.1 1910.302-308/1926.402-408 Design Safety Standards For Electrical Systems

This section contains design safety regulations for all the electrical equipment and installations used to provide power and light to employee workplaces. The articles listed are outlined in the following sections.

4.1.2 1910.302/1926.402 Electric Utilization Systems

This article identifies the scope of the standard. Listings are included to show which electrical installations and equipment are covered under the standard, and which installations and equipment are not covered under the standard. Furthermore, certain sections of the standard apply only to utilization equipment installed after March 15, 1972, and some apply only to equipment installed after April 16, 1981. Article 1910.302 (1926.402) addresses these oddities and provides guidance to clarify them.

4.1.3 1910.303/1926.403 General Requirements

This article covers topics that mostly concern equipment installation clearances, identification, and examination. Some of the major subjects addressed in this article are:

- Equipment installation examinations
- Splicing
- Marking
- Identification of disconnecting means
- Workspace around electrical equipment

4.1.4   1910.304/1926.404 Wiring Design And Protection

This article covers the application, identification, and protection requirements of grounding conductors, outside conductors, service conductors, and equipment enclosures. Some of the major topics discussed are:

- Grounded conductors
- Outside conductors
- Service conductors
- Overcurrent protection
- System grounding requirements

4.1.5   1910.305/1926.405 Wiring Methods, Components, And Equipment For General Use

In general, this article addresses the wiring method requirements of raceways, cable trays, pull and junction boxes, switches, and switchboards; the application requirements of temporary wiring installations; the equipment and conductor requirements for general wiring; and the protection requirements of motors, transformers, capacitors, and storage batteries. Some of the major topics are:

- Wiring methods
- Cabinets, boxes, and fittings
- Switches
- Switchboards and panelboards
- Enclosures for damp or wet locations
- Conductors for general wiring
- Flexible cords and cables
- Portable cables
- Equipment for general use

4.1.6   1910.306/1926.406 Specific Purpose Equipment And Installations

This article addresses the requirements of special equipment and installations not covered in other articles. Some of the major types of equipment and installations found in this article are:

- Electric signs and outline lighting
- Cranes and hoists
- Elevators, dumbwaiters, escalators, and moving walks
- Electric welders
- Data processing systems

- X-ray equipment
- Induction and dielectric heating equipment
- Electrolytic cells
- Electrically driven or controlled irrigation machines
- Swimming pools, fountains, and similar installations

4.1.7 1910.307/1926.407 Hazardous (Classified) Locations

This article covers the requirements for electric equipment and wiring in locations that are classified because they contain flammable vapors, liquids, and/or gases, or combustible dust or fibers; and the likelihood that a flammable or combustible concentration or quantity is present. Some of the major topics covered in this article are:

- Scope
- Electrical installations in hazardous locations
- Conduit
- Equipment in Division 2 locations

The OSHA standard for hazardous (classified) locations is included in *Appendix A* at the back of this module.

4.1.8 1910.308/1926.408 Special Systems

This article covers the wiring methods, grounding, protection, identification, and other general requirements of special systems not covered in other articles. Some of the major subtopics found in this article are:

- Systems over 600V nominal
- Emergency power systems
- Class 1, 2, and 3 remote control, signaling, and power-limited circuits
- Fire-protective signaling systems
- Communications systems

4.1.9 1910.331/1926.416 Scope

This article serves as an overview of the following articles and also provides a summary of the installations that this standard allows qualified and unqualified persons to work on or near, as well as the installations that this standard does *not* cover.

4.1.10 1910.332 Training

The training requirements contained in this article apply to employees who face a risk of electric shock. Some of the topics that appear in this article are:

- Content of training
- Additional requirements for unqualified persons
- Additional requirements for qualified persons
- Type of training

4.1.11 1910.333/1926.416-417 Selection And Use Of Work Practices

This article covers the implementation of safety-related work practices necessary to prevent electrical shock and other related injuries to the employee. Some of the major topics addressed in this article are listed below:

- General work practices
- Working on or near exposed deenergized parts
- Working on or near exposed energized parts

4.1.12 1910.334/1926.431 Use Of Equipment

This article was added to reinforce the regulations pertaining to portable electrical equipment, test equipment, and load break switches. Major topics include:

- Portable electric equipment
- Electric power and lighting circuits
- Test instruments and equipment

4.1.13 1910.335/1926.416 Safeguards For Personnel Protection

This article covers the personnel protection requirements for employees in the vicinity of electrical hazards. It addresses regulations that protect personnel working on equipment as well as personnel working nearby. Some of the major topics are:

- Use of protective equipment
- Alerting techniques

Now that background topics have been covered and an overview of the OSHA electrical safety standards has been provided, it is time to move on to topics related directly to safety. As we discuss these topics, we will continually refer to the OSHA standards to identify the requirements that govern them.

OSHA 1926 Subpart K also addresses electrical safety requirements that are necessary for the practical safeguarding of employees involved in construction work.

4.2.0 SAFETY PHILOSOPHY AND GENERAL SAFETY PRECAUTIONS

The most important piece of safety equipment in performing work in an electrical environment is common sense. All areas of electrical safety precautions and practices draw upon common sense and attention to detail. One of the most dangerous conditions in an electrical work area is a poor attitude toward safety.

WARNING! Only qualified individuals may work on electrical equipment. Your employer will determine who is qualified. Remember, your employer's safety rules must always be followed.

As stated in CFR 1910.333(a)/1926.403, safety-related work practices shall be employed to prevent electric shock or other injuries resulting from either direct or indirect electrical contact when work is performed near or on equipment or circuits that are or may be energized. The specific safety-related work practices shall be consistent with the nature and extent of the associated electrical hazards. The following are considered some of the basic and necessary attitudes and electrical safety precautions that lay the groundwork for a proper safety program. Before going on any electrical work assignment, these safety precautions should be reviewed and adhered to.

- *All work on electrical equipment should be done with circuits deenergized and cleared or grounded* – It is obvious that working on energized equipment is much more dangerous than working on equipment that is deenergized. Work on energized electrical equipment should be avoided if at all possible. CFR 1910.333(a)(1)/1926.403 states that live parts to which an employee may be exposed shall be deenergized before the employee works on or near them, unless the employer can demonstrate that deenergizing introduces additional or increased hazards or is not possible because of equipment design or operational limitations. Live parts that operate at less than 50 volts to ground need not be deenergized if there will be no increased exposure to electrical burns or to explosion due to electric arcs.
- *All conductors, buses, and connections should be considered energized until proven otherwise* – As stated in 1910.333(b)(1)/1926.417, conductors and parts of electrical equipment that have not been locked out or tagged out in accordance with this section should be considered energized. Routine operation of the circuit breakers and disconnect switches contained in a power distribution system can be hazardous if not approached in the right manner. Several basic precautions that can be observed in switchgear operations are:
  - Wear proper clothing made of 100% cotton or fire-resistant fabric.
  - Eye, face, and head protection should be worn. Turn your head away whenever closing devices.
  - Whenever operating circuit breakers in low-voltage or medium-voltage systems, always stand off to the side of the unit.

ELECTRICAL SAFETY — TRAINEE MODULE 33107

- Always try to operate disconnect switches and circuit breakers under a no-load condition.
- Never intentionally force an interlock on a system or circuit breaker.
- Always verify what you are closing a device into; you could violate a lockout or close into a hard fault.

Often, a circuit breaker or disconnect switch is used for providing lockout on an electrical system. To ensure that a lockout is not violated, perform the following procedures when using the device as a lockout point:

- Breakers must always be locked out and tagged out whenever you are working on a circuit that is tied to an energized breaker. Breakers capable of being opened and racked out to the disconnected position should have this done. Afterward, approved safety locks must be installed. The breaker may be removed from its cubicle completely to prevent unexpected mishaps. Always follow the standard rack-out and removal procedures that were supplied with the switchgear. Once removed, a sign must be hung on the breaker identifying its use as a lockout point, and approved safety locks must be installed when the breaker is used for isolation. In addition, the closing springs should be discharged.

- Some of the circuit breakers used are equipped with keyed interlocks for protection during operation. These locks are generally called *kirklocks* and are relied upon to ensure proper sequence of operation only. These are not to be used for the purpose of locking out a circuit or system. Where disconnects are installed for use in isolation, they should never be opened under load. When opening a disconnect manually, it should be done quickly with a positive force. Again, lockouts should be used when the disconnects are open.

- Whenever performing switching or fuse replacements, always use the protective equipment necessary to ensure personnel safety. *Never* make the assumption that because things have gone fine the last 999 times, they will not go wrong this time. Always prepare yourself for the worst case accident when performing switching.

- Whenever reenergizing circuits following maintenance or removal of a faulted component, extreme care should be used. Always verify that the equipment is in a condition to be reenergized safely. All connections should be insulated and all covers should be installed. Have all personnel stand clear of the area for the initial reenergization. *Never* assume everything is in perfect condition. Verify the conditions!

The following procedure is provided as a guideline for ensuring that equipment and systems will not be damaged by reclosing low-voltage circuit breakers into faults. If a low-voltage circuit breaker has opened for no apparent reason, perform the following:

Step 1 Verify that the equipment being supplied is not physically damaged and shows no obvious signs of overheating or fire.

Step 2 Make all appropriate tests to locate any faults.

Step 3   Reclose the feeder breaker. Stand off to the side when closing the breaker.

Step 4   If the circuit breaker trips again, do not attempt to reclose the breaker. In a plant environment, the Electrical Engineering Department should be notified, and the cause of the trip should be isolated and repaired.

The same general procedure should be followed for fuse replacement, with the exception of transformer fuses. If a transformer fuse blows, the transformer and feeder cabling should be inspected and tested before reenergizing. A blown fuse to a transformer is very significant because it normally indicates an internal fault. Transformer failures are catastrophic in nature and can be extremely dangerous. If applicable, contact the in-plant Electrical Engineering Department prior to commencing any effort to reenergize a transformer.

Power must always be removed from a circuit when removing and installing fuses. The air break disconnects (or quick disconnects) provided on the upstream side of a large transformer must be opened prior to removing the transformer's fuses. Otherwise, severe arcing will occur as the fuse is removed. This arcing can result in personnel injury and equipment damage.

To replace fuses servicing circuits below 600V:
- Secure power to fuses or ensure all downstream loads have been disconnected.
- Always use a positive force to remove and install fuses.

4.3.0   ELECTRICAL REGULATIONS

OSHA has certain regulations that apply to job site electrical safety. These regulations include the following:

- All electrical work shall be in compliance with the latest NEC and OSHA standards.

Note: OSHA may not recognize the current edition of the NEC, which can sometimes cause problems; however, OSHA typically will *not* cite for any differences.

- The non-current-carrying metal parts of fixed, portable, and plug-connected equipment shall be grounded. It is best to choose **grounded tools**. However, portable tools and appliances protected by an approved system of double insulation need not be grounded. *Figure 3* shows an example of a **double-insulated/ungrounded tool**.
- Extension cords shall be the three-wire type, shall be protected from damage, and shall not be fastened with staples or hung in a manner that could cause damage to the outer jacket or insulation. Never run an extension cord through a doorway or window that can pinch the cord. Also, never allow vehicles or equipment to drive over cords.
- Exposed lamps in temporary lights shall be guarded to prevent accidental contact, except where lamps are deeply recessed in the reflector. Temporary lights shall not be suspended, except in accordance with their listed labeling.

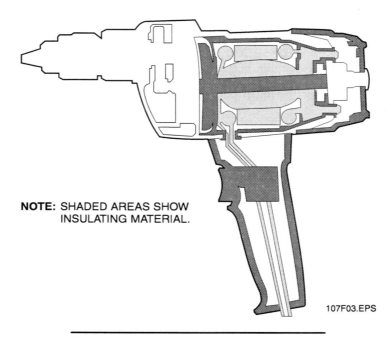

NOTE: SHADED AREAS SHOW INSULATING MATERIAL.

Figure 3. Double-Insulated Electric Drill

- Receptacles for attachment plugs shall be of an approved type and properly installed. Installation of the receptacle will be in accordance with the listing and labeling for each receptacle and shall be GFCI-protected if the setting is a temporarily wired construction site. If permanent receptacles are used with extension cords, then you must use GFCI protection.
- Each disconnecting means for motors and appliances, and each service feeder or branch circuit at the point where it originates, shall be legibly marked to indicate its purpose and voltage.
- Flexible cords shall be used in continuous lengths (no splices) and shall be of a type listed in *NEC Table 400-4*.
- Personnel safety ground fault protection for temporary construction is required. There are two methods for accomplishing this: an assured grounding program (this method is still used but should be used in conjunction with a GFCI program) or ground fault protection receptacles or breakers. Each employer will set the standard and method to be used. *Figure 4* shows a typical ground-fault circuit interrupter. The OSHA standard for ground fault protection on construction sites is included in *Appendix A* at the back of this module.

4.3.1 OSHA Lockout/Tagout Rule

OSHA released the 29 CFR 1926 lockout/tagout rule in December 1991. This rule covers the specific procedure to be followed for the servicing and maintenance of machines and equipment in which the unexpected energization or startup of the machines or equipment, or releases of stored energy, could cause injury to employees. This standard establishes minimum performance requirements for the control of such hazardous energy.

Figure 4. Typical GFCI

WARNING! This procedure is provided for your information only. The OSHA procedure provides only the minimum requirements for lockouts/tagouts. Consult the lockout/tagout procedure for your company and the plant or job site at which you are working. Remember that your life could depend on the lockout/tagout procedure. It is critical that you use the correct procedure for your site.

The purpose of the OSHA procedure is to ensure that equipment is isolated from all potentially hazardous energy (e.g., electrical, mechanical, hydraulic, chemical, or thermal), and tagged and locked out before employees perform any servicing or maintenance activities where the unexpected energization, startup, or release of stored energy could cause injury. All employees shall be instructed in the lockout/tagout procedure.

CAUTION: Although 99% of your work may be electrical, be aware that you may also need to lock out mechanical equipment.

The following is an example of a lockout/tagout procedure. Make sure to use the procedure that is specific to your employer or job site.

I. Introduction

A. This lockout/tagout procedure has been established for the protection of personnel from potential exposure to hazardous energy sources during construction, installation, service, and maintenance of electrical energy systems.

B. This procedure applies to and must be followed by all personnel who may be potentially exposed to the unexpected startup or release of hazardous energy (e.g., electrical, mechanical, pneumatic, hydraulic, chemical, or thermal).

Exception: This procedure does not apply to process and/or utility equipment or systems with cord and plug power supply systems when the cord and plug are the only source of hazardous energy, are removed from the source, and remain under the exclusive control of the authorized employee.

Exception: This procedure does not apply to troubleshooting (diagnostic) procedures and installation of electrical equipment and systems when the energy source cannot be deenergized because continuity of service is essential or shutdown of the system is impractical. Additional personal protective equipment for such work is required and the safe work practices identified for this work must be followed.

II. Definitions

- *Affected employee* – Any person working on or near equipment or machinery when maintenance or installation tasks are being performed by others during lockout/tagout conditions.
- *Appointed authorized employee* – Any person appointed by the job site supervisor to coordinate and maintain the security of a group lockout/tagout condition.
- *Authorized employee* – Any person authorized by the job site supervisor to use lockout/tagout procedures while working on electrical equipment.
- *Authorized supervisor* – The assigned job site supervisor who is in charge of coordination or procedures and maintenance of security of all lockout/tagout operations at the job site.
- *Energy isolation device* – An approved electrical disconnect switch capable of accepting approved lockout/tagout hardware for the purpose of isolating and securing a hazardous electrical source in an open or safe position.
- *Lockout/tagout hardware* – A combination of padlocks, danger tags, and other devices designed to attach to and secure electrical isolation devices.

III. Training

A. Each authorized supervisor, authorized employee, and appointed authorized employee shall receive initial and as-needed user-level training in lockout/tagout procedures.

B. Training is to include recognition of hazardous energy sources, the type and magnitude of energy sources in the workplace, and the procedures for energy isolation and control.

C. Retraining will be conducted on an as-needed basis whenever lockout/tagout procedures are changed or there is evidence that procedures are not being followed properly.

IV. Protective Equipment and Hardware

A. Lockout/tagout devices shall be used exclusively for controlling hazardous electrical energy sources.

B. All padlocks must be numbered and assigned to one employee only.

C. No duplicate or master keys will be made available to anyone except the site supervisor.

D. A current list with the lock number and authorized employee's name must be maintained by the site supervisor.

E. Danger tags must be of the standard white, red, and black *DANGER—DO NOT OPERATE* design and shall include the authorized employee's name, the date, and the appropriate network company (use permanent markers).

F. Danger tags must be used in conjunction with padlocks, as shown in *Figure 5*.

Figure 5. Lockout/Tagout Device

V. Procedures

A. Preparation for lockout/tagout:
1. Check the procedures to ensure that no changes have been made since you last used a lockout/tagout.
2. Identify all authorized and affected employees involved with the pending lockout/tagout.

B. Sequence for lockout/tagout:
1. Notify all authorized and affected personnel that a lockout/tagout is to be used and explain the reason why.
2. Shut down the equipment or system using the normal OFF or STOP procedures.

3. Lock out energy sources and test disconnects to be sure they cannot be moved to the ON position and open the control cutout switch. If there is not a cutout switch, block the magnet in the switch open position before working on electrically operated equipment/apparatus such as motors, relays, etc. Remove the control wire.
4. Lock and tag the required switches in the open position. Each authorized employee must affix a separate lock and tag. An example is shown in *Figure 6*.

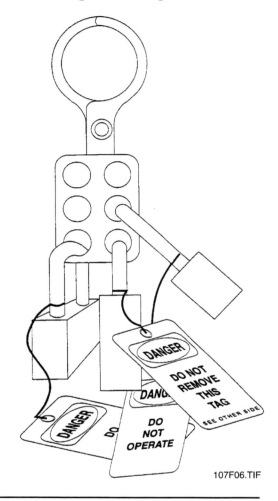

Figure 6. Multiple Lockout/Tagout Device

5. Dissipate any stored energy by attaching the equipment or system to ground.
6. Verify that the test equipment is functional via a known power source.
7. Confirm that all switches are in the open position and use test equipment to verify that all parts are deenergized.
8. If it is necessary to temporarily leave the area, upon returning, retest to ensure that the equipment or system is still deenergized.

C. Restoration of energy:
  1. Confirm that all personnel and tools, including shorting probes, are accounted for and removed from the equipment or system.

2. Completely reassemble and secure the equipment or system.
3. Replace and/or reactivate all safety controls.
4. Remove locks and tags from isolation switches. Authorized employees must remove their own locks and tags.
5. Notify all affected personnel that the lockout/tagout has ended and the equipment or system is energized.
6. Operate or close isolation switches to restore energy.

VI. Emergency Removal Authorization

A. In the event a lockout/tagout device is left secured, and the authorized employee is absent, or the key is lost, the authorized supervisor can remove the lockout/tagout device.
B. The authorized employee must be informed that the lockout/tagout device has been removed.
C. Written verification of the action taken, including informing the authorized employee of the removal, must be recorded in the job journal.

4.4.0 OTHER OSHA REGULATIONS

There are other OSHA regulations which you need to be aware of on the job site. For example:

- OSHA requires the posting of hard hat areas. Be alert to those areas and always wear your hard hat properly, with the bill in front. Hard hats should be worn whenever overhead hazards exist or there is the risk of exposure to electric shock or burns.
- You should wear safety shoes on all job sites. Keep them in good condition.
- Do not wear clothing with exposed metal zippers, buttons, or other metal fasteners. Avoid wearing loose-fitting or torn clothing.
- Protect your eyes. Your eyesight is threatened by many activities on the job site. Always wear safety glasses with full side shields. In addition, the job may also require protective equipment such as face shields or goggles.

4.4.1 Testing For Voltage

OSHA also requires that you inspect or test existing conditions before beginning work on electrical equipment or lines. Usually, you will use a voltmeter/sensor or voltage tester to do this. You should assume that all electrical equipment and lines are energized until you have determined that they are not. Do not proceed to work on or near energized parts until the operating voltage is determined.

After the electrical equipment to be worked on has been locked and tagged out, the equipment must be verified as deenergized before work can proceed. This section sets the requirements that must be met before any circuits or equipment can be considered deenergized. First, and most importantly, only qualified persons may verify that a circuit or piece of equipment is deenergized. Before approaching the equipment to be worked on, the qualified person shall operate the equipment's normal operating controls to check that the proper energy sources have been disconnected.

Upon opening a control enclosure, the qualified person shall note the presence of any components that may store electrical energy. Initially, these components should be avoided.

To verify that the lockout was adequate and the equipment is indeed deenergized, a qualified person must use appropriate test equipment to check for power, paying particular attention to induced voltages and unrelated feedback voltage.

Ensure that your testing equipment is working properly by performing the *live-dead-live* check before each use. To perform this test, first check your voltmeter on a known live voltage source. This known source must be in the same range as the electrical equipment you will be working on. Next, without changing scales on your voltmeter, check for the presence of power in the equipment you have locked out. Finally, to ensure that your voltmeter did not malfunction, check it again on the known live source. Performing this test will assure you that your voltage testing equipment is reliable.

In accordance with OSHA section 1910.333(b)(2)(iv)/1926.417(d)(4)(ii), if the circuit to be tested normally operates at more than 600 volts, the live-dead-live check must be performed.

Once it has been verified that power is not present, stored electrical energy that might endanger personnel must be released. A qualified person must use the proper devices to release the stored energy, such as using a shorting probe to discharge a capacitor.

5.0.0 LADDERS AND SCAFFOLDING

Ladders and scaffolding account for about half of the injuries from workplace electrocutions. The involuntary recoil which can occur when a person is shocked can cause them to be thrown from a ladder or high place.

5.1.0 LADDERS

Many job site accidents involve the misuse of ladders. Make sure to follow these general rules every time you use any ladder. Following these rules can prevent serious injury or even death.

- Before using any ladder, inspect it. Look for loose or missing rungs, cleats, bolts, or screws, and check for cracked, broken, or badly worn rungs, cleats, or side rails.
- If you find a ladder in poor condition, do not use it. Report it and tag it for repair or disposal.
- Never modify a ladder by cutting it or weakening its parts.
- Do not set up ladders where they may be run into by others, such as in doorways or walkways. If it is absolutely necessary to set up a ladder in such a location, protect the ladder with barriers.

- Do not increase a ladder's reach by standing it on boxes, barrels, or anything other than a flat surface.
- Check your shoes for grease, oil, or mud before climbing a ladder. These materials could make you slip.
- Always face the ladder and hold on with both hands when climbing up or down.
- Never lean out from the ladder. Keep your belt buckle centered between the rails. If something is out of reach, get down and move the ladder.

WARNING!   When performing electrical work, always use ladders made of nonconductive material.

5.1.1 Straight And Extension Ladders

There are some specific rules to follow when working with straight and extension ladders:

- Always place a straight ladder at the proper angle. The distance from the ladder feet to the base of the wall or support should be about one-fourth the working height of the ladder (see *Figure 7*).

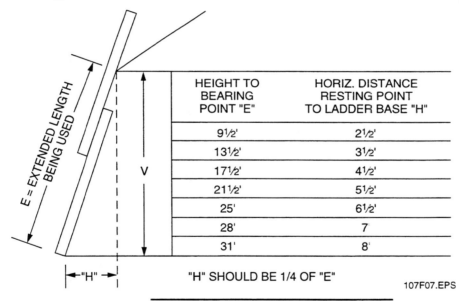

HEIGHT TO BEARING POINT "E"	HORIZ. DISTANCE RESTING POINT TO LADDER BASE "H"
9½'	2½'
13½'	3½'
17½'	4½'
21½'	5½'
25'	6½'
28'	7
31'	8'

"H" SHOULD BE 1/4 OF "E"

Figure 7. Straight Ladder Positioning

- Secure straight ladders to prevent slipping. Use ladder shoes or hooks at the top and bottom. Another method is to secure a board to the floor against the ladder feet. For brief jobs, someone can hold the straight ladder.
- Side rails should extend above the top support point by at least 36" (see *Figure 7*).
- It takes two people to safely extend and raise an extension ladder. Extend the ladder only after it has been raised to an upright position.
- Never carry an extended ladder.

- Never use two ladders spliced together.
- Ladders should not be painted because paint can hide defects.

5.1.2 Step Ladders

There are also a few specific rules to use with a step ladder:

- Always open the step ladder all the way and lock the spreaders to avoid collapsing the ladder accidentally.
- Use a step ladder that is high enough for the job so that you do not have to reach. Get someone to hold the ladder if it is more than 10' high.
- Never use a step ladder as a straight ladder.
- Never stand on or straddle the top two rungs of a step ladder.
- Ladders are not shelves.

WARNING!   Do not leave tools or materials on a step ladder.

Sometimes you will need to move or remove protective equipment, guards, or guardrails to complete a task using a ladder. Remember, always replace what you moved or removed before leaving the area.

5.2.0 SCAFFOLDING

Working on scaffolding also involves being safe and alert to hazards. In general, keep scaffold platforms clear of unnecessary material or scrap. These can become deadly tripping hazards or falling objects. Carefully inspect each part of the scaffolding as it is erected. Your life may depend on it! Makeshift scaffolding has caused many injuries and deaths on job sites. Use only scaffolding and planking materials designed and marked for their specific use. When working on scaffolding, follow the established specific requirements set by OSHA for the use of fall protection. When appropriate, wear an approved harness with a lanyard properly anchored to the structure.

Note: The following requirements represent a compilation of the more stringent requirements of both CFR 1910 and CFR 1926.

The following are some of the basic OSHA rules for working safely on scaffolding:

- Scaffolding must be erected on sound, rigid footing that can carry the maximum intended load.
- Guardrails and toe boards must be installed on the open sides and ends of platforms which are higher than 6' above the ground or floor.
- There must be a screen of ½" maximum openings between the toe board and the midrail where persons are required to work or pass under the scaffold.

- Scaffold planks must extend over their end supports not less than 6" nor more than 12" and must be properly blocked.
- If the scaffold does not have built-in ladders which meet the standard, then it must have an attached ladder access.
- All employees must be trained to erect, dismantle, and use scaffolding.
- Unless it is impossible, fall protection must be worn while building or dismantling all scaffolding.
- Work platforms must be completely decked for use by employees.
- Your hard hat is the first line of protection from falling objects. Your hard hat, however, cannot protect your shoulders, arms, back, or feet from the danger of falling objects. The person working below depends on those working above. When you are working above the ground, be careful so that material, including your tools, cannot fall from your work site. Use trash containers or other similar means to keep debris from falling and never throw or sweep material from above.

6.0.0 LIFTS, HOISTS, AND CRANES

On the job, you may be working in the operating area of lifts, hoists, or cranes. The following safety rules are for those who are working in the area with overhead equipment but are not directly involved in its operation:

- Stay alert and pay attention to the warning signals from operators.
- Never stand or walk under a load, regardless of whether it is moving or stationary.
- Always warn others of moving or approaching overhead loads.
- Never attempt to distract signal persons or operators of overhead equipment.
- Obey warning signs.
- Do not use equipment that you are not qualified to operate.

7.0.0 LIFTING

Back injuries cause many lost working hours every year. That is in addition to the misery felt by the person with the hurt back! Learn how to lift properly and size up the load. To lift, first stand close to the load. Then, squat down and keep your back straight. Get a firm grip on the load and keep the load close to your body. Lift by straightening your legs. Make sure that you lift with your legs and not your back. Do not be afraid to ask for help if you feel the load is too heavy. See *Figure 8* for an example of proper lifting.

Figure 8. Proper Lifting

8.0.0 BASIC TOOL SAFETY

When using any tools for the first time, read the operator's manual to learn the recommended safety precautions. If you are not certain about the operation of any tool, ask the advice of a more experienced worker. Before using a tool, you should know its function and how it works.

Always use the right tool for the job. Incorrectly using tools is one of the leading causes of job site injury. Using a hammer as a pry bar, or a screwdriver as a chisel, can cause damage to the tool and injure you in the process.

9.0.0 CONFINED SPACE ENTRY PROCEDURES

Occasionally, you may be required to do your work in a manhole or vault. If this is the case, there are some special safety considerations that you need to be aware of. For details on the subject of working in manholes and vaults, refer to 1910.146/1926.21(a)(6)(i) and (ii) and the *National Electrical Safety Code*. The general precautions are listed in the following paragraphs.

9.1.0 GENERAL GUIDELINES

A confined space includes (but is not limited to) any of the following: a manhole, boiler, tank, trench (4' or deeper), tunnel, hopper, bin, sewer, vat, pipeline, vault, pit, air duct, or vessel.

A confined space is identified as follows:

- It has limited entry and exit.
- It is not intended for continued human occupancy.
- It has poor ventilation.
- It has the potential for entrapment/engulfment.
- It has the potential for accumulating a dangerous atmosphere.
- Entry into a confined space occurs when any part of the body crosses the plane of entry. No employee shall enter a confined space unless the employee has been trained in confined space entry procedures.
- All hazards must be eliminated or controlled before a confined space entry is made.
- All appropriate personal protective equipment shall be worn at all times during confined space entry and work. The minimum required equipment includes a hard hat, safety glasses, full body harness, and life line.
- Ladders used for entry must be secured.
- A rescue retrieval system must be in use when entering confined spaces and while working in permit-required confined spaces (discussed later). Each employee must be capable of being rescued by the retrieval system.
- Only no-entry rescues will be performed by company personnel. Entry rescues will be performed by trained rescue personnel identified on the entry permit.
- The area outside the confined space must be properly barricaded, and appropriate warning signs must be posted.
- Entry permits can be issued and signed by job site supervisors only. Permits must be kept at the confined space while work is being conducted. At the end of the shift, the entry permits must be made part of the job journal and retained for one year.

9.2.0 CONFINED SPACE HAZARD REVIEW

Before determining the proper procedure for confined space entry, a hazard review shall be performed. The hazard review shall include, but not be limited to, the following conditions:

- The past and current uses of the confined space
- The physical characteristics of the space including size, shape, air circulation, etc.
- Proximity of the space to other hazards
- Existing or potential hazards in the confined space, such as:
  - Atmospheric conditions (oxygen levels, flammable/explosive levels, and/or toxic levels)
  - Presence/potential for liquids
  - Presence/potential for particulates
- Potential for mechanical/electrical hazards in the confined space (including work to be done)

Once the hazard review is completed, the supervisor, in consultation with the project managers and/or safety manager, shall classify the confined space as one of the following:

- A nonpermit confined space
- A permit-required confined space controlled by ventilation
- A permit-required confined space

Once the confined space has been properly classified, the appropriate entry and work procedures must be followed.

9.3.0 ENTRY AND WORK PROCEDURES

Nonpermit spaces – A hazard review checklist must be completed before a confined space is designated as a *nonpermit space*. The checklist must be made part of the job journal, and a copy of the checklist must be sent to the safety office. A nonpermit confined space must meet the following criteria:

- There is no actual or potential atmospheric hazard.

Note: Using ventilation to clear the atmosphere does not meet this criteria.

- There are no actual or potential physical, electrical, or mechanical hazards capable of causing harm or death.

Documentation using the hazards checklist and entry permit forms and verifying that the confined space is hazard-free must be made available to employees and maintained at the confined space while work is conducted. If it is necessary to enter the space to verify it is hazard-free or to eliminate hazards, entry must be made under the requirements of a permit-required space.

An employee may enter the confined space using the minimum fall protection of harness and anchored life line. Once in the space, the employee may disconnect the life line and reconnect it before exiting.

If the work being done creates a hazard, the space must be reclassified as a permit-required space. If any other atmospheric, physical, electrical, or mechanical hazards arise, the space is to be evacuated immediately and reclassified as a permit-required entry space.

Permit-required spaces controlled by ventilation – A hazard review checklist must be completed before a confined space is designated as a *permit-required space controlled by ventilation*. The checklist must be made part of the job journal and a copy of the checklist must be sent to the safety office. A permit-required confined space controlled by ventilation must meet the following criteria:

- The only hazard in the confined space is an actual/potential atmospheric hazard.
- Continuous forced-air ventilation maintains a safe atmosphere (i.e., within the limits designated on the entry permit).

- Inspection and monitoring data are documented.
- No other physical, electrical, or mechanical hazard exists.

An entry permit must be issued and signed by the job site supervisor and be kept at the confined space while work is being conducted.

Atmospheric testing must be conducted before entry into the confined space and in the following order:

- Oxygen content
- Flammable gases and vapors
- Toxic contaminants

Unacceptable atmospheric conditions must be eliminated with forced air ventilation. If continuous forced air ventilation is required to maintain an acceptable atmosphere, employees may not enter until forced air ventilation has eliminated any hazardous atmosphere.

Periodic atmospheric testing must be conducted during the work shift to ensure that the atmosphere remains clear. Periodic monitoring must be documented on the entry permit. If atmospheric conditions change, employees must exit the confined space immediately, and atmospheric conditions must be re-evaluated. Continuous communication must be maintained with the employees working in the confined space.

If *hot work* is to be performed, a hot work permit is required, and the hazard analysis must document that the hot work does not create additional hazards that are not controlled by ventilation only. Hot work is defined as any work that produces arcs, sparks, flames, heat, or other sources of ignition.

A rescue plan using trained rescue personnel must be in place prior to the start of work in the confined space. All employees should be aware of the rescue plan and how to activate it.

Permit-required confined spaces – A hazard review checklist must be completed before a confined space is designated as a *permit-required confined space*. The checklist must be made part of the job journal and a copy must be sent to the safety office. A permit-required space meets the following criteria:

- There are actual/potential hazards other than a hazardous atmosphere.
- Ventilation alone does not eliminate atmospheric hazards.
- Conditions in and around the confined space must be continually monitored.

An entry permit must be issued and signed by the job site supervisor. The permit is to be kept at the confined space while work is being performed in the space.

ELECTRICAL SAFETY — TRAINEE MODULE 33107

Atmospheric testing must be conducted before entry into the confined space and in the following order:

- Oxygen content
- Flammable gases and contaminants
- Toxic contaminants

Unacceptable atmospheric conditions must be eliminated/controlled prior to employee entry. Methods of elimination may include isolation, purging, flushing, or ventilating. Continuous atmospheric monitoring must be conducted while employees are in the confined space. Triggering of a monitoring alarm means employees should evacuate the confined space immediately. Any other physical hazards must be eliminated or controlled by engineering and work practice controls before entry. Additional personal protective equipment should be used as a follow-up to the above methods. An attendant, whose job it is to monitor conditions in and around the confined space and to maintain contact with the employees in the space, must be stationed outside the confined space for the duration of entry operations.

A rescue plan using trained rescue personnel must be in place before confined space entry. The attendant should be aware of the rescue plan and have the means to activate it.

10.0.0 FIRST-AID

You should be prepared in case an accident does occur on the job site or anywhere else. First-aid training that includes certification classes in CPR and artificial respiration could be the best insurance you and your fellow workers ever receive. Make sure that you know where first aid is available at your job site. Also, make sure you know the accident reporting procedure. Each job site should also have a first-aid manual or booklet giving easy-to-find emergency treatment procedures for various types of injuries. Emergency first-aid telephone numbers should be readily available to everyone on the job site. Refer to CFR 1910.151/1926.23 and 1926.50 for specific requirements.

11.0.0 SOLVENTS AND TOXIC VAPORS

The solvents that are used on a job site may give off vapors that are toxic enough to make people temporarily ill or even cause permanent injury. Many solvents are skin and eye irritants. Solvents can also be systemic poisons when they are swallowed or absorbed through the skin.

Solvents in spray or aerosol form are dangerous in another way. Small aerosol particles or solvent vapors mix with air to form a combustible mixture with oxygen. The slightest spark could cause an explosion in a confined area because the mix is perfect for fast ignition.

There are procedures and methods for using, storing, and disposing of most solvents and chemicals. These procedures are normally found in the Material Safety Data Sheets (MSDSs) available at your facility.

An MSDS is required for all materials that could be hazardous to personnel or equipment. These sheets contain information on the material, such as the manufacturer and chemical makeup. As much information as possible is kept on the hazardous material to prevent a dangerous situation. In the event of a dangerous situation, the information is used to rectify the problem in as safe a manner as possible. See *Figure 9* for an example of procedures you may find on the job.

Section VII — Precautions for Safe Handling and Use

Steps to Be Taken in Case Material is Released or Spilled
Isolate from oxidizers, heat, sparks, electric equipment, and open flames.

Waste Disposal Method
Recycle or incinerate observing local, state and federal health, safety and pollution laws.

Precautions to Be Taken in Handling and Storing
Store in a cool dry area. Observe label cautions and instructions.

Other Precautions
SEE ATTACHMENT PARA #3

Section VIII — Control Measures

Respiratory Protection (Specify Type)
Suitable for use with organic solvents

Ventilation	Local Exhaust: preferable	Special: none
	Mechanical (General): acceptable	Other: none

Protective Gloves: recommended (must not dissolve in solvents)
Eye Protection: goggles

Other Protective Clothing or Equipment: none

Work/Hygienic Practices: Use with adequate ventilation. Observe label cautions.

Figure 9. Portion Of An MSDS

11.1.0 PRECAUTIONS WHEN USING SOLVENTS

It is always best to use a nonflammable, nontoxic solvent whenever possible. However, any time solvents are used, it is essential that your work area be adequately ventilated and that you wear the appropriate personal protective equipment:

- A chemical face shield with chemical goggles should be used to protect the eyes and skin from sprays and splashes.

- A chemical apron should be worn to protect your body from sprays and splashes. Remember that some solvents are acid-based. If they come into contact with your clothes, solvents can eat through your clothes to your skin.
- A paper filter mask does not stop vapors; it is used only for nuisance dust. In situations where a paper mask does not supply adequate protection, chemical cartridge respirators might be needed. These respirators can stop many vapors if the correct cartridge is selected. In areas where ventilation is a serious problem, a self-contained breathing apparatus (SCBA) must be used. Make sure that you have been given a full medical evaluation and that you are properly trained in using respirators at your site.

11.2.0 RESPIRATORY PROTECTION

Protection against high concentrations of dust, mist, fumes, vapors, gases, and/or oxygen deficiency is provided by appropriate respirators.

Appropriate respiratory protective devices should be used for the hazardous material involved and the extent and nature of the work performed.

An air-purifying respirator is, as its name implies, a respirator that removes contaminants from air inhaled by the wearer. The respirators may be divided into the following types: particulate-removing (mechanical filter), gas-removing and vapor-removing (chemical filter), and a combination of particulate-removing and gas-removing and vapor-removing.

Particulate-removing respirators are designed to protect the wearer against the inhalation of particulate matter in the ambient atmosphere. They may be designed to protect against a single type of particulate, such as pneumoconiosis-producing and nuisance dust, toxic dust, metal fumes or mist, or against various combinations of these types.

Gas and vapor-removing respirators are designed to protect the wearer against the inhalation of gases or vapors in the ambient atmosphere. They are designated as gas masks, chemical cartridge respirators (nonemergency gas respirators), and self-rescue respirators. They may be designed to protect against a single gas such as chlorine; a single type of gas, such as acid gases; or a combination of types of gases, such as acid gases and organic vapors.

If you are required to use a respiratory protective device, you must be evaluated by a physician to ensure that you are physically fit to use a respirator. You must then be fitted and thoroughly instructed in the respirator's use.

WARNING! Do not use any respirator unless you have been fitted for it and thoroughly understand its use. As with all safety rules, follow your employer's respiratory program and policies.

Any employee whose job entails having to wear a respirator must keep their face free of facial hair in the seal area.

Respiratory protective equipment must be inspected regularly and maintained in good condition. Respiratory equipment must be properly cleaned on a regular basis and stored in a sanitary, dustproof container.

12.0.0 ASBESTOS

Asbestos is a mineral-based material that is resistant to heat and corrosive chemicals. Depending on the chemical composition, asbestos fibers may range in texture from coarse to silky. The properties that make asbestos fibers so valuable to industry are its high tensile strength, flexibility, heat and chemical resistance, and good frictional properties.

Asbestos fibers enter the body by inhalation of airborne particles or by ingestion and can become embedded in the tissues of the respiratory or digestive systems. Years of exposure to asbestos can cause numerous disabling or fatal diseases. Among these diseases are asbestosis, an emphysema-like condition; lung cancer; mesothelioma, a cancerous tumor that spreads rapidly in the cells of membranes covering the lungs and body organs; and gastrointestinal cancer.

12.1.0 MONITORING

Employers who have a workplace or work operation covered by OSHA 3096 (*Asbestos Standard for the Construction Industry*) must perform initial monitoring to determine the airborne concentrations of asbestos to which employees may be exposed. If employers can demonstrate that employee exposures are below the action level and/or excursion limit by means of objective or historical data, initial monitoring is not required. If initial monitoring indicates that employee exposures are below the action level and/or excursion limit, then periodic monitoring is not required. Within regulated areas, the employer must conduct daily monitoring unless all workers are equipped with supplied-air respirators operated in the positive-pressure mode. If daily monitoring by statistically reliable measurements indicates that employee exposures are below the action level and/or excursion limit, then no further monitoring is required for those employees whose exposures are represented by such monitoring. Employees must be given the chance to observe monitoring, and affected employees must be notified as soon as possible following the employer's receipt of the results.

12.2.0 REGULATED AREAS

The employer must establish a regulated area where airborne concentrations of asbestos exceed or can reasonably be expected to exceed the locally determined exposure limit, or when certain types of construction work are performed, such as cutting asbestos-cement sheets and removing asbestos-containing floor tiles. Only authorized personnel may enter regulated areas. All persons entering a regulated area must be supplied with an appropriate respirator.

No smoking, eating, drinking, or applying cosmetics is permitted in regulated areas. Warning signs must be displayed at each regulated area and must be posted at all approaches to regulated areas. These signs must bear the following information:

> **DANGER**
>
> **ASBESTOS**
>
> **CANCER AND LUNG DISEASE HAZARD**
>
> **AUTHORIZED PERSONNEL ONLY**
>
> **RESPIRATORS AND PROTECTIVE CLOTHING**
>
> **ARE REQUIRED IN THIS AREA**

Where feasible, the employer shall establish negative-pressure enclosures before commencing asbestos removal, demolition, and renovation operations. The setup and monitoring requirements for negative-pressure enclosures are as follows:

- A competent person shall be designated to set up the enclosure and ensure its integrity and supervise employee activity within the enclosure.
- Exemptions are given for small-scale, short duration maintenance or renovation operations.
- The employer shall conduct daily monitoring of the exposure of each employee who is assigned to work within a regulated area. Short-term monitoring is required whenever asbestos concentrations will not be uniform throughout the workday and where high concentrations of asbestos may reasonably be expected to be released or created in excess of the local limit.

In addition, warning labels must be affixed on all asbestos products and to all containers of asbestos products, including waste containers, that may be in the workplace. The label must include the following information:

> **DANGER**
>
> **CONTAINS ASBESTOS FIBERS**
>
> **AVOID CREATING DUST**
>
> **CANCER AND LUNG DISEASE HAZARD**

12.3.0 METHODS OF COMPLIANCE

To the extent feasible, engineering and work practice controls must be used to reduce employee exposure to within the permissible exposure limit (PEL). The employer must use one or more of the following control methods to achieve compliance:

- Local exhaust ventilation equipped with high-efficiency particulate air (HEPA) filter dust collection systems
- General ventilation systems
- Vacuum cleaners equipped with HEPA filters
- Enclosure or isolation of asbestos dust-producing processes
- Use of wet methods, wetting agents, or removal encapsulants during asbestos handling, mixing, removal, cutting, application, and cleanup
- Prompt disposal of asbestos-containing wastes in leak-tight containers

Prohibited work practices include the following:

- The use of high-speed abrasive disc saws that are not equipped with appropriate engineering controls
- The use of compressed air to remove asbestos-containing materials, unless the compressed air is used in conjunction with an enclosed ventilation system

Where engineering and work practice controls have been instituted but are insufficient to reduce employee exposure to a level that is at or below the PEL, respiratory protection must be used to supplement these controls.

13.0.0 BATTERIES

Working around batteries can be dangerous if the proper precautions are not taken. Batteries often give off hydrogen gas as a byproduct. When hydrogen mixes with air, the mixture can be explosive in the proper concentration. For this reason, smoking is strictly prohibited in battery rooms, and only insulated tools should be used. Proper ventilation also reduces the chance of explosion in battery areas. Follow your company's procedures for working near batteries. Also, ensure that your company's procedures are followed for lifting heavy batteries.

CAUTION: Batteries cannot be thrown into the regular trash. The disposal of batteries is regulated by the Environmental Protection Agency (EPA) and must be done through a regulated disposal company in accordance with EPA regulations and local laws. This requirement applies to all types of batteries.

13.1.0 ACIDS

Batteries also contain acid, which will eat away human skin and many other materials. Appropriate personal protective equipment for battery work typically includes chemical aprons, sleeves, gloves, face shields, and goggles to prevent acid from contacting skin and eyes. Follow your site procedures for dealing with spills of these materials. Also, know the location of first aid when working with these chemicals.

13.2.0 WASH STATIONS

Because of the chance that battery acid may contact someone's eyes or skin, wash stations are located near battery rooms. Do not connect or disconnect batteries without proper supervision. Everyone who works in the area should know where the nearest wash station is and how to use it. Battery acid should be flushed from the skin and eyes with large amounts of water or with a neutralizing solution.

CAUTION: If you come in contact with battery acid, report it immediately to your supervisor.

14.0.0 PCBs

Polychlorinated biphenyls (PCBs) are chemicals that were marketed under various trade names as a liquid insulator/cooler in older transformers. In addition to being used in older transformers, PCBs are also found in some large capacitors and in the small ballast transformers used in street lighting and ordinary fluorescent light fixtures. Disposal of these materials is regulated by the EPA and must be done through a regulated disposal company; use extreme caution and follow your facility procedures.

WARNING! Do not come into contact with PCBs. They present a variety of serious health risks, including lung damage and cancer.

15.0.0 FALL PROTECTION

15.1.0 FALL PROTECTION PROCEDURES

Fall protection must be used when employees are on a walking or working surface that is 6' or more above a lower level and has an unprotected edge or side. The areas covered include, but are not limited to:

- Finished and unfinished floors or mezzanines
- Temporary or permanent walkways/ramps
- Finished or unfinished roof areas

- Elevator shafts and hoist-ways
- Floor, roof, or walkway holes
- Working 6' or more above dangerous equipment

Exception: If the dangerous equipment is unguarded, fall protection must be used at all heights regardless of the fall distance.

Note: Walking/working surfaces do not include ladders, scaffolding, vehicles, or trailers. Also, an unprotected edge or side is an edge/side where there is no guardrail system at least 39 inches high.

Fall protection is not required during inspection, investigation, or assessment of job site conditions before or after construction work.

These fall protection guidelines do not apply to the following areas. Fall protection for these areas is located in the subparts cited in parenthesis.

- Cranes and derricks (1926 subpart N/1910 subpart N)
- Scaffolding (1926 subpart L/1910 subpart D)
- Electrical power transmission and distribution (1926 subpart V/1910 subpart R)
- Stairways and ladders (1926 subpart X/1910 subpart D)
- Excavations (1926 subpart P)

Fall protection must be selected in order of preference as listed below. Selection of a lower-level system (e.g., safety nets) must be based only on feasibility of protection. The list includes, but is not limited to, the following:

- Guardrail systems and hole covers
- Personal fall arrest systems
- Safety nets

These fall protection procedures are designed to warn, isolate, restrict, or protect workers from a potential fall hazard.

15.2.0 TYPES OF FALL PROTECTION SYSTEMS

The type of system selected shall depend on the fall hazards associated with the work to be performed. First, a hazard analysis shall be conducted by the job site supervisor prior to the start of work. Based on the hazard analysis, the job site supervisor and project manager, in consultation with the safety manager, will select the appropriate fall protection system. All employees will be instructed in the use of the fall protection system before starting work.

SUMMARY

Safety must be your concern at all times so that you do not become either the victim of an accident or the cause of one. Safety requirements and safe work practices are provided by OSHA and your employer. It is essential that you adhere to all safety requirements and follow your employer's safe work practices and procedures. Also, you must be able to identify the potential safety hazards of your job site. The consequences of unsafe job site conduct can often be expensive, painful, or even deadly. Report any unsafe act or condition immediately to your supervisor. You should also report all work-related accidents, injuries, and illnesses to your supervisor immediately. Remember, proper construction techniques, common sense, and a good safety attitude will help to prevent accidents, injuries, and fatalities.

References

For advanced study of topics covered in this task module, the following books are suggested:

29 CFR Parts 1900 – 1910, Standards for General Industry, Occupational Safety and Health Administration, US Department of Labor.

29 CFR Part 1926, Standards for the Construction Industry, Occupational Safety and Health Administration, US Department of Labor.

National Electrical Code Handbook, Latest Edition, National Fire Protection Association, Quincy, MA.

National Electrical Safety Code, Latest Edition, National Fire Protection Association, Quincy, MA.

REVIEW/PRACTICE QUESTIONS

1. How much electrical current does it take before a person would begin to feel a painful shock?
    a. Less than 1mA
    b. 20mA to 50mA
    c. 50mA to 200mA
    d. 1A to 5A

2. If a person's heart begins to fibrillate due to an electrical shock, the solution is to _____.
    a. leave the person alone until the fibrillation stops
    b. administer heart massage
    c. use the Heimlich maneuver
    d. have a qualified person use emergency defibrillation equipment

3. The majority of injuries due to electrical shock are caused by _____.
    a. electrically operated hand tools
    b. contact with low-voltage conductors
    c. contact with high-voltage conductors
    d. lightning

4. The government agency whose purpose is to ensure safe and healthful working conditions is known as the _____.
    a. PCB
    b. EPA
    c. OSHA
    d. GFCI

5. A kirklock is a _____.
    a. keyed circuit breaker interlock
    b. church key
    c. padlock used to secure an electrical disconnect switch
    d. type of fuse puller

6. Which of these statements correctly describes a double-insulated power tool?
    a. There is twice as much insulation on the power cord.
    b. It can safely be used in place of a grounded tool.
    c. It is made entirely of plastic or other nonconducting material.
    d. The entire tool is covered in rubber.

ELECTRICAL SAFETY — TRAINEE MODULE 33107

7. Which of the following applies in a lockout/tagout procedure?
   a. Only the supervisor can install lockout/tagout devices.
   b. If several employees are involved, the lockout/tagout equipment is applied only by the first employee to arrive at the disconnect.
   c. Lockout/tagout devices applied by one employee can be removed by another employee as long as it can be verified that the first employee has left for the day.
   d. Lockout/tagout devices are installed by every authorized employee involved in the work.

8. What is the proper distance from the feet of a straight ladder to the wall?
   a. One-fourth the working height of the ladder
   b. One-half the height of the ladder
   c. Three feet
   d. One-fourth of the square root of the height of the ladder

9. What are the minimum and maximum distances (in inches) that a scaffold plank can extend beyond its end support?
   a. 4; 8
   b. 6; 10
   c. 6; 12
   d. 8; 12

10. Which of these conditions applies to a permit-required confined space but not to a permit-required space controlled by ventilation?
    a. A hazard review checklist must be completed.
    b. An attendant, whose job is to monitor the space, must be stationed outside the space.
    c. Unacceptable atmospheric conditions must be eliminated.
    d. Atmospheric testing must be conducted.

notes

ANSWERS TO REVIEW/PRACTICE QUESTIONS

Answer	Section Reference
1. b	2.0.0
2. d	2.0.0
3. b	2.1.0
4. c	4.0.0
5. a	4.2.0
6. b	4.3.0
7. d	4.3.1
8. a	5.1.1
9. c	5.2.0
10. b	9.3.0

APPENDIX A

SUMMARY OF OSHA CONSTRUCTION STANDARDS

CONSTRUCTION SAFETY AND
HEALTH OUTREACH PROGRAM

U.S. Department of Labor
OSHA Office of Training and Education

ELECTRICAL STANDARDS FOR CONSTRUCTION

INTRODUCTION

Electricity has long been recognized as a serious workplace hazard, exposing employees to such dangers as electric shock, electrocution, fires, and explosions.

Experts in electrical safety have traditionally looked toward the widely used *National Electrical Code* (NEC) for help in the practical safeguarding of persons from these hazards. The Occupational Safety and Health Administration (OSHA) recognized the important role of the NEC in defining basic requirements for safety in electrical installations by including the entire 1971 NEC by reference in Subpart K of 29 *Code of Federal Regulations* Part 1926 (Construction Safety and Health Standards).

In a final rule dated July 11, 1986, OSHA updated, simplified, and clarified Subpart K, 29 CFR 1926. The revisions serve these objectives:

- NEC requirements that directly affect employees in construction workplaces have been placed in the text of the OSHA standard, eliminating the need for the NEC to be incorporated by reference.

- Certain requirements that supplemented the NEC have been integrated in the new format.

- Performance language is utilized and superfluous specifications omitted and changes in technology accommodated.

In addition, the standard is easier for employers and employees to use and understand. Also, the OSHA revision of the electrical standards has been made more flexible, eliminating the need for constant revision to keep pace with the NEC, which is revised every three years.

SUBPART K

The NEC provisions directly related to employee safety are included in the body of the standard itself - making it unnecessary to continue the adoption by reference of the NEC. Subpart K is divided into four major groups plus a general definitions section:

- Installation Safety Requirements
  [29 CFR 1926.402 - 1926.415]

- Safety-Related Work Practices
  [29 CFR 1926.416 - 1926.430]

- Safety-Related Maintenance and Environmental Considerations
  [29 CFR 1926.431 - 1926.440]

- Safety Requirements for Special Equipment
  [29 CFR 1926.441 - 1926.448]

- Definitions
  [29 CFR 1926.449]

I. INSTALLATION SAFETY REQUIREMENTS

Part I of the standard is very comprehensive. Only some of the major topics and brief summaries of these requirements are included in this discussion.

Sections 29 CFR 1926.402 through 1926.408 contain installation safety requirements for electrical equipment and installations used to provide electric power and light at the jobsite. These sections apply to installations, both temporary and permanent, used on the jobsite; but they *do not* apply to existing permanent installations that were in place before the construction activity commenced.

If an installation is made in accordance with the 1984 *National Electrical Code*, it will be considered to be in compliance with Sections 1926.403 through 1926.408, except for:

1926.404(b)(1)	Ground-fault protection for employees
1926.405(a)(2)(ii)(E)	Protection of lamps on temporary wiring
1926.405(a)(2)(ii)(F)	Suspension of temporary lights by cords
1926.405(a)(2)(ii)(G)	Portable lighting used in wet or conductive locations
1926.405(a)(2)(ii)(J)	Extension cord sets and flexible cords

Approval

The electrical conductors and equipment used by the employer must be approved.

Examination, Installation, and Use of Equipment

The employer must ensure that electrical equipment is free from recognized hazards that are likely to cause death or serious physical harm to employees. Safety of equipment must be determined by the following:

- Suitability for installation and use in conformity with the provisions of the standard. Suitability of equipment for an identified purpose may be evidenced by a listing, by labeling, or by certification for that identified purpose.

- Mechanical strength and durability. For parts designed to enclose and protect other equipment, this includes the adequacy of the protection thus provided.

- Electrical insulation.

- Heating effects under conditions of use.

- Arcing effects.

- Classification by type, size, voltage, current capacity, and specific use.

- Other factors that contribute to the practical safeguarding of employees who use or are likely to come in contact with the equipment.

Guarding

Live parts of electric equipment operating at 50 volts or more must be guarded against accidental contact. Guarding of live parts must be accomplished as follows:

- Location in a cabinet, room, vault, or similar enclosure accessible only to qualified persons.

- Use of permanent, substantial partitions or screens to exclude unqualified persons.

- Location on a suitable balcony, gallery, or platform elevated and arranged to exclude unqualified persons.

- Elevation of eight feet or more above the floor.

Entrance to rooms and other guarded locations containing exposed live parts must be marked with conspicuous warning signs forbidding unqualified persons to enter.

Electric installations that are over 600 volts and that are open to unqualified persons must be made with metal-enclosed equipment or enclosed in a vault or area controlled by a lock. In addition, equipment must be marked with appropriate caution signs.

Overcurrent Protection

The following requirements apply to overcurrent protection of circuits rated 600 volts, nominal, or less.

- Conductors and equipment must be protected from overcurrent in accordance with their ability to safely conduct current and the conductors must have sufficient current-carrying capacity to carry the load.

- Overcurrent devices must not interrupt the continuity of the grounded conductor unless all conductors of the circuit are opened simultaneously, except for motor-running overload protection.

- Overcurrent devices must be readily accessible and not located where they could create an employee safety hazard by being exposed to physical damage or located in the vicinity of easily ignitable material.

- Fuses and circuit breakers must be so located or shielded that employees will not be burned or otherwise injured by their operation, e.g., arcing.

Grounding of Equipment Connected by Cord and Plug

Exposed noncurrent-carrying metal parts of cord- and plug-connected equipment that may become energized must be grounded in the following situations:

- When in a hazardous (classified) location.

- When operated at over 150 volts to ground, except for guarded motors and metal frames of electrically heated appliances if the appliance frames are permanently and effectively insulated from ground.

- When one of the types of equipment listed below. But see Item 6 for exemption.

    1. Hand held motor-operated tools.

    2. Cord- and plug-connected equipment used in damp or wet locations or by employees standing on the ground or on metal floors or working inside metal tanks or boilers.

    3. Portable and mobile X-ray and associated equipment.

    4. Tools likely to be used in wet and/or conductive locations.

    5. Portable hand lamps.

    6. [Exemption] Tools likely to be used in wet and/or conductive locations need not be grounded if supplied through an isolating transformer with an ungrounded secondary of not over 50 volts. Listed or labeled portable tools and appliances protected by a system of double insulation, or its equivalent, need not be grounded. If such a system is employed, the equipment must be distinctively marked to indicate that the tool or appliance uses a system of double insulation.

II. Safety-Related Work Practices

Protection of Employees

The employer must not permit an employee to work near any part of an electric power circuit that the employee could contact in the course of work, unless the employee is protected against shock by de-energizing the circuit and grounding it or by guarding it effectively by insulation or other means.

Where the exact location of underground electric power lines is unknown, employees using jack hammers or hand tools that may contact a line must be provided with insulated protective gloves.

Even before work is begun, the employer must determine by inquiry, observation, or instruments where any part of an exposed or concealed energized electric power circuit is located. This is necessary because a person, tool or machine could come into physical or electrical contact with the electric power circuit.

The employer is required to advise employees of the location of such lines, the hazards involved, and protective measures to be taken as well as to post and maintain proper warning signs.

Passageways and Open Spaces

The employer must provide barriers or other means of guarding to ensure that workspace for electrical equipment will not be used as a passageway during the time when energized parts of electrical equipment are exposed. Walkways and similar working spaces must be kept clear of electric cords. Other standards cover load ratings, fuses, cords, and cables.

Lockout and Tagging of Circuits

Tags must be placed on controls that are to be deactivated during the course of work on energized or de-energized equipment or circuits. Equipment or circuits that are de-energized must be rendered inoperative and have tags attached at all points where such equipment or circuits can be energized.

III. Safety-Related Maintenance and Environmental Considerations

Maintenance of Equipment

The employer must ensure that all wiring components and utilization equipment in hazardous locations are maintained in a dust-tight, dust-ignition-proof, or explosion-proof condition without loose or missing screws, gaskets, threaded connections, seals, or other impairments to a tight condition.

Environmental Deterioration of Equipment

Unless identified for use in the operating environment, no conductors or equipment can be located:

- In damp or wet locations.

- Where exposed to gases, fumes, vapors, liquids, or other agents having a deteriorating effect on the conductors or equipment.

- Where exposed to excessive temperatures.

Control equipment, utilization equipment, and busways approved for use in dry locations only must be protected against damage from the weather during building construction.

For protection against corrosion, metal raceways, cable armor, boxes, cable sheathing, cabinets, elbows, couplings, fittings, supports, and support hardware must be of materials appropriate for the environment in which they are installed.

Ground-Fault Circuit Interrupter

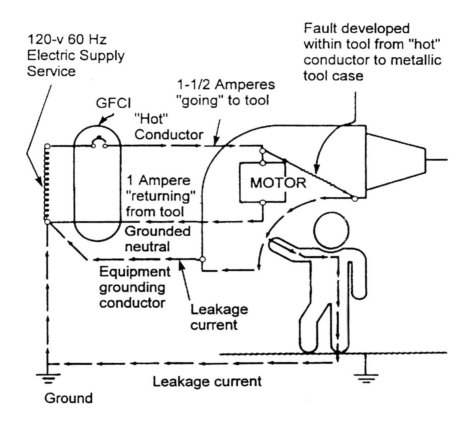

GFCI monitors the difference in current flowing into the "hot" and out to the grounded neutral conductors. The difference (1/2 ampere in this case) will flow back through any available path, such as the equipment grounding conductor, and through a person holding the tool, if the person is in contact with a grounded object.

However, the GFCI will not protect the employee from line-to-line contact hazards (such as a person holding two "hot" wires or a hot and a neutral wire in each hand). It does provide protection against the most common form of electrical shock hazard--the ground fault. It also provides protection against fires, overheating, and destruction of insulation on wiring.

WHAT ARE THE HAZARDS?

With the wide use of portable tools on construction sites, the use of flexible cords often becomes necessary. Hazards are created when cords, cord connectors, receptacles, and cord- and plug-connected equipment are improperly used and maintained.

Generally, flexible cords are more vulnerable to damage than is fixed wiring. Flexible cords must be connected to devices and to fittings so as to prevent tension at joints and terminal screws. Because a cord is exposed, flexible, and unsecured, joints and terminals become more vulnerable. Flexible cord conductors are finely stranded for flexibility, but the strands of one conductor may loosen from under terminal screws and touch another conductor, especially if the cord is subjected to stress or strain.

A flexible cord may be damaged by activities on the job, by door or window edges, by staples or fastenings, by abrasion from adjacent materials, or simply by aging. If the electrical conductors become exposed, there is a danger of shocks, burns, or fire. A frequent hazard on a construction site is a cord assembly with improperly connected terminals.

When a cord connector is wet, hazardous leakage can occur to the equipment grounding conductor and to humans who pick up that connector if they also provide a path to ground. Such leakage is not limited to the face of the connector but also develops at any wetted portion of it.

When the leakage current of tools is below 1 ampere, and the grounding conductor has a low resistance, no shock should be perceived. However, should the resistance of the equipment grounding conductor increase, the current through the body also will increase. Thus, if the resistance of the equipment grounding conductor is significantly greater than 1 ohm, tools with even small leakages become hazardous.

PREVENTING AND ELIMINATING HAZARDS

GFCIs can be used successfully to reduce electrical hazards on construction sites. Tripping of GFCIs--interruption of current flow--is sometimes caused by wet connectors and tools. It is good practice to limit exposure of connectors and tools to excessive moisture by using watertight or sealable connectors. Providing more GFCIs or shorter circuits can prevent tripping caused by the cumulative leakage from several tools or by leakages from extremely long circuits.

EMPLOYER'S RESPONSIBILITY

OSHA ground-fault protection rules and regulations have been determined necessary and appropriate for employee safety and health. Therefore, it is the employer's responsibility to provide either: (a) ground-fault circuit interrupters on construction sites for receptacle outlets in use and not part of the permanent wiring of the building or structure; or (b) a scheduled and recorded assured equipment grounding conductor program on construction sites, covering all cord sets, receptacles which are not part of the permanent wiring of the building or structure, and equipment connected by cord and plug which are available for use or used by employees.

GROUND-FAULT CIRCUIT INTERRUPTERS

The employer is required to provide approved ground-fault circuit interrupters for all 120-volt, single-phase, 15- and 20-ampere receptacle outlets on construction sites which are not a part of the permanent wiring of the building or structure and which are in use by employees. Receptacles on the ends of extension cords are not part of the permanent wiring and, therefore, must be protected by GFCIs whether or not the extension cord is plugged into permanent wiring. These GFCIs monitor the current-to-the-load for leakage to ground. When this leakage exceeds 5 mA ± 1 mA, the GFCI interrupts the current. They are rated to trip quickly enough to prevent electrocution. This protection is required in addition to, not as a substitute for, the grounding requirements of OSHA safety and health rules and regulations, 29 CFR 1926. The requirements which employers must meet, if they choose the GFCI option, are stated in 29 CFR 1926.404(b)(1)(ii). (See appendix.)

ASSURED EQUIPMENT GROUNDING CONDUCTOR PROGRAM

The assured equipment grounding conductor program covers all cord sets, receptacles which are not a part of the permanent wiring of the building or structure, and equipment connected by cord and plug which are available for use or used by employees. The requirements which the program must meet are stated in 29 CFR 1926.404(b)(1)(iii), but employers may provide additional tests or procedures. (See appendix.) OSHA requires that a written description of the employer's assured equipment grounding conductor program, including the specific procedures adopted, be kept at the jobsite. This program should outline the employer's specific procedures for the required equipment inspections, tests, and test schedule.

The required tests must be recorded, and the record maintained until replaced by a more current record. The written program description and the recorded tests must be made available, at the jobsite, to OSHA and to any affected employee upon request. The employer is required to designate one or more **competent persons** to implement the program.

Electrical equipment noted in the assured equipment grounding conductor program must be visually inspected for damage or defects before each day's use. Any damaged or defective equipment must not be used by the employee until repaired.

Two tests are required by OSHA. One is a continuity test to ensure that the equipment grounding conductor is electrically continuous. It must be performed on all cord sets, receptacles which are not part of the permanent wiring of the building or structure, and on cord- and plug-connected equipment which is required to be grounded. This test may be performed using a simple continuity tester, such as a lamp and battery, a bell and battery, an ohmmeter, or a receptacle tester.

The other test must be performed on receptacles and plugs to ensure that the equipment grounding conductor is connected to its proper terminal. This test can be performed with the same equipment used in the first test.

These tests are required before first use, after any repairs, after damage is suspected to have occurred, and at 3-month intervals. Cord sets and receptacles which are essentially fixed and not exposed to damage must be tested at regular intervals not to exceed 6 months. Any equipment which fails to pass the required tests shall not be made available or used by employees.

SUMMARY

This discussion provides information to help guide employers and employees in protecting themselves against 120-volt electrical hazards on the construction site, through the use of ground-fault circuit interrupters or through an assured equipment grounding conductor program.

When planning your program, remember to use the OSHA rules and regulations as a guide to ensure employee safety and health. Following these rules and regulations will help reduce the number of injuries and accidents from electrical hazards. Work disruptions should be minor, and the necessary inspections and maintenance should require little time.

An effective safety and health program requires the cooperation of both the employer and employees.

If you need additional information planning your program, contact the OSHA office nearest you.

CONSTRUCTION SAFETY AND HEALTH OUTREACH PROGRAM

U.S. Department of Labor
OSHA Office of Training and Education

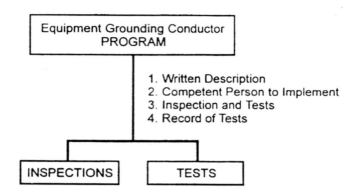

Visual inspection of following:

1. cord sets
2. cap, plug and receptacle of cord sets
3. equipment connected by cord and plug

Exceptions:

- receptacles and cord sets which are fixed and not exposed to damage

Frequency of Inspections:

- before each day's use

Conduct tests for:

1. continuity of equipment grounding conductor
2. proper terminal connection of equipment grounding conductor

Frequency of Tests:

- before first use
- after repair, and before placing back in service
- before use, after suspected damage
- every 3 months, except that cord sets and receptacles that are fixed and not exposed to damage must be tested at regular intervals not to exceed 6 months.

APPENDIX

Construction Safety and Health Regulations Part 1926 Subpart K (Partial)

§1926.404 Wiring design and protection.

(b) Branch circuits--(1) Ground-fault protection--(i) General.
The employer shall use either ground-fault circuit interrupters as specified in paragraph (b)(l)(ii) of this section or an assured equipment grounding conductor program as specified in paragraph (b)(l)(iii) of this section to protect employees on construction sites. These requirements are in addition to any other requirements for equipment grounding conductors.

(ii) Ground-fault circuit interrupters. All 120-volt, single-phase, 15- and 20-ampere receptacle outlets on construction sites, which are not a part of the permanent wiring of the building or structure and which are in use by employees, shall have approved ground-fault circuit interrupters for personnel protection. Receptacles on a two-wire, single-phase portable or vehicle-mounted generator rated not more than 5kW, where the circuit conductors of the generator are insulated from the generator frame and all other grounded surfaces, need not be protected with ground-fault circuit interrupters.

(iii) Assured equipment grounding conductor program. The employer shall establish and implement an assured equipment grounding conductor program on construction sites covering cord sets, receptacles which are not a part of the building or structure, and equipment connected by cord and plug which are available for use or used by employees. This program shall comply with the following minimum requirements:

(A) A written description of the program, including the specific procedures adopted by the employer, shall be available at the jobsite for inspection and copying by the Assistant Secretary and any affected employee.

(B) The employer shall designate one or more competent persons [as defined in §1926.32(f)] to implement the program.

(C) Each cord set, attachment cap, plug and receptacle of cord sets, and any equipment connected by cord and plug, except cord sets and receptacles which are fixed and not exposed to damage, shall be visually inspected before each day's use for external defects, such as deformed or missing pins or insulation damage, and for indications of possible internal damage. Equipment found damaged or defective shall not be used until repaired.

(D) The following tests shall be performed on all cord sets, receptacles which are not a part of the permanent wiring of the building or structure, and cord- and plug-connected equipment required to be grounded:

(1) All equipment grounding conductors shall be tested for continuity and shall be electrically continuous.

(2) Each receptacle and attachment cap or plug shall be tested for correct attachment of the equipment grounding conductor. The equipment grounding conductor shall be connected to its proper terminal.

(E) All required tests shall be performed:

(1) Before first use;

(2) Before equipment is returned to service following any repairs;

(3) Before equipment is used after any incident which can be reasonably suspected to have caused damage (for example, when a cord set is run over); and

(4) At intervals not to exceed 3 months, except that cord sets and receptacles which are fixed and not exposed to damage shall be tested at intervals not exceeding 6 months.

(F) The employer shall not make available or permit the use by employees of any equipment which has not met the requirements of this paragraph (b)(l)(iii) of this section.

(G) Tests performed as required in this paragraph shall be recorded. This test record shall identify each receptacle, cord set, and cord- and plug-connected equipment that passed the test and shall indicate the last date it was tested or the interval for which it was tested. This record shall be kept by means of logs, color coding, or other effective means and shall be maintained until replaced by a more current record. The record shall be made available on the jobsite for inspection by the Assistant Secretary and any affected employee.

NCCER CRAFT TRAINING USER UPDATES

The NCCER makes every effort to keep these manuals up-to-date and free of technical errors. We appreciate your help in this process. If you have an idea for improving this manual, or if you find an error, a typographical mistake, or an inaccuracy in the NCCER's Craft Training Manuals, please write us, using this form or a photocopy. Be sure to include the exact module number, page number, a description of the problem, and the correction, if possible. Your input will be brought to the attention of the Technical Review Committee. Thank you for your assistance.

Instructors – If you found that additional materials were necessary in order to teach this module effectively, please let us know so that we may include them in the Equipment/Materials list in the Instructor's Guide.

Write: Curriculum Revision and Development Department
National Center for Construction Education and Research
P.O. Box 141104
Gainesville, FL 32614-1104
Fax: 352-334-0932

Craft _____ Module Name _____

Copyright Date _____ Module Number _____ Page Number(s) _____

Description of Problem
_____
_____
_____
_____

(Optional) Correction of Problem
_____
_____
_____

(Optional) Your Name and Address
_____
_____
_____

Low-Voltage Cabling

Module 33108

Electronic Systems Technician Trainee Task Module 33108

NATIONAL
CENTER FOR
CONSTRUCTION
EDUCATION AND
RESEARCH

LOW-VOLTAGE CABLING

OBJECTIVES

Upon completion of this module, the trainee will be able to:

1. Explain the various sizes and gauges of wire in accordance with the American Wire Gauge (AWG) standards and determine the proper gauge for an application.
2. Read and identify markings on conductors and cables.
3. Describe the different materials from which conductors are made.
4. Describe the different types of conductor insulation.
5. Describe the color coding of insulation.
6. Identify selected NEC low-voltage cable classifications.
7. Plan and set up for a cable pull.
8. Properly install a pull line for a cable pulling operation.
9. Prepare the ends of conductors for pulling.
10. Safely pull cable through conduit in vertical and horizontal pathways.
11. Wrap, tie, fasten, label, and protect cable, and explain the importance of maintaining the proper slack.
12. Describe the installation of cables in cable trays.
13. Describe and/or demonstrate a residential low-voltage cable installation.
14. State the restrictions imposed by the NEC on the uses of various types of cable.

Prerequisites

Successful completion of the following Task Modules is recommended before beginning study of this Task Module: Core Curricula; Electronic Systems Technician Level One, Modules 33101 through 33107.

Required Trainee Materials

1. Trainee Task Module
2. Appropriate Personal Protective Equipment
3. Copy of the Latest Edition of the *National Electrical Code*

Copyright © 1999 National Center for Construction Education and Research, Gainesville, FL 32614-1104. All rights reserved. No part of this work may be reproduced in any form or by any means, including photocopying, without written permission of the publisher.

Note: The designations "National Electrical Code," "NE Code," and "NEC," where used in this document, refer to the *National Electrical Code®*, which is a registered trademark of the National Fire Protection Association, Quincy, MA. *All National Electrical Code (NEC) references in this module refer to the 1999 edition of the NEC.*

COURSE MAP

This course map shows all of the modules in the first level of the Electronic Systems Technician curricula. The suggested training order begins at the bottom and proceeds up. Skill levels increase as a trainee advances on the course map. The training order may be adjusted by the local Training Program Sponsor.

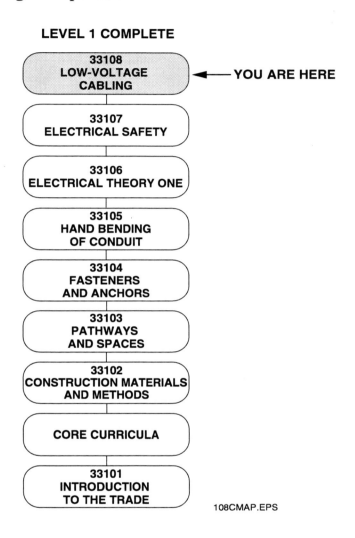

TABLE OF CONTENTS

Section	Topic	Page
	Course Map	3
1.0.0	Introduction	8
2.0.0	Low-Voltage Cable Conductors And Insulation	9
2.1.0	Conductor Wire Size	9
2.2.0	Conductor Material	10
2.3.0	Insulation	10
2.4.0	Conductor Voltage Drop	11
3.0.0	Optical Fiber Cable Signal Conductor And Sheathing	13
4.0.0	Low-Voltage And Optical Fiber Cables	15
4.1.0	NEC Classifications And Ratings	15
4.2.0	PTLC, Fire Alarm, And Class 2/3 Cable Styles And Construction	24
4.3.0	Communication Cable Styles And Construction	24
4.3.1	Unshielded Twisted-Pair Cable (UTP)	26
4.3.2	Unshielded Twisted-Pair Patch Cords	28
4.3.3	Undercarpet Telecommunication Cable (UTC)	28
4.3.4	Screened Twisted-Pair (ScTP) Cable and Patch Cord	29
4.3.5	Shielded Twisted-Pair (STP) Cable, Enhanced Shielded Twisted-Pair (STP-A) Cable, And STP Patch Cord	30
4.3.6	Coaxial Cable	30
4.3.7	Optical Multi-Fiber Cable	32
5.0.0	Commercial Low-Voltage Cable Installation	34
5.1.0	Planning The Installation	36
5.1.1	Cabling Pathways	38
5.1.2	Pulling Location	40
5.1.3	Pathway Cable Pull Operations	41
5.2.0	Setting Up For Cable Pulling	42
5.2.1	Setting Up The Cable Reels Or Boxes	43
5.2.2	Preparing Conduit Pathways For Cables	45
5.2.3	Installing A Pull Line In Conduit Or Inner Duct	47
5.2.4	Installing A Pull Line In Open Ceilings	49
5.2.5	Preparing Cable Ends For Pulling	49
5.2.6	Types Of Pulling Lines	52
5.3.0	Cable Pulling Equipment	52
5.3.1	Pulling Safety	55
5.4.0	Vertical And Horizontal Pathway Cable Pulls	55
5.4.1	Vertical Backbone Cable Pulls From The Top Down	56
5.4.2	Vertical Backbone Cable Pulls From The Bottom Up	59
5.4.3	Horizontal Backbone Cable Pulls	62
5.4.4	Optical Fiber Backbone Cable Pulls	67

TABLE OF CONTENTS (Continued)

Section	Topic	Page
5.4.5	Horizontal Work Area Cable Pulls	69
5.4.6	Conduit Fill For Backbone Cable	71
6.0.0	Residential Low-Voltage Cable Installation	72
6.1.0	Residential Unit Communication/Data Cabling Requirements And Grades	72
6.1.1	Residential Unit Grade 1 Service Cabling	73
6.1.2	Residential Unit Grade 2 Service Cabling	73
6.1.3	Residential Unit Communication/Data Cable Types	73
6.2.0	Understanding The Job	73
6.3.0	Residential Cable Installation Requirements/Considerations	74
6.4.0	Drilling And Fishing Cable In Existing Construction	77
7.0.0	Interior Low-Voltage Cabling Installation Requirements	78
7.1.0	Class 1 Circuits	78
7.1.1	Conductors Of Different Circuits In The Same Cable, Enclosure, Or Raceway	78
7.2.0	Class 2 and 3 Circuits	79
7.2.1	Separation Of Class 2 And 3 Circuits From Power Circuits	79
7.2.2	Separation In Hoistways	80
7.2.3	Separation In Other Applications	80
7.2.4	Support Of Conductors	80
7.2.5	Installation In Plenums, Risers, Cable Trays, And Hazardous Locations	80
7.3.0	Instrumentation Tray Cable Circuits	80
7.4.0	Nonpower-Limited Fire Alarm Circuits	81
7.4.1	Conductors Of Different Circuits In The Same Cable, Enclosure, Or Raceway	81
7.5.0	Power-Limited Fire Alarm Circuits	82
7.5.1	Separation Of Power-Limited Fire Alarm Circuits From Power Circuits	82
7.5.2	Separation In Hoistways	83
7.5.3	Separation In Other Applications	83
7.5.4	Support Of Conductors	83
7.5.5	Current-Carrying Continuous Line-Type Fire Detectors	83
7.5.6	Plenums And Risers	84
7.6.0	Optical Fiber Cable	84
7.7.0	Hybrid Cable	85
7.8.0	Communication Circuits Within Buildings	85
7.8.1	Separation of Communication Circuits From Power Circuits	85

TABLE OF CONTENTS (Continued)

Section	Topic	Page
7.8.2	Separation In Other Applications	85
7.8.3	Support Of Conductors	86
7.8.4	Cable Trays	86
7.8.5	Plenums And Risers	86
7.8.6	Plenums, Risers, And General-Purpose Raceways	86
7.9.0	Coaxial CATV Cable Installation Within Buildings	86
7.9.1	Separation In Raceways And Boxes	86
7.9.2	Separation In Other Applications	87
7.9.3	Support Of Conductors	87
7.10.0	Network-Powered Broadband Communication System Installation Within Buildings	87
7.10.1	Separation In Raceways And Boxes	87
7.10.2	Separation In Other Applications	88
7.10.3	Support Of Conductors	88
8.0.0	Electromagnetic Interference Considerations	88
8.1.0	EMI Guidelines	89
	Summary	91
	Review/Practice Questions	92
	Answers To Review/Practice Questions	96
	Appendix A	97
	Appendix B	101
	Appendix C	108

Trade Terms Introduced In This Module

Ampacity: The amount of current, in amperes, that a conductor is permitted to carry continuously without exceeding its temperature rating.

Attenuation: A decrease in the magnitude of a signal. It represents a loss of signal power between two points and is measured in decibels (dB).

Basket grip: A flexible steel mesh grip used on the ends of cable and conductors for attaching the pulling rope. The more force exerted by the pull, the tighter the grip wraps around the cable.

Capstan: The turning drum of a cable puller on which the rope is wrapped and pulled. An increase in the number of wraps increases the pulling force of the cable puller.

CATV: Abbreviation for community access television.

Clevis: A device used in cable pulls to facilitate connecting the pulling rope to the cable grip.

Conduit piston: A foam cylinder that fits inside the conduit and is then propelled by compressed air or vacuumed through the conduit run to pull a line, rope, or measuring tape. Also called a *mouse*.

Cross-connect: A facility for connecting cable runs, subsystems, and equipment using patch cords or jumper cables.

Crosstalk: Unwanted coupling of signals from circuit to circuit.

Drain wires: Wires running parallel to and in contact with a foil cable shield. They are used to connect the shield to a grounding point.

Electromagnetic interference (EMI): An electrical phenomenon in which undesirable random electrical energy from a source such as adjacent wiring, equipment, etc. is induced or picked up by wiring and subsequently causes problems with equipment connected to the wiring. Also known as *radio frequency interference (RFI), noise,* and *radio interference*.

Fish line: A light cord used in conjunction with vacuum/blower power fishing systems that attaches to the conduit piston to be pushed or pulled through the conduit. Once through, a pulling rope is attached to one end and pulled back through the conduit for use in pulling conductors.

Fish tape: A flat iron wire or fiber cord used to pull conductors or a pulling rope through conduit.

Head end: The central distribution and/or control equipment that is used in residential installations.

Impedance: A measurement of the opposition to the flow of alternating current (AC).

Messenger: A strong support member such as a steel strand used to carry the weight of cables and wiring.

Multiplex: Combining two or more signals into a single wave from which either of the two original signals can be recovered.

Nonpower-limited circuit: A circuit that complies with *NEC Chapters 1 through 4* and is less than 600V.

Patch cord: A connecting cable used with cross-connect equipment and work area cables.

Power-limited circuit: A circuit that has a power source limited to 1,000VA or that is self-limiting to 1,000VA or less.

Sheave: A pulley-like device used in cable pulls in both conduit and cable tray systems.

Snub: Wrapping a rope around a post or other secure object to check the movement of the rope.

Soap: Slang for wire-pulling lubricant.

Thicknet (10 base 5): A specification for a network using a thick 50Ω coaxial cable.

Thinnet (10 base 2): A specification for a network using a thin 50Ω coaxial cable.

Token ring: A network topology in which a token must be passed to a terminal or workstation by the network controller before it can transmit.

Tracer: A strip or band of color that is different than the insulation color on a wire. The band or strip is used for identification purposes. In older wire, the tracer was a colored thread encased by the insulation.

Twisted-pair cable: A multi-conductor cable consisting of two or more copper conductors twisted in a manner designed to cancel electrical interference.

1.0.0 INTRODUCTION

As an electronic systems technician, you will be required to pull various types of cable through conduit and wireways, or run it through walls, over ceilings, and under floors in order to terminate it at desired locations for a particular job. In some instances, you may be required to select the proper type of cabling to be used for the job. This module covers the general procedures for installing cable in residential and commercial structures. It also covers the various types and ratings of common low-voltage cable, as well as fiber optic cable used for nonpower-limited or power-limited circuits. Materials and procedures are given for remote control, signaling, fire alarm, and communication circuits.

Manufacturers produce a wide variety of cables that meet the requirements for various types of installations. The cable construction and rating of each manufacturer may differ, even though the cables may meet the classification type established by the NEC or other standards. The manufacturer's specifications and installation recommendations must always be followed when selecting or installing cable.

2.0.0 LOW-VOLTAGE CABLE CONDUCTORS AND INSULATION

The term *conductor* is used two ways. It is usually used to describe individual insulation-covered wires within a cable. It is also used to define the current-carrying material of a wire, either insulated or uninsulated. In this module, it will be used to describe individual insulation-covered wires, unless otherwise stated. Conductors are identified by size and insulation material. The size refers to the physical American Wire Gauge (AWG) size of the current-carrying wire of the conductor.

2.1.0 CONDUCTOR WIRE SIZE

The AWG system uses numbers to identify different sizes of drawn, solid wire. The numbers represent the number of progressively smaller dies that the wire must be drawn through to reach a desired size. The larger the number, the smaller the cross-sectional area of the wire (*Figure 1*). The smaller the cross-sectional area, the smaller the current that the wire can conduct. The AWG numbers range from 50 (smallest) to 1; then 0, 00, 000, and 0000 (largest). Low-voltage cable typically contains conductors with solid wire sizes ranging anywhere from No. 26 to No. 12 AWG. Stranded wire of the same gauge as solid wire is somewhat larger in diameter, but has the same cross-sectional conducting material area as its solid counterpart. Unfortunately, its DC resistance per foot is higher and at high data transfer frequencies, stranding can cause up to 20% more signal **attenuation**. For conductor wire sizes larger than No. 16 AWG, the size is usually marked on the insulation. For smaller size conductors, a standard wire gauge (*Figure 2*) can be used to check the size of the solid wires.

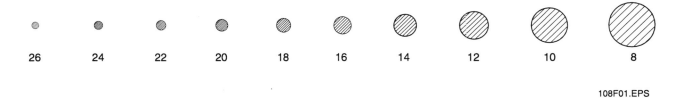

108F01.EPS

Figure 1. Wire Sizes (Enlarged) Showing Increasing Wire Diameter Versus Decreasing AWG Numbers

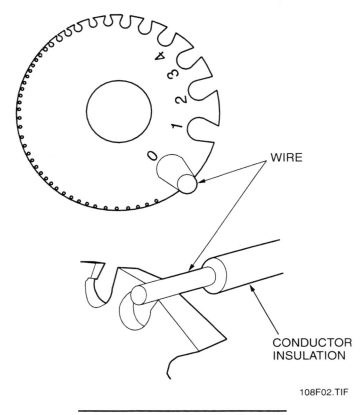

Figure 2. Standard Wire Gauge

2.2.0 CONDUCTOR MATERIAL

The standard material used for the current-carrying wire of conductors used in low-voltage cable is annealed (soft) solid copper. In some cases, copper-coated (copper-clad) steel or copper alloys such as cadmium-chromium copper or zirconium copper are used to increase the strength of the conductors. Unfortunately, copper-clad steel or copper alloys have inferior current-carrying ability.

2.3.0 INSULATION

Thermoplastics are widely used for conductor insulation materials and cable jackets. Some of the most popular and effective types are listed below. Multiple-wire cable should use the same insulation types on all wires, otherwise signal propagation speed on various pairs of wires may be an issue in high-speed applications.

- *Polyvinyl chloride (PVC)* – The most common material used for conductor insulation and cable jackets. Available as plenum or non-plenum rated.
- *Polyethylene (PE)* – An excellent weatherproofing material used primarily for insulation of control and communication wiring and for underground cabling.
- *Fluorinated ethylene propylene (FEP) or ethylene chlorotrifluoroethylene (ECTFE)* – Provides better smoke and flame resistance and improved transmission performance for high-performance cable.

- *Cross-linked polyethylene (XLP)* – An improved PE with superior heat and moisture resistance.
- *Nylon* – Primarily used as cable jacketing material.
- *Teflon™* – A high-temperature insulation that is widely used for communication wiring in plenum-rated cable.
- *Rubber compounds* – Typically used for outdoor cabling.

2.4.0 CONDUCTOR VOLTAGE DROP

In installations where cabling or wiring for fire alarm circuits or other types of control or communications circuits may be required to supply current over long distances, excessive voltage drop may occur due to the conductor sizes selected. Normally, the maximum voltage drop allowable at the load end of the wiring run should be no more than 3% of the source voltage. To check the voltage drop, either one of the following equations may be used to determine the DC or single-phase AC voltage drop for a specific size of copper conductor:

$$VD = \frac{2 \times L \times K \times I}{CM} \quad \text{or} \quad VD = \frac{2 \times L \times R \times I}{1,000}$$

Where:

VD = voltage drop

L = one way length of wiring run

K = constant for copper (12.9)

I = current in amperes required by the load at the rated source voltage

CM = cross-sectional area of the conductor in circular mils

R = DC resistance of the conductor per 1,000'

For conductors that are No. 18 AWG or larger, the circular mil area or DC resistance of the conductor can be found in tables included in the NEC.

For selecting conductors with light loads over long distances at low voltages, load current wire selection tables similar to *Tables 1* and *2* are very common and can save calculation time. The numbers within the 12V or 24V tables at the intersection of various wire lengths and current requirements represent the AWG conductor size that can support the current load with no more than a 3% voltage drop.

For example, to select a wire size adequate for a 24V control circuit to a load requiring 1A at 150', refer to *Table 2* and scan down the 1A column to the 150' row. At the intersection of the column and row, note that a No. 16 AWG wire is specified for the stated distance and load.

Total One Way Length of Wire Run	¼A	½A	¾A	1A	1¼A	1½A	2A	3A
100'	20	18	16	14	14	12	12	10
150'	18	16	14	12	12	12	10	–
200'	16	14	12	12	10	10	–	–
250'	16	14	12	10	10	10	–	–
300'	16	12	12	10	10	–	–	–
400'	14	12	10	–	–	–	–	–
500'	14	10	10	–	–	–	–	–
750'	12	10	–	–	–	–	–	–
1,000'	10	–	–	–	–	–	–	–
1,500'	10	–	–	–	–	–	–	–

Table 1. Wire Size Selection For Load Current At 12V

Total One Way Length of Wire Run	¼A	½A	¾A	1A	1¼A	1½A	2A	3A
100'	24	20	18	18	16	16	14	12
150'	22	18	16	16	14	14	12	10
200'	20	18	16	14	14	12	12	10
250'	18	16	14	14	12	12	12	10
300'	18	16	14	12	12	12	10	–
400'	18	14	12	12	10	10	–	–
500'	16	14	12	10	10	–	–	–
750'	14	12	10	10	–	–	–	–
1,000'	14	10	10	–	–	–	–	–
1,500'	12	10	–	–	–	–	–	–

Table 2. Wire Size Selection For Load Current At 24V

3.0.0 OPTICAL FIBER CABLE SIGNAL CONDUCTOR AND SHEATHING

Optical fiber cable is used to conduct a light signal instead of an electrical current. *Figure 3* shows a typical simplex (single) optical fiber cable with its fiber optic light conductor and sheathing. The components shown are described below.

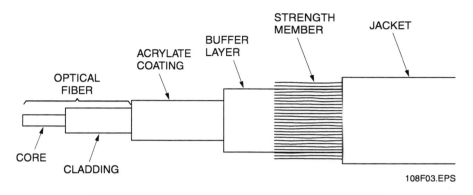

Figure 3. Typical Simplex Optical Fiber Cable

- *Jacket* – An outer layer that may be constructed of various types of materials, from flexible metal to soft PVC. The jacket coloring usually identifies one of two types of fiber used inside.

- *Strength member* – Normally, these are nonmetallic fiber strands under the jacket that provide strength and flexibility to the cable. They are usually made of aramid (Kevlar®) yarn to which certain types of connectors can be attached. They can also be used to pull the cable into position at installation.

- *Buffer layer* – An intermediate layer that is usually 900 microns in diameter. A micron is one millionth of a meter and is also known a *micrometer* (µm). This layer can be one of two types, tight or loose. For indoor cable, the layer is a soft plastic that is tight against the layer below it and is known as a *tight buffer*. For outdoor cable, the layer consists of a plastic tube filled with a gel (called a *loose-tube buffer*) that protects the optical fiber from impact damage, damage from temperature extremes, or water infiltration damage. Either type of buffer contributes to the strength and flexibility of the cable.

- *Acrylate coating* – This layer is a clear coating, also known as the *acrylate buffer*, that is about 250 microns in diameter. It provides strength and flexibility to the optical fiber and improves the handling of the fiber.

- *Cladding* – The cladding and the core make up the optical fiber that transmits the light signal. The cladding surrounds the core and is made of glass or plastic that is purer than the core glass. The difference in purity between the core and cladding glass or plastic results in a reflective boundary where the core and cladding meet. This reflective surface keeps the light signals traveling in the core from escaping the core.

- *Core* – This part of the fiber is made of very pure glass or plastic and carries the information-modulated light signal(s) being transmitted through the fiber. Light travels through the core as a result of total internal reflection. As the light travels through the core, it is reflected off the boundary between the core and cladding at a shallow angle that allows the light to continue through the core. Light travels in the core on a path called a *mode*. The number of modes depends on the core diameter. Fiber is classified as either *single-mode* or *multi-mode* fiber. In any installation where optical fiber cables will be joined together, it is essential that cables with the same size core and cladding be used. Joining cables with different cores or cladding will result in unacceptable signal losses.
  - Single-mode core fiber *(Figure 4)* has a core diameter of 8 to 9 microns and carries one light wave at a time. The cladding diameter is usually 125 microns. Cable with this type of fiber is usually used for long-distance transmission outdoors up to 9,840' (3,000m) due to its low loss. The light source for this type of fiber is usually a laser-emitting wavelength of 1,310ηm (nanometers) and 1,550ηm. The jacket of single-mode core cable is usually yellow.

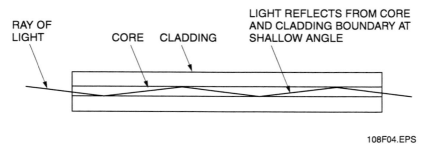

Figure 4. Single-Mode Core Fiber

  - Multi-mode core fiber *(Figure 5)* is required to transmit many modes of light at one time down the fiber and, as a result, has a much bigger core than single-mode core cable. The two most common core diameters are 50 microns or 62.5 microns with a cladding diameter of 125 microns. Cables with these fibers are called *50/125µm* or *62.5/125µm* cables. Core diameters at 100 microns with 140 micron cladding, 200 microns with 230 micron cladding, and others are also available. Multi-mode core cable is typically used only for short distances (under 2 kilometers) because the losses in this type of cable are higher than single-mode core cable. The light source is usually light-emitting diodes (LEDs).

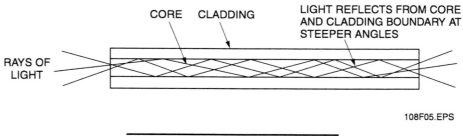

Figure 5. Multi-Mode Core Fiber

4.0.0 LOW-VOLTAGE AND OPTICAL FIBER CABLES

It is necessary to select the proper type and rating of low-voltage and optical fiber cables to meet NEC safety requirements, local codes, and equipment manufacturer requirements for the system being installed. With the exception of optical fiber or communication cables that enter a structure and are terminated in an enclosure within 50' of the entrance, the NEC requires that low-voltage and optical fiber cables used for installation be listed and marked with an appropriate classification code. Various styles of cables are available that meet the various classification codes. The NEC classification codes, along with a description of the typical cable styles available, are provided in the following paragraphs.

4.1.0 NEC CLASSIFICATIONS AND RATINGS

Low-voltage and optical fiber cables are type-classified, rated, and listed for use in various areas of a structure in accordance with the following:

- *NEC Article 725*, Remote Control, Signaling, and Power-Limited Circuits (Class 1, 2, and 3 Circuits)
- *NEC Article 727*, Instrumentation Tray Cable
- *NEC Article 760*, Fire Alarm Systems
- *NEC Article 770*, Optical Fiber Cables And Raceways
- *NEC Article 780*, Closed-Loop And Programmed Power Distribution
- *NEC Article 800*, Communications Circuits
- *NEC Article 820*, Community Antenna Television and Radio Distribution Systems
- *NEC Article 830*, Network-Powered Broadband Communications Systems

These classifications and ratings are summarized in *Table 3*. All cable conforming to the requirements of the NEC is normally marked by the cable manufacturer with the appropriate classification code. Cable types listed for each article are arranged in ascending order of fire resistance. This means that any cable within the listing for a NEC article may be substituted for any cables listed prior to it in the table. For more detail, refer to the latest edition of the NEC.

Note: *NEC Article 310* covers single-conductor power wire used for installation of Class 1 circuits under *NEC Article 725* and nonpower-limited fire alarm circuits under *NEC Article 760*. Both types of circuits shall be installed in accordance with *NEC Chapter 3*. Insulation of single-conductor power wire shall be rated at 600V and shall comply with *NEC Article 310* for sizes larger than No. 16 AWG. For No. 18 and No. 16 AWG, the insulation shall be Type KF-2, KFF-2, PAFF, PF, PFF, PGF, PGFF, PTFF, RFH-2, RFHH-2, RFHH-3, SF-2, SFF-2, TF, TFF, TFFN, TFN, ZF, or ZFF for nonpower-limited fire alarm or Class 1 single-conductor wire. In addition, insulation Type FFH-2, PAF, or PTF is allowed for Class 1 single-conductor wire.

NEC Reference	Cable Type Marking	Listed Voltage Rating	Listed Usage
NEC Article 725	CL2 or CL3	300V	Power-limited Class 2 or 3 cable for general-purpose use except in environmental plenums, risers, and other environmental air spaces. Resistant to the spread of fire.
NEC Article 725	CL2X or CL3X	300V	Power-limited Class 2 or 3 limited-use cable for installation only in dwellings and raceways. Flame-retardant.
NEC Article 725	CL2R or CL3R	300V	Power-limited Class 2 or 3 riser cable suitable for use in vertical shafts or in runs from floor to floor. Resistant to fire to the extent that flames will not be carried from floor to floor.
NEC Article 725	CL2P or CL3P	300V	Power-limited Class 2 or 3 plenum cable suitable for use in environmental plenums and other environmental air spaces. Has good fire resistance and low smoke-producing characteristics.
NEC Article 725	PLTC	300V	Power-limited tray cable (PLTC) for use in cable trays. Consists of two or more insulated conductors in a nonmetallic jacket that may be covered with a metallic jacket and an additional nonmetallic jacket. Individual insulated conductors can range from No. 22 to No. 12 AWG. Conductor material may be solid or stranded copper. Conductors can be parallel or twisted or a combination of both. They can be individually shielded or shielded as one or more groups or as a combination of both. Class 2 thermocouple extension wire may also be included. Cable resists flame spread, sunlight, and moisture. Can be used in hazardous locations where permitted, direct buried if marked for that use, and as open limited-length wiring in certain industrial applications.
NEC Article 727	(Per **NEC Section 310-11**)	300V	Instrumentation tray cable (ITC) that can be used only in certain industrial applications. Consists of two or more insulated copper and/or thermocouple alloy conductors covered with a nonmetallic sheath or armor. Can be installed in cable trays, raceways, certain hazardous locations, as aerial cable on a messenger, direct buried if marked for that use, and under raised floors if protected. Conductors can range from No. 22 to No. 12 AWG. Shielding is permitted. Cannot be used on circuits operating at more than 150V or more than 5A; No. 22 AWG is limited to 3A.

Table 3. Cable Classifications, Ratings, And Usages (1 Of 7)

NEC Reference	Cable Type Marking	Listed Voltage Rating	Listed Usage
NEC Article 760	NPLF	600V	Nonpower-limited fire alarm cable suitable for general-purpose use except in plenums, risers, and other environmental air spaces. Resistant to the spread of fire. Insulated solid or stranded copper multi-conductors are No. 18 AWG or larger.
NEC Article 760	NPLFR	600V	Nonpower-limited fire alarm riser cable suitable for use in vertical shafts or in runs from floor to floor. Resistant to fire to the extent that flames will not be carried from floor to floor. Insulated solid or stranded copper multi-conductors are No. 18 AWG or larger.
NEC Article 760	NPLFP	600V	Nonpower-limited fire alarm plenum cable suitable for use in plenums and other environmental air spaces. Has good fire resistance and low smoke-producing characteristics. Insulated solid or stranded copper multi-conductors are No. 18 AWG or larger.
NEC Article 760	FPL-CI FPLR-CI FPLP-CI PLTC	—	Circuit integrity (CI) marked cable is nonpower-limited fire alarm, fire alarm riser, or fire alarm plenum cable rated for survivability for a specified length of time under fire conditions and used for critical fire alarm signal and/or control circuits.
NEC Article 760	FPL	300V	Power-limited fire alarm cable and/or insulated, continuous, line-type fire detector cable suitable for general-purpose use except in plenums, risers, and other environmental air spaces. Resistant to the spread of fire. Insulated solid or stranded copper multi-conductors are No. 26 AWG or larger. Coaxial cables with 30% conductivity, copper-clad center conductors are also permitted.
NEC Article 760	FPLR	300V	Power-limited fire alarm riser cable and/or insulated, continuous, line-type fire detector cable suitable for use in vertical shafts or in runs from floor to floor. Resistant to fire to the extent that flames will not be carried from floor to floor. Insulated solid or stranded copper multi-conductors are No. 26 AWG or larger. Coaxial cables with 30% conductivity, copper-clad center conductors are also permitted.

Table 3. Cable Classifications, Ratings, And Usages (2 Of 7)

NEC Reference	Cable Type Marking	Listed Voltage Rating	Listed Usage
NEC Article 760	FPLP	300V	Power-limited fire alarm plenum cable and/or insulated, continuous, line-type fire detector cable suitable for use in plenums and other environmental air spaces. Has good fire resistance and low smoke-producing characteristics. Insulated solid or stranded copper multi-conductors are No. 26 AWG or larger. Coaxial cables with 30% conductivity, copper-clad center conductors are also permitted.
NEC Article 760	FPL-CI FPLR-CI FPLP-CI	300V	Circuit integrity (CI) marked cable is power-limited fire alarm, fire alarm riser, or fire alarm plenum cable or an equivalent line-type fire detector rated for survivability for a specified length of time under fire conditions and used for critical fire alarm signal and/or control circuits.
NEC Article 770	OFN or OFC	—	Optical fiber nonconductive (N) or conductive (C) cable for general-purpose use except in risers and other environmental air spaces. Consists of one or more optical fibers. Conductive cable uses metallic strength members and/or sheathing. Resistant to the spread of fire. Composite cables containing optical fibers and electrical conductors permitted, but must be listed as the appropriate electrical cable type. May also be installed in undercarpet protective carrier devices.
NEC Article 770	OFNG or OFCG	—	Optical fiber nonconductive or conductive cable for general-purpose use except in risers. Consists of one or more optical fibers. Conductive cable uses metallic strength members and/or sheathing. Resistant to the spread of fire. Composite cables containing optical fibers and electrical conductors permitted, but must be listed as the appropriate electrical cable type.
NEC Article 770	OFNR or OFCR	—	Optical fiber nonconductive or conductive riser cable suitable for use in vertical shafts or in runs from floor to floor. Resistant to fire to the extent that flames will not be carried from floor to floor. Consists of one or more optical fibers. Conductive cable uses metallic strength members and/or sheathing. Composite cables containing optical fibers and electrical conductors permitted, but must be listed as the appropriate electrical cable type.

Table 3. Cable Classifications, Ratings, And Usages (3 Of 7)

NEC Reference	Cable Type Marking	Listed Voltage Rating	Listed Usage
NEC Article 770	OFNP or OFCP	—	Optical fiber nonconductive or conductive plenum cable suitable for use in plenums and other environmental air spaces. Has good fire resistance and low smoke-producing characteristics. Consists of one or more optical fibers. Conductive cable uses metallic strength members and/or sheathing. Composite cables containing optical fibers and electrical conductors permitted, but must be listed as the appropriate electrical cable type.
NEC Article 780	Hybrid cable [Per **NEC Section 800-51(i)**]	—	Hybrid cable for closed-loop and programmed power distribution use consisting of power, communications, and/or signaling conductors all in one jacket, with the power conductors separated from the other conductors. An additional optional outer jacket may be used. Individual conductors are marked in accordance with the appropriate listed use. Signaling conductors may be No. 24 AWG or larger and for voltages of 24V or less, currents may not exceed 1A.
NEC Article 800	CMUC	300V	Undercarpet communications wire or cable suitable for use under carpets (for walking weight only). Resistant to flame spread. Insulated solid copper conductors are normally used (typically No. 26 to No. 20 AWG). Typically a flat ribbon cable with twisted pairs. Coaxial cable with copper-clad center conductor is also permitted.
NEC Article 800	CMX	300V	Limited-use communications cable suitable for use in dwellings and raceways. Resistant to flame spread. Insulated solid copper conductors are normally used (typically No. 26 to No. 18 AWG). Coaxial cables with copper-clad center conductors are also permitted.
NEC Article 800	CM	300V	Communications wire and cable for general-purpose use except in plenums, risers, and other environmental air spaces. Resistant to the spread of fire. Insulated solid copper conductors are normally used (typically No. 26 to No. 18 AWG). Coaxial cables with copper-clad center conductors are also permitted.
NEC Article 800	Hybrid cable [Per **NEC Section 800-51(i)**]	600V	Hybrid cable for power and communications use with conductors all in one jacket, with the power conductors separated from the other conductors. Power cables are Type NM or NM-B and communications cables are Type CM. Hybrid cable is resistant to flame spread.

Table 3. Cable Classifications, Ratings, And Usages (4 Of 7)

NEC Reference	Cable Type Marking	Listed Voltage Rating	Listed Usage
NEC Article 800	NP	300V	Multi-purpose communications wire or cable for general use except in plenums, risers, and other environmental air spaces. Resistant to the spread of fire. Insulated solid copper conductors are normally used (typically No. 26 to No. 18 AWG). Coaxial cables with copper-clad center conductors are also permitted.
NEC Article 800	CMG	300V	Communications cable for general-purpose use except in plenums, risers, and other environmental air spaces. Resistant to the spread of fire. Insulated solid copper conductors are normally used (typically No. 26 to No. 18 AWG). Coaxial cables with copper-clad center conductors are also permitted.
NEC Article 800	MPG	300V	Multi-purpose cable for general-purpose communications or power-limited fire alarm use. Resistant to the spread of fire. Insulated solid copper conductors are normally used (typically No. 26 to No. 18 AWG). Coaxial cables with copper-clad center conductors are also permitted.
NEC Article 800	CMR	300V	Communications riser cable suitable for use in vertical shafts or in runs from floor to floor. Resistant to fire to the extent that flames will not be carried from floor to floor. Insulated solid copper conductors are normally used (typically No. 26 to No. 18 AWG). Coaxial cables with copper-clad center conductors are also permitted.
NEC Article 800	MPR	300V	Multi-purpose riser cable for communications or power-limited fire alarm use in vertical shafts or in runs from floor to floor. Resistant to fire to the extent that flames will not be carried from floor to floor. Insulated solid copper conductors are normally used (typically No. 26 to No. 18 AWG). Coaxial cables with copper-clad center conductors are also permitted.
NEC Article 800	MPP	300V	Communications plenum cable suitable for use in plenums and other environmental air spaces. Has good fire resistance and low smoke-producing characteristics. Insulated solid copper conductors are normally used (typically No. 26 to No. 18 AWG). Coaxial cables with copper-clad center conductors are also permitted.

Table 3. Cable Classifications, Ratings, And Usages (5 Of 7)

NEC Reference	Cable Type Marking	Listed Voltage Rating	Listed Usage
NEC Article 800	MPP	300V	Multi-purpose plenum cable for communications or power-limited fire alarm use in environmental plenums and other environmental air spaces. Has good fire resistance and low smoke-producing characteristics. Insulated solid copper conductors are normally used (typically No. 26 to No. 18 AWG). Coaxial cables with copper-clad center conductors are also permitted.
NEC Article 820	CATVX	—	Limited-use community antenna television coaxial cable that is suitable for interior use in dwellings and raceways. Resistant to the spread of fire.
NEC Article 820	CATV	—	Community antenna television coaxial cable that is suitable for general-purpose interior CATV use except in risers and plenums. Resistant to the spread of fire.
NEC Article 820	CATVR	—	Community antenna television coaxial cable suitable for interior use in vertical shafts or in runs from floor to floor. Resistant to fire to the extent that flames will not be carried from floor to floor.
NEC Article 820	CATVP	—	Community antenna television coaxial cable suitable for interior use in plenums and other environmental air spaces. Has good fire resistance and low smoke-producing characteristics.
NEC Article 830	BMU	300V	Medium power, multimedia network cable rated for circuit voltages up to 150V and 250VA. May consist of a jacketed factory-assembled coaxial cable, a combination of coaxial and individual conductors, or a combination of optical fiber cable and multiple conductors. Type BMU is suitable for underground use.
NEC Article 830	BM	300V	Medium power, multimedia network cable rated for circuit voltages up to 150V and 250VA. May consist of a jacketed factory-assembled coaxial cable, a combination of coaxial and individual conductors, or a combination of optical fiber cable and multiple conductors. Type BM is suitable for general-purpose use except in risers and plenums. Resistant to the spread of fire.

Table 3. Cable Classifications, Ratings, And Usages (6 Of 7)

NEC Reference	Cable Type Marking	Listed Voltage Rating	Listed Usage
NEC Article 830	BMR	300V	Medium power, multimedia network cable rated for circuit voltages up to 150V and 250VA. May consist of a jacketed factory-assembled coaxial cable, a combination of coaxial and individual conductors, or a combination of optical fiber cable and multiple conductors. Type BMR is suitable for use in vertical shafts or in runs from floor to floor. Resistant to fire to the extent that flames will not be carried from floor to floor.
NEC Article 830	BLU	300V	Low power, multimedia network cable rated for circuit voltages up to 100V and 250VA. May consist of a jacketed factory-assembled coaxial cable, a combination of coaxial and individual conductors, or a combination of optical fiber cable and multiple conductors. Type BLU is suitable for underground use.
NEC Article 830	BLX	300V	Low power, multimedia network cable rated for circuit voltages up to 100V and 250VA. May consist of a jacketed factory-assembled coaxial cable, a combination of coaxial and individual conductors, or a combination of optical fiber cable and multiple conductors. Type BLX is a limited-use cable suitable for use outside, in dwellings, and in raceways. Resistant to flame spread.
NEC Article 830	BLP	300V	Low power, multimedia network cable rated for circuit voltages up to 100V and 250VA. May consist of a jacketed factory-assembled coaxial cable, a combination of coaxial and individual conductors, or a combination of optical fiber cable and multiple conductors. Type BLP cables are suitable for use in risers, plenums, and other environmental air spaces. Has good fire resistance and low smoke-producing characteristics.

Table 3. Cable Classifications, Ratings, And Usages (7 Of 7)

In addition to the cable substitutions allowed within the various cable types for any one classification, cables for other classifications may also be substituted (*Figures 6, 7,* and *8*).

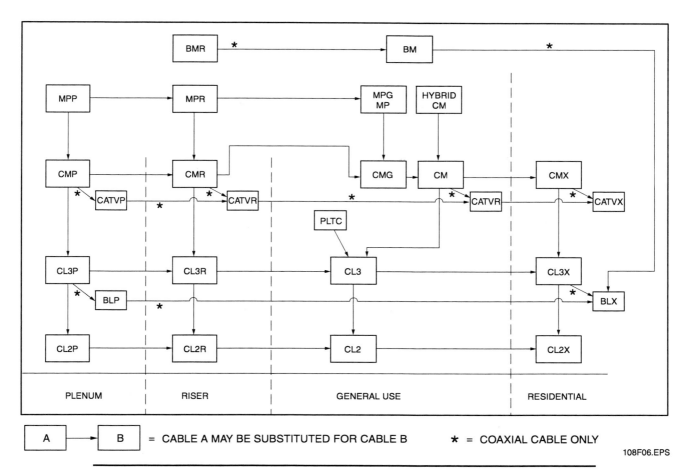

Figure 6. Permitted Communication, Class 2, and Class 3 Cable Substitutions

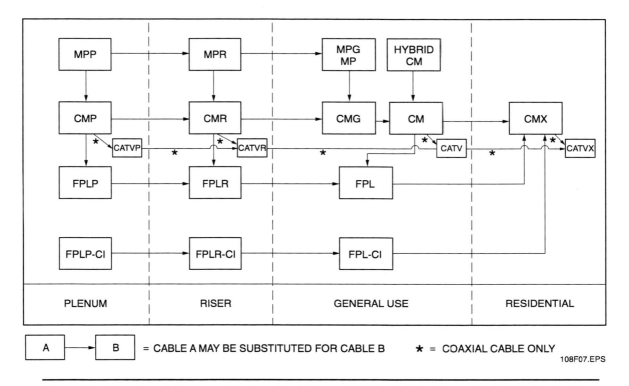

Figure 7. Permitted Communication And Power-Limited Fire Alarm Cable Substitutions

LOW-VOLTAGE CABLING — TRAINEE TASK MODULE 33108

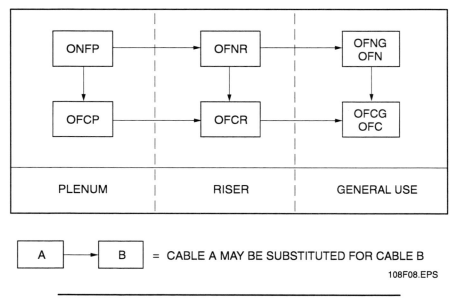

Figure 8. Permitted Optical Fiber Cable Substitutions

4.2.0 PTLC, FIRE ALARM, AND CLASS 2/3 CABLE STYLES AND CONSTRUCTION

Figure 9 depicts several styles available for PLTC, fire alarm, and Class 2/3 cable. Conductor sizes range from No. 12 to No. 24 AWG, depending on the number of conductors in the cable. Conductors can be tinned or bare, solid or stranded, and twisted or parallel. Shielding, if used, is usually an aluminum-coated Mylar® film with one or more **drain wires** (ground wires) inside a jacket that, along with the conductor insulation, is rated for the applicable usage. To be effective, the shielding drain wire(s) must be grounded. The installation loop diagrams or instructions for the system equipment will indicate if all cabling shields are connected together and grounded at only one point, or whether each cable is grounded at both ends or only one end. If only one end is grounded, the drain wire at the other end is folded back and taped to the cable. This is called *floating the ground*. If problems ever develop with the drain wire at the other end, the taped drain wire can be untaped and used. Color coding of the conductors is generally dictated by the manufacturer. The outer jackets of fire alarm cable supplied from some manufacturers may be colored red to identify the cable as a fire alarm circuit cable.

4.3.0 COMMUNICATION CABLE STYLES AND CONSTRUCTION

Some of the communication cables described in the following paragraphs are available with jackets/conductors that are labeled by the manufacturer with feet and/or inch markings to help with the installation of the cable/conductor (ANSI/TIA/EIA-56A and EIA-570). In other cases, the cables are available and marked for use as direct-buried cable or for use as aerial cable supported by a steel wire **messenger**. Some cables may have thin outer jackets that are color coded (usually yellow) to identify them as communication cable.

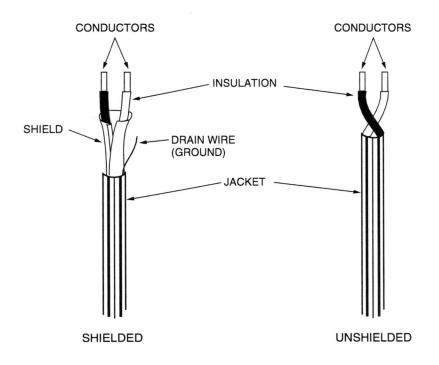
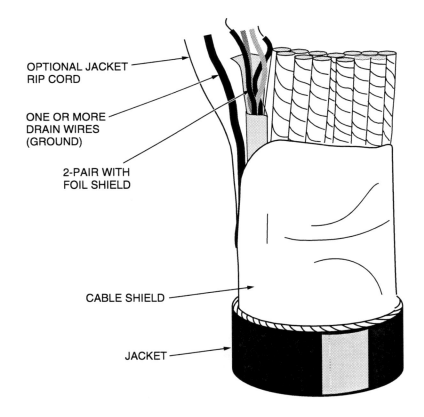

Figure 9. Typical PLTC, Fire Alarm, And Class 2/3 Cables

4.3.1 Unshielded Twisted-Pair Cable (UTP)

Unshielded **twisted-pair cable** (*Figure 10*) has been used for many years for both voice and data transmission. Currently, it is the most widely used type of cable in these applications. It consists of from one pair to as many as 1,800 pairs of solid copper conductors twisted together. The conductors range in size from No. 24 to No. 22 AWG. In cables exceeding 600 pairs, an overall aluminum-steel shield is used to enclose the cable conductors. Each pair of conductors has a nominal **impedance** of 100 ohms (Ω). They are currently available as Category 3 (transmission rates to 16 MHz), Category 4 (20 MHz), and Category 5 or 5e (100 MHz) cables. Category 1 and 2 cables (known as *Level 1* and *Level 2* cables) are not used for new installations; however, they may exist in old Bell System private branch exchange (PBX) installations or in residential installations for voice applications. They range from three-wire to four-wire twisted cable (1 to 2 twists per foot) to very large, parallel, multi-pair cables using the old Bell System conductor color codes. They can only be used for analog phone systems.

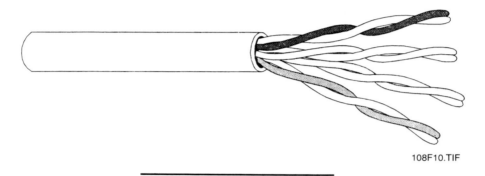

Figure 10. Typical UTP Cable

Category 3 cable is the minimum grade acceptable for new installations; however, Category 5 is recommended. When properly installed, Category 5 cable can support high-speed data communications.

UTP cable pairs consist of conductors referred to as the *tip and ring conductors*. The tip conductor is normally connected to a positive DC signal voltage and the ring conductor is connected to the return or negative signal voltage. Originally, these terms came from early telephone systems where operators used **patch cords** to connect calls. The patch cord plug had three conductive portions separated by insulators. The very end of the plug was the tip. This was followed by a ring section that was, in turn, followed by a grounded sleeve section. The original colors used for the tip, ring, and ground (sleeve) wires of telephone equipment were green for the tip (L1), red for the ring (L2), and yellow (G) for the ground. Many old three-wire residential Bell System installations and telephones are wired using this color code. In new communication system installations, only the tip and ring conductors are required.

UTP cable tip and ring conductors are usually identified with the color code shown in *Table 4*, although other codes may be used. This color code uses five colors for the tip conductor: white, red, black, yellow, and violet (in that order). There are up to five pairs for each tip color as defined by five different ring colors: blue, orange, green, brown, and slate (in that order).

Pair			Binder Group	
Number	Tip	Ring	Color	Pair Count
1	White	Blue	White-Blue	001 – 025
2	White	Orange	White-Orange	026 – 050
3	White	Green	White-Green	051 – 075
4	White	Brown	White-Brown	076 – 100
5	White	Slate	White-Slate	101 – 125
6	Red	Blue	Red-Blue	126 – 150
7	Red	Orange	Red-Orange	151 – 175
8	Red	Green	Red-Green	176 – 200
9	Red	Brown	Red-Brown	201 – 225
10	Red	Slate	Red-Slate	226 – 250
11	Black	Blue	Black-Blue	251 – 275
12	Black	Orange	Black-Orange	276 – 300
13	Black	Green	Black-Green	301 – 325
14	Black	Brown	Black-Brown	326 – 350
15	Black	Slate	Black-Slate	351 – 375
16	Yellow	Blue	Yellow-Blue	376 – 400
17	Yellow	Orange	Yellow-Orange	401 – 425
18	Yellow	Green	Yellow-Green	426 – 450
19	Yellow	Brown	Yellow-Brown	451 – 475
20	Yellow	Slate	Yellow-Slate	476 – 500
21	Violet	Blue	Violet-Blue	501 – 525
22	Violet	Orange	Violet-Orange	526 – 550
23	Violet	Green	Violet-Green	551 – 575
24	Violet	Brown	Violet-Brown	576 – 600
25	Violet	Slate	No Binder	n/a

Table 4. Typical Color Codes For Wire Pairs And Binder Groups

The color code allows the definition of up to 25 pairs of conductors in a cable without repeating a color combination. Some manufacturers add a **tracer** to the tip conductor that matches the corresponding ring conductor color to improve pair identification. For cables with over 25 pairs, each group of 25 pairs up to the 24th group (600 pairs) are wrapped with a color-coded binder tape or thread that identifies the sequential order of each set of 25 pairs, as shown in the table. The last set of 25 has no binder. For some smaller cables (125 pairs or less), only the ring color may be used since the white tip color can be assumed. For cables over 600 pairs, each group of 600 is wrapped with a super-binder. Each super-binder is identified by a tip color in sequential order: 1 to 600 is white, 601 to 1,200 is red, 1,201 to 1,800 is black, etc. Multi-pair UTP cable is generally used for communication system backbone cable. Four-pair UTP cable is employed as horizontal cable for connection to work area outlets. Some manufacturers may use an outer yellow jacket to denote the cable as part of a communication circuit.

4.3.2 Unshielded Twisted-Pair Patch Cords

Unshielded twisted-pair patch cords use four, twisted-pair (100Ω), stranded copper conductors for flexibility and can exhibit up to 20% more attenuation than solid conductors. They are usually equipped with 8-position, 8-contact (8P8C) connectors on the ends. The nominal impedance is 100Ω. Typical color coding for UTP patch cords is given in *Table 5*.

Pair	Identification	Conductor Color Code – Option 1	Conductor Color Code – Option 2
Pair 1	Tip Ring	White-Blue (W-BL) Blue-White (BL-W)	Green (G) Red (R)
Pair 2	Tip Ring	White-Orange (W-O) Orange-White (O-W)	Black (BK) Yellow (Y)
Pair 3	Tip Ring	White-Green (W-G) Green-White (G-W)	Blue (BL) Orange (O)
Pair 4	Tip Ring	White-Brown (W-BR) Brown-White (BR-W)	Brown (BR) Slate (S)

Table 5. Typical Color Codes For UTP Patch Cords

4.3.3 Undercarpet Telecommunication Cable (UTC)

Another form of twisted-pair cable is undercarpet telecommunication cable. See *Figures 11* and *12*. This cable is available up through Category 5 with nominal impedances of 100Ω and can be either:

- Shielded or unshielded, round, twisted, four-pair cable or coaxial cable enclosed or integrated in a support device
- Shielded or unshielded, twisted, four-pair, flat ribbon cable

Optical fiber cable is also being used as undercarpet cabling. Unfortunately, undercarpet cable is often the last choice for new installations due to its susceptibility to damage, its limited reconfiguration ability, and its aesthetically poor appearance due to witness lines that show through carpeting. However, it does offer an effective solution for difficult renovation jobs. When installing undercarpet cabling, high traffic areas, heavy furniture locations, and undercarpet power cables must be avoided.

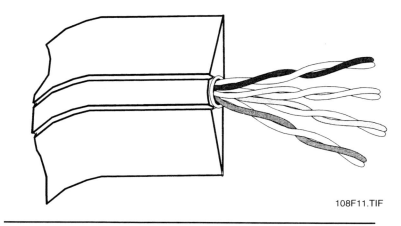

Figure 11. Typical Undercarpet Cable And Support Device

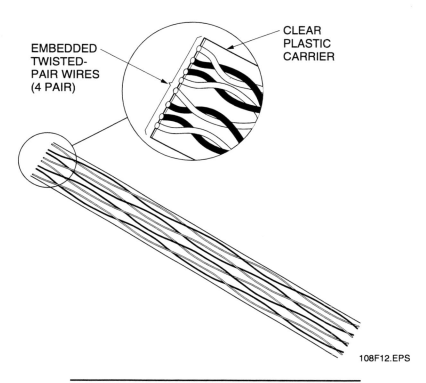

Figure 12. Typical Undercarpet Ribbon Cable

4.3.4 Screened Twisted-Pair (ScTP) Cable and Patch Cord

Screened twisted-pair cable (*Figure 13*) provides very good rejection of high-frequency electrical noise and interference. It is more expensive than UTP and, with the exception of the foil shield and a drain wire that must be grounded, is essentially the same as UTP cable. It is available up to and including Category 5 cable. ScTP patch cords with four twisted-pair stranded conductors, along with the shield and drain wire, are also available. Like UTP patch cords, they have up to 20% more attenuation than solid conductors.

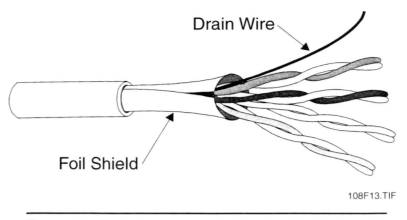

Figure 13. Typical Screened Twisted-Pair (ScTP) Cable

4.3.5 Shielded Twisted-Pair (STP) Cable, Enhanced Shielded Twisted-Pair (STP-A) Cable, And STP Patch Cord

Enhanced shielded twisted-pair cable (*Figure 14*) has two separate foil-shielded twisted pairs and an impedance of 150Ω per pair, with the highest bandwidth (300 MHz) of any ANSI/TIA/EIA-568-A approved cable. The foil shields reduce high-frequency interference and provide a positive attenuation-to-**crosstalk** ratio (ACR) throughout the bandwidth. The exterior braid reduces low-frequency interference and provides immunity to other **electromagnetic interference (EMI)**. STP cable has only the exterior braid and a bandwidth of about 20 MHz. Conductors for both types are No. 22 AWG solid copper and have a specific color code of green/red for one pair and black/orange for the other pair. STP patch cords use No. 26 AWG stranded copper conductors and exhibit up to 50% more attenuation than solid No. 22 AWG conductors. STP-A cable is used in high-speed data systems, including **token ring** installations, that do not require more than two pairs.

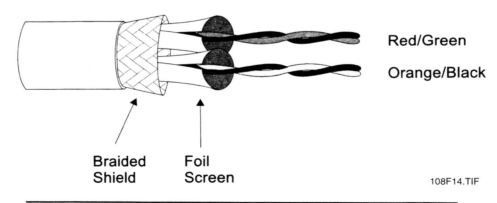

Figure 14. Typical Enhanced Shielded Twisted-Pair (STP-A) Cable

4.3.6 Coaxial Cable

Figure 15 shows several examples of coaxial cable that are or have been used for data communication. The 50Ω cables are no longer being installed in data communication systems, but can be found in many existing installations.

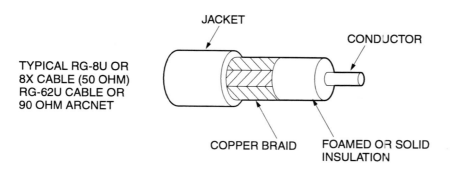
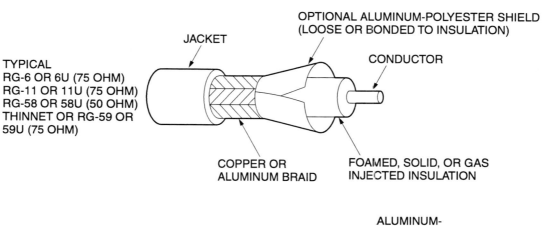
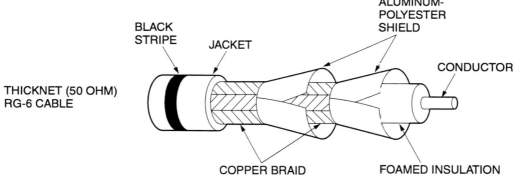
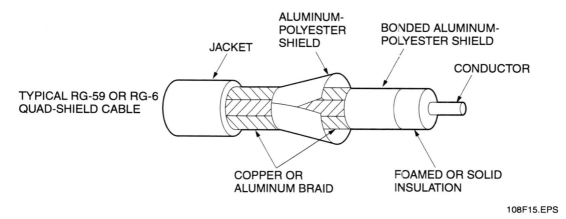

Figure 15. Typical Shielded Coaxial Cables

- *RG-59, RG-6, and RG-11 coaxial cable* – These types of coaxial cable are used primarily in **CATV**, video systems, and security systems. Quad-shield is recommended for all CATV applications. These types are available with 60% to 95% copper or aluminum braid outer shields. Coaxial cable provides a much higher bandwidth and much better protection against EMI than twisted-pair conductors. The aluminum-shielded versions of these cables are less expensive than the copper-braided versions. RG-59, RG-6, and RG-11 cable, also referred to as *Series 59, Series 6,* and *Series 11*, respectively, all exhibit the same 75Ω impedance. However, RG-59 has the greatest attenuation and is limited to run distances of less than 200' (61m) before amplification is required. RG-6 has less attenuation and its maximum run distance is about 500' (152m). Of the three, RG-11 has the lowest loss and can be run for much longer distances.
- *Series 7* – This coaxial cable has higher performance characteristics than Series 6, but is less expensive than Series 11, and is often used in CATV applications.
- *Special RG-8 (IEEE 802.3 Thicknet) and RG-58 (IEEE 802.3 Thinnet)* – These types of coaxial cable are no longer installed for new data systems. They were used in **thicknet (10 base 5)** and **thinnet (10 base 2)** bus-type networks. RG-8 cable has a maximum segment length of 1,640' (500m). The maximum segment length for RG-58 cable is 606' (185m). Cable marked RG-8 or RG-58 should also be marked *IEEE 802.3* or either *thicknet* or *thinnet* (as applicable) to ensure that the cable is compatible for that use.

4.3.7 Optical Multi-Fiber Cable

Like individual fiber cables, optical multi-fiber cable is supplied as tight-buffered (*Figure 16*) or loose-tube (*Figure 17*) cable. The tight-buffered cable is used mostly indoors and the loose-tube cable is normally used outdoors. The gel buffer in loose-tube cable allows the fiber to expand and contract with changes in temperature. It also protects the fiber from any external impact damage to the cable.

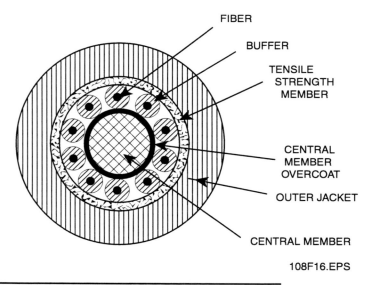

Figure 16. Tight-Buffered Optical Multi-Fiber Cable

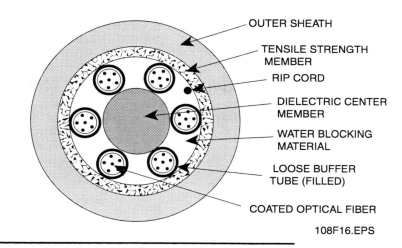

Figure 17. Loose-Tube Optical Multi-Fiber Cable

Either type of cable can be single-mode or multi-mode fiber. Backbone (riser or plenum) or horizontal cable can be multi-mode or single-mode (the choice depends on the equipment connected to it). Some cable jackets are colored yellow to identify the cable as communication cable.

ANSI/TIA/EIA-598-A defines the color coding of optical multi-fiber cable strands, as shown in *Table 6*. If included, strands 13 through 24 are the same as strands 1 through 12 with the addition of a black tracer (except for the black strand, which has a yellow tracer). The tracer may be a dashed or solid line.

Fiber Number	Color	Fiber Number	Color
1	Blue	13	Blue/Black Tracer
2	Orange	14	Orange/Black Tracer
3	Green	15	Green/Black Tracer
4	Brown	16	Brown/Black Tracer
5	Slate	17	Slate/Black Tracer
6	White	18	White/Black Tracer
7	Red	19	Red/Black Tracer
8	Black	20	Black/Yellow Tracer
9	Yellow	21	Yellow/Black Tracer
10	Violet	22	Violet/Black Tracer
11	Rose	23	Rose/Black Tracer
12	Aqua	24	Aqua/Black Tracer

Table 6. Optical Multi-Fiber Cable Strand Color Code

5.0.0 COMMERCIAL LOW-VOLTAGE CABLE INSTALLATION

In most cases, the installation of cables in pathway systems is fairly routine. However, there are certain practices that can reduce labor and materials, and also help to prevent damage to the cables. The use of modern equipment, such as vacuum **fish tape** systems, is one way to reduce labor during this phase of the wiring installation.

There are three types of manual fish tape: steel, nylon, and fiberglass. They also come in different weights for various applications. The proper size and length of the fish tape, as well as the type, should be one of the first considerations. When longer runs are encountered, a pull line is fished through the conduit using a fish tape enclosed in a metal or plastic fish tape reel. This way, the fish tape can be rewound on the reel as the pull line is drawn through the conduit. This avoids having an excessive length of tape lying around on the floor or deck (*Figure 18*).

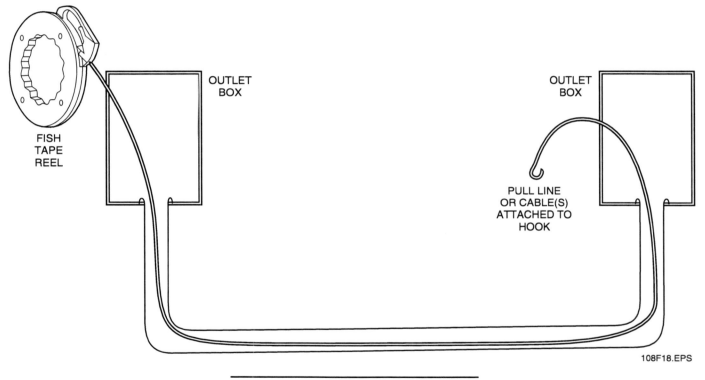

Figure 18. Use Of A Manual Fish Tape

WARNING! Never fish a steel tape through or into enclosures or pathways containing an energized conductor.

When several bends are present in the pathway system, the insertion of the fish tape may be made easier by using flexible fish tape leaders on the end of the fish tape.

Combination blower and vacuum fish tape systems are ideal for use on long runs and can save much time. Basically, the system consists of a tank and air pump with accessories (*Figure 19*). A technician can vacuum or blow a pull line in any size conduit from ½" through 4", or even up to 6" conduit with optional accessories.

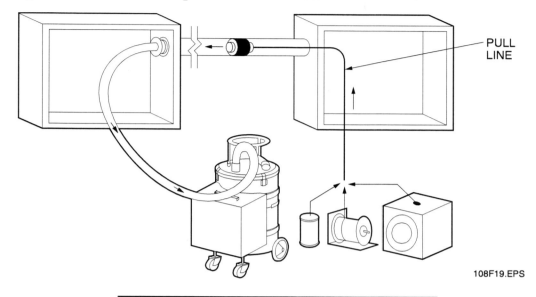

Figure 19. Power Fishing System (Vacuum Mode)

After the fish tape or pull line is inserted in the pathway system, the cables must be firmly attached by some approved means. On short runs where only a few cables are involved, all that is necessary is to strip the jacket from the ends of the cables, bend the exposed wires around the hook in the fish tape or a loop in the end of the pull line, and securely tape them in place. However, this method may not be appropriate for high-speed data cable. Where several cables are to be pulled together, the cables should be staggered and securely taped at the point of attachment so that the overall diameter of the group of cables is not increased any more than is absolutely necessary.

Basket grips (*Figure 20*) are available in many sizes for almost any size and combination of conductors. They are designed to hold the conductors firmly to the fish tape or pull line and can save much time and trouble that would be required when taping wires.

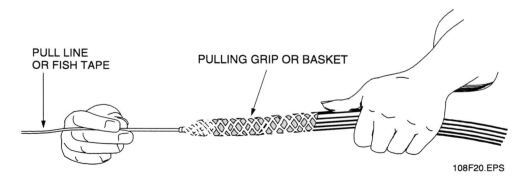

Figure 20. Typical Basket Or Pulling Grip

LOW-VOLTAGE CABLING — TRAINEE TASK MODULE 33108

In all but very short runs, the wires should be lubricated with wire lubricant prior to attempting the pull, and also during the pull. Some of this lubricant should also be applied to the inside of the conduit.

Wire dispensers are helpful for keeping the conductors straight and facilitating the pull. Many different types of wire dispensers are now marketed that handle virtually any spool size. Some of the smaller dispensers can handle up to ten spools of cable; the larger ones can handle a lot more. These dispensers are sometimes called *cable caddies* or *cable trees* (see *Figure 21*).

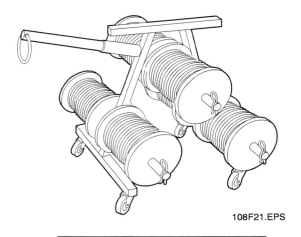

Figure 21. Typical Cable Caddy

5.1.0 PLANNING THE INSTALLATION

The importance of planning any wire pulling installation cannot be over-stressed. Proper planning will make the work go easier and much labor will be saved.

Large cables are usually shipped on reels, involving considerable weight and bulk. Consequently, setting up these reels for the pull, measuring cable run lengths, and similar preliminary steps will often involve a relatively large amount of the total cable installation time. Therefore, consideration must be given to reel setup space, proper equipment, and moving the cable reels into place.

Whenever possible, the cables should be pulled directly from the shipping reels without pre-handling them. This can usually be done through proper coordination of the ordering of the conductors with the job requirements. While doing so requires extremely close checking of the drawings and on-the-job measurements (allowing for extra cable for pull boxes, elbows, troughs, connections, splices, service loops, and outlet service slack), the extra effort is well worth the time.

When the lengths of cable have been established, the length of cable per reel can be ordered so that the total length per reel will be equal to the total of a given number of pathway lengths, and the reel so identified.

The individual cables of the proper length for a given number of runs are reeled separately onto two or more reels at the factory.

With individual cables shipped on separate reels, it is necessary to set up for the same number of reels as the number of conductors to be pulled into a given run, as shown in *Figure 22*. Smaller and/or shorter cables are available on spools or in boxes with up to 1,000' in a box.

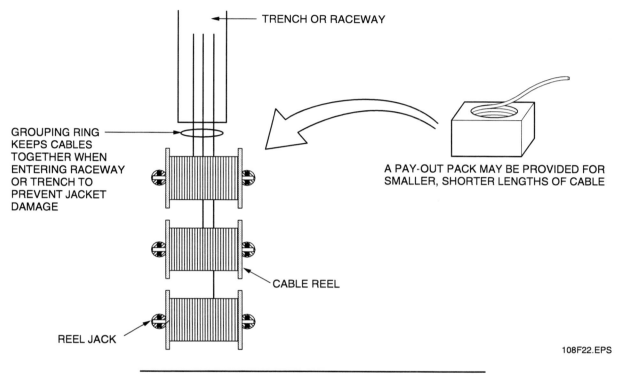

Figure 22. Staggered Reels For A Multiple Cable Pull

Several styles of boxes are used. One is called a *pay-out pack* and another is called a *reel-in-a-box*. The reel-in-a-box types can be stacked and the cable is unreeled through a hole in the end of the box. The pay-out packs are used for shorter lengths of cable and cannot be stacked. The cable is dispensed from a large hole in the box, similar to electrical cable. Like electrical cable boxes, this type of box imparts a twist to the cable as the cable is dispensed.

As an extra precaution against error when calculating the lengths of the cables involved, it is a good idea to actually measure all runs with a fish tape before starting the cable pull, adding for makeup to reach the terminations, slack to create service loops at each end, and accounting for discarding the cable underneath the pulling sleeve or pulling wrap, as it will have been excessively stressed during the pull. Check these totals against the totals indicated on the reels. Under normal cable delivery schedules, when the pathways have been installed at a relatively early stage of the overall building construction, it may not delay the final completion of the installation to delay ordering the cables until the pathways can actually be measured.

When pulling conductors directly from reels in boxes, care must be taken to ensure that each given run is cut off from the reel so that there is a minimum amount of waste. This is to avoid the possibility that the final run of cable taken from the reel will end up being too short for that run.

5.1.1 Cabling Pathways

Communication pathways for a typical commercial building are shown in *Figure 23*. Only the first floor and the Nth floor are shown. In actual practice, a separate set of backbone cables would exit the main **cross-connect** in the equipment room in a star fashion and be routed up to an individual telecommunications closet on each floor of the building. In this example, two types of backbone cable are shown; one is optical fiber for data and the other is UTP for voice traffic. For this example, the fiber optic cable is connected to signal conversion equipment located in the telecommunications closet. This equipment converts optical signals to electrical signals that are routed to the horizontal run outlets and vice versa. The equipment **multiplexes** a number of horizontal electrical runs into one pair of fibers due to the high capacity of optical fiber. Thus, the number of fibers in each backbone optical fiber cable depends on the number of multiplexed outlets on each floor plus any fibers included for future system expansion. The number of conductor pairs in the UTP backbone cable depends on the number of outlets on each floor plus any conductor pairs included for future system expansion.

A typical vertically-aligned telecommunications closet is shown in *Figure 24*. When closets are aligned one above the other on each floor of a building, sleeves and/or slots are normally used to route vertical backbone cable (*Figure 25*).

In some cases, cable shafts are used when large groups of cables must be routed near the top of the building. In cases where the closets are not vertically aligned, pathways or conduits are used. Horizontal runs can be through conduits or pathways in walls, floors, and ceilings or through plenums or other environmental air spaces above false ceilings. Elevator shafts are not used for routing communication cables.

For runs between buildings and other structures, cables that are rated for direct burial should be trench-laid without crossovers and slightly snaked to allow for possible earth settlement, movement, or heaving due to frost action. They should be buried to a depth specified by ***NEC Table 300-5*** or local code and should be placed on, and covered with, cushions of sand or screened fill to protect the cable jackets from damage due to sharp objects.

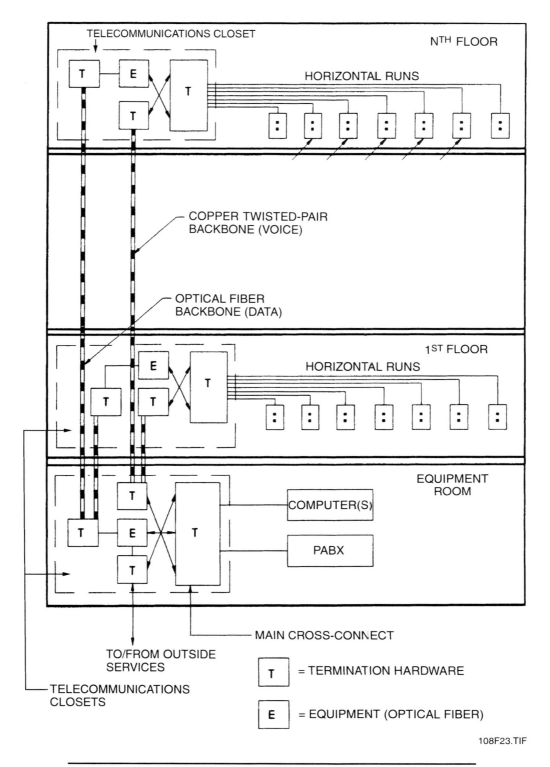

Figure 23. Example Pathways In A Typical Commercial Building

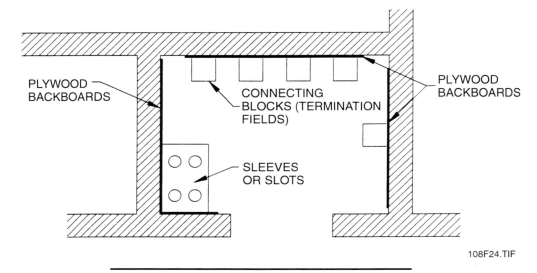

Figure 24. Typical Telecommunications Closet

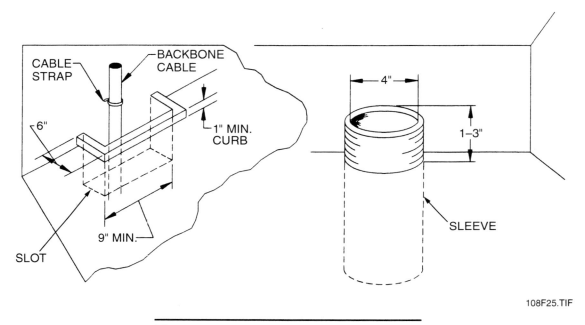

Figure 25. Typical Sleeve And Slot Sizes

5.1.2 Pulling Location

Each job will have to be judged separately as to the best location for pulling setups. The number of setups should be reduced to a minimum in line with the best direction of the pull. For example, it is usually best to pull cables downward rather than up to avoid having to pull the total weight of the cable at the final stages of the pull and also the possibility of injury to workers should the conductors break loose from the pulling line on long vertical pulls. On the other hand, it may not be practical to hoist the cable reels and setup equipment to the upper locations in the building. Also, a separate setup might have to be made at the top of each rise, whereas a single setup might be made at a ground floor pull box location from which several floors are served with the same size and number of cables.

The equipment required depends on the direction of the pull. For instance, if a large cable or a bundle of cables is pulled up vertically or horizontally, a tugger, similar to a winch, may be needed to aid in applying pulling force to a pull line. If pulling down vertically, reel brakes may be needed to prevent the weight of a large, heavy cable or bundle of cables from causing an uncontrolled unreeling of cable called a *runaway*. Two-way communication devices that allow co-workers to keep each other apprised of the pulling progress are essential in every cable pull. Everyone must be alert to inform the person pulling the cable to halt if the cable twists, kinks, or binds in some way.

The location of the pulling equipment determines the number of workers required for the job. A piece of equipment that can be moved in and set up on the first floor by four workers in an hour's time may require six workers working two hours when set up in the basement or the top floor of a building.

It is a simple operation for a few workers to roll cable reels from a loading platform to a first-floor setup, whereas moving them to upper floors involves much handling and usually requires a crane or other hoists. After the pulling operation is completed, reels, jacks, tuggers, cable brakes, etc., all have to be removed.

5.1.3 Pathway Cable Pull Operations

The following operations are performed in almost all cable pulls with larger sizes of cables:

Step 1   Measure or recheck runs and establish communication between both ends of the pull. Secure the area with barricades and/or caution tape.

Step 2   If needed, set up and anchor pulling equipment and/or reel brakes.

Step 3   Move the cable to the setup point using a cable caddy or by moving reels or boxes.

Step 4   If necessary, move the reel jacks and mandrel to the setup point and mount the reels.

Step 5   Prepare the cable ends and label the cable(s).

Step 6   Install the fish tape, pulling line, or rope.

Step 7   Connect the fish tape, pulling line, or rope to the cable(s).

Step 8   If the cable is being pulled through conduit, lubricate the conduit.

Step 9   Pull and secure the cable(s).

Step 10   Disconnect the pulling line.

Step 11   Create the service loops, then label the cable(s) at the reel or box end and cut the cable(s).

Step 12   Remove the reels or boxes and pulling equipment.

Step 13   Terminate the cables.

Step 14   Check and test.

Note: In some instances, the following additional operations are involved, depending upon the exact details of the project. Other items of importance will be discussed later in this module.

- Remove the lagging or other protective covering from the reels.
- Unreel the cable and cut it to length.
- Re-reel the cable for pulling.
- Replace the lagging on the reels.
- Operate such auxiliary equipment as cable brakes, guide-through cabinets or pull boxes, and signal systems between reel and pulling setups.

Each of the pulling operations is discussed in detail in this module.

5.2.0 SETTING UP FOR CABLE PULLING

As mentioned previously, much planning is required for pulling the larger sizes of cables or cable bundles in pathways. There are several preliminary steps required before the actual pull begins.

The proper use of appropriate equipment is crucial to a successful cable installation. The equipment needed for most installations is shown in the checklist in *Figure 26*. Some projects may require all of these items, while others may require only some of them. Each cable pulling project must be taken on an individual basis and analyzed accordingly; seldom will two pulls require identical procedures.

EQUIPMENT CHECKLIST

- ❏ PORTABLE ELECTRIC GENERATOR (IF NEEDED)
- ❏ EXTENSION CORDS WITH GFCI
- ❏ WARNING FLAGS, SIGNS, BARRICADES
- ❏ RADIOS OR TELEPHONES
- ❏ GLOVES
- ❏ FISH TAPE OR STRING BLOWER/VACUUM
- ❏ PULL LINE OR ROPES
- ❏ CAPSTAN-TYPE ROPE PULLER
- ❏ SHORT ROPES FOR TEMPORARY TIE-OFFS
- ❏ SWIVELS
- ❏ BASKET-GRIP PULLERS
- ❏ REEL ANCHOR
- ❏ REEL JACKS
- ❏ REEL BRAKES
- ❏ CABLE CUTTERS
- ❏ CABLE-PULLING LUBRICANTS
- ❏ 50' MEASURING TAPE
- ❏ BULLWHEELS
- ❏ PULLEYS
- ❏ TELESCOPING POLE WITH HOOK
- ❏ REMOTE-CONTROLLED TOY

Figure 26. Cable Pulling Equipment Checklist

5.2.1 Setting Up The Cable Reels Or Boxes

When reels of cable arrive at the job site, it is best to move them directly to the setup location whenever possible. This prevents having to handle the reels more than necessary. However, if this is not practical, arrangements must be made for storage until the cable is needed. The exact method of handling reels of cable depends upon their size and the available tools and equipment.

In many cases, the reels may be rolled to the pulling location by one or more workers. Smaller reels or spools can be mounted on and transported to the setup site on a cable reel transporter, as shown in *Figure 27*.

(A) CABLE REEL TRANSPORTER (B) FORKLIFT

Figure 27. Two Methods Of Transporting Cable Reels

Payout packs or reel-in-a-box containers can be stacked on a pallet and moved by forklift or stacked on a hand truck and manually wheeled to the setup site. For reels up to 24" wide × 40" in diameter, a cable reel transporter can be used to transport the cable reel; it also acts as a dispenser during the pulling operation. When available, a forklift is ideal for lifting and transporting cable reels. See *Figure 27*.

However, for reels 14" or more in diameter, a crane or similar hoisting apparatus is usually necessary for lifting the reels onto reel jacks supported by jack stands to acquire the necessary height. *Figure 28* shows a summary of proper and improper ways to transport reels of cable on the job site.

LOW-VOLTAGE CABLING — TRAINEE TASK MODULE 33108

ON THE RIMS OF THE SPOOL (MOVING EQUIPMENT DOES NOT COME INTO CONTACT WITH CABLE)

ON THE FLAT SIDE OF THE SPOOL OR ON THE CABLE (MOVING EQUIPMENT COMPRESSES INSULATION AND MAY DAMAGE CABLE)

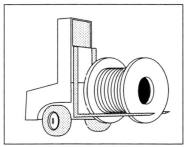 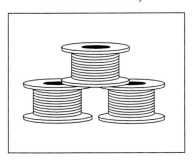
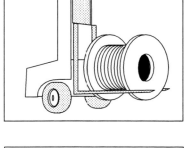 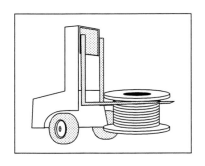
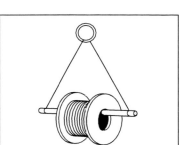 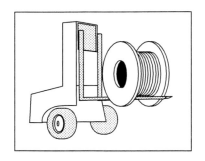
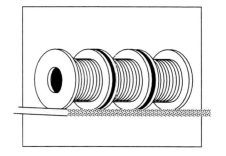 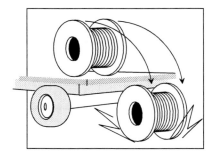

108F28.EPS

Figure 28. Proper And Improper Ways Of Transporting Cable Reels

Figure 29 shows several types of reel stands, including the spindle. For a complete setup, two stands and a spindle are required for each reel. The stands are available in various sizes from 13" to 54" high to accommodate reel diameters up to 96". Extension stands used in conjunction with reel stands can accommodate 8", 10", 12", and 14" reels.

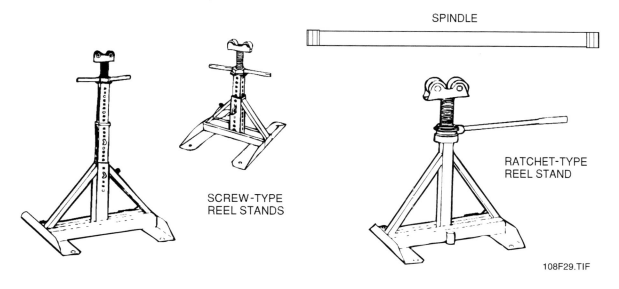

Figure 29. Typical Reel Stands

Reel-stand spindles are commonly available in diameters from 2⅜" to 3½" and from 59" to 100" in length for carrying reel loads up to 7,500 pounds. However, some heavy-duty spindles are rated for loads up to 15,000 pounds.

5.2.2 Preparing Conduit Pathways For Cables

A preliminary step prior to pulling cables in conduit systems is to inspect the conduit itself. Few things are more frustrating than to pull cables through a conduit and find out when the pull is almost done that the conduit is blocked or damaged. Such a situation usually requires pulling the cables back out, repairing the fault, and starting all over again.

Figure 30 shows several devices used to inspect conduit systems, as well as to prepare the conduit for easier and safer cable pulls. Go and no-go steel and aluminum mandrels are available for pulling through runs of conduit before the cable installation. Mandrels should be approximately 80% of the conduit size (twice the 40% fill factor).

Figure 30. Devices Used To Inspect, Clean, And Lubricate Conduit Systems

LOW-VOLTAGE CABLING — TRAINEE TASK MODULE 33108

A test pull will detect any hidden obstructions in the conduit prior to the pull. If any are found, they can be corrected before wasting time on an installation which might result in cable damage and the possibility of having to re-pull the cable.

The conduit swab in *Figure 30* can be used to swab out water and debris from the pathway and spread a uniform film of pulling compound inside the conduit for easier pulling.

One final step before starting the pull is to measure the length of the conduit, including all turns in junction boxes and other devices. A fish tape may be pushed through the conduit system and a piece of tape used to mark the end. When it is pulled back out, a tape measure may be used to measure the exact length. An easier way, however, is to use a power fishing system to push or pull a measuring tape through the conduit run. Details of this operation are explained in the next section.

When measuring the conduit run length, be sure to allow sufficient room where measurements are made through a pull box. Cables should enter and leave pull boxes in such a manner as to allow the greatest possible sweep for the cables. Large cables are especially difficult to bend, but with proper planning, you can simplify the feeding of these cables from one conduit to another.

For example, if a conduit run makes a right-angle bend through a pull box, the conduit for a given cable should come into the box at the lower left-hand corner and leave diagonally opposite at the upper right-hand corner, as shown in *Figure 31*. This gives the cable the greatest possible sweep, eliminating sharp bends and consequent damage to the cable insulation.

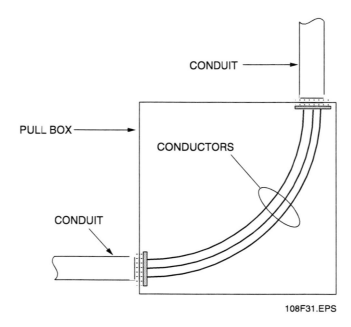

Figure 31. Obtaining The Greatest Possible Cable Sweep In A Pull Box

Runs should also be calculated to allow for terminations, service loops, and end wastage due to pulling.

In some cases, inner duct (*Figure 32*) is installed inside conduit, on cable trays, or other supports to simplify cable pulling for initial cable installations and subsequent additions or alterations. Inner duct is a nonmetallic electrical tubing available in either plenum or nonplenum forms and in a variety of colors used to identify the type of circuits contained within the duct. Basically, red is used for fire protection circuits, blue for power circuits, and yellow for signal and communication circuits. Inner duct is also available with or without pull lines installed.

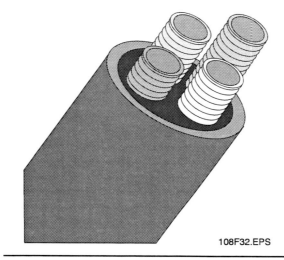

Figure 32. Conduit With Inner Duct Installed

5.2.3 Installing A Pull Line In Conduit Or Inner Duct

At one time, pull-in lines were frequently placed in conduit runs as they were installed. However, in recent times, with modern cable-fishing equipment, this practice is seldom used except in the case of inner duct, where pre-installed pull lines can be ordered.

In conduit or inner duct without pull lines, pull lines or ropes are sometimes manually fished through the conduit using a steel fish tape, but much time can be saved by using a blower/vacuum fish tape system. In general, a **conduit piston**—sometimes referred to as a *mouse* or *missile*—is blown with air pressure or vacuumed through the run. The foam piston is sized to the conduit and has a loop on both ends. In most cases, **fish line** or measuring tape is attached to the piston as it is blown or vacuumed through the conduit run. The measuring tape serves two purposes: it provides an accurate measurement of the conduit run and the tape is used to pull the cable-draw pulling line or rope into the conduit run. In some cases, if the run is suitable, the pulling line or rope is attached directly to the piston and vacuumed into the run. *Figure 33* shows a blower/vacuum fish tape system being used to vacuum a pull line in a conduit, while *Figure 34* shows the same apparatus blowing the piston through the conduit. Most of these units provide enough pressure to clean dirt or water from the conduit during the fishing operation.

LOW-VOLTAGE CABLING — TRAINEE TASK MODULE 33108

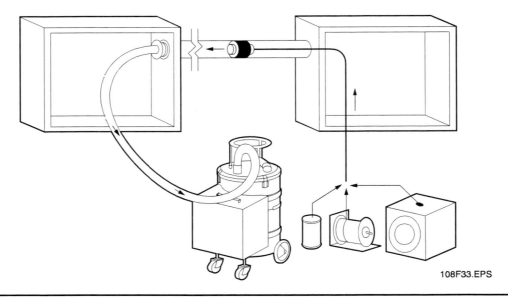

Figure 33. Blower/Vacuum Fish Tape System Used To Vacuum A Pull Line In Conduit

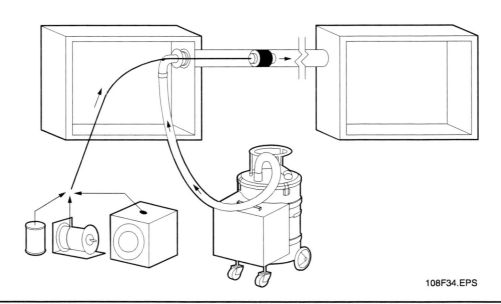

Figure 34. Blower/Vacuum Fish Tape System Used To Blow A Pull Line In Conduit

WARNING!   Before blowing anything through a conduit, find out what is at the other end of the conduit run. Make sure that no one is looking into or near the other end of the conduit. Also, ensure that no live electrical wiring is in the pull or termination box at the other end of the conduit.

Ensure that the tensile strength of the pulling line or rope is sufficient for the cable run to be pulled. Any rope used must be specially designed so that it does not stretch appreciably under tension and can withstand the friction generated by a power tugger.

There are certain precautions that should be taken when using a power fish tape system:

- Read and understand all instructions and warnings before using the tool.
- Never attempt to fish runs that might contain live power.
- Be prepared for the unexpected. Make sure that your footing and body position are such that you will not lose your balance in any unexpected event.
- Use blower/vacuum systems only for specified light fishing and exploring the pathway system.
- Never use pliers or other devices that are not designed to pull a fish tape. They can kink or nick the tape, creating a weak spot.

5.2.4 Installing A Pull Line In Open Ceilings

Installing cables in open ceilings requires that a pull line be fished across supports or through plenums above the ceiling. Cables may be supported by beam clamps that hold bridle rings, D-rings, wire hooks, or J-hooks (preferred). The distance between the supports as well as the supported cable weight must comply with all applicable codes. Ensure that any changes in direction for the cables are gradual enough that the minimum bend radius for the cables is not exceeded. High-performance cables are not normally supported by bridle rings, D-rings, or wire hooks. As an alternative, the cable can be pulled through the structural supports for the floor or ceiling above. If extensive cable will be pulled across open ceilings, cable trays or ladder racks should be installed.

Like conduit pulls, a pull line with adequate tensile strength or a pull tape may be used to pull cable in open ceiling runs. The pull line can be threaded when the supports are installed or it can be threaded over ducts and through supports or plenums using one or more of the following methods:

- Line attached to a rubber ball (tennis ball) and thrown over supports
- Remote-controlled toy
- Telescoping pole (up to 25' long) with a hook to lift the line and place in supports

The use of a telescoping pole reduces the number of ceiling panels that must be removed to place the pull line. Each 90° change of direction constitutes a new pull point. A technician must be stationed at each pull point or a pulley must be used to assist in pulling the cable.

5.2.5 Preparing Cable Ends For Pulling

The pulling line or rope must be attached to the cable in such a manner that it cannot part from the cable during the pull. Four common methods include:

- Direct connection with the copper cable conductors (core hitches) of the cable/cable bundle. *Figure 35* shows two different core hitch methods. These methods may not be applicable with some types of high-performance cable.

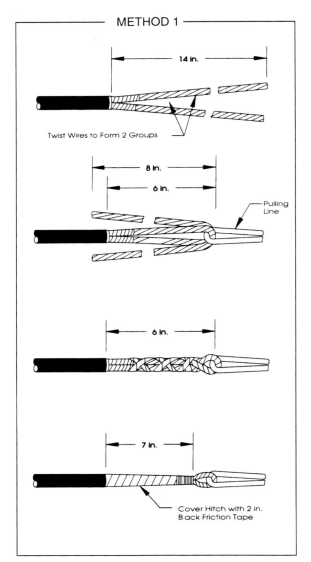 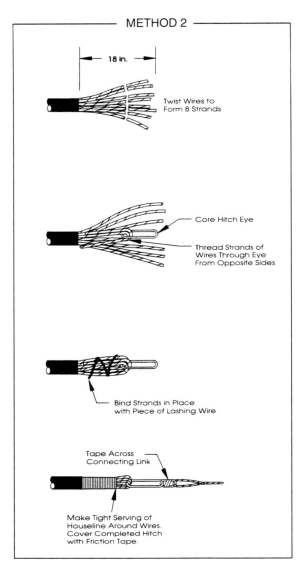

NOTE 1: There is a maximum pulling tension for each type of cable. For example, the pulling tension may not exceed 25 pounds on high-speed data cable. Check the manufacturer's instructions for the type of cable you are pulling.

NOTE 2: The methods shown above may not be applicable to certain types of high-performance cable.

Figure 35. Core Hitches

- Connection by means of a pulling grip or cable basket placed over the cable/cable bundles and taped to the cable/cable bundle to prevent slippage. See *Figure 36*.
- Connection of a pull line or rope to the cable/cable bundle using a rolling hitch knot. See *Figure 37*.
- For optical fiber cables without pre-installed connectors, connection to the aramid yarn strength member. See *Figure 38*.

In some cases, manufacturers can supply large, multi-pair backbone cable with a factory-installed pulling eye.

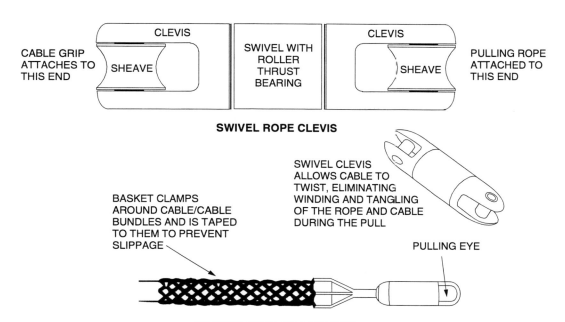

NOTE: There is a maximum pulling tension for each type of cable. For example, the pulling tension may not exceed 25 pounds on high-speed data cable. Check the manufacturer's instructions for the type of cable you are pulling.

Figure 36. Typical Pulling Grip And Clevis Used During Cable Installation

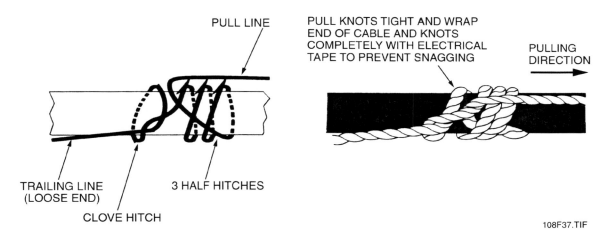

Figure 37. Rolling Hitch Knot

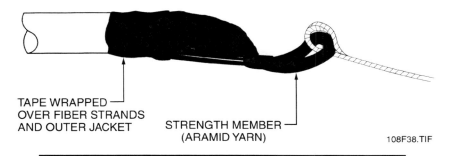

Figure 38. Connection To Optical Fiber Strength Member

LOW-VOLTAGE CABLING — TRAINEE TASK MODULE 33108

Most pulling lines or ropes have a **clevis** as an integral component. However, when using pulling grips or baskets, a clevis that is part of the basket is normally used to facilitate connecting the pulling rope to the wire grip. The clevis allows the pulling line or rope on the cable(s) to twist, eliminating any tangling of the rope and cable(s) during the pull.

For high-speed data cable, the pulling tension must not exceed 25 foot-pounds per cable. In these cases, a breakaway clevis or link that is also rated at 25 foot-pounds can be used to protect the cable.

5.2.6 Types Of Pulling Lines

The type of pulling line or rope selected will depend mainly on the pulling load, which includes the weight of the cable, the length of the pull, and the total resistance to the pull. For example, Greenlee's multiplex cable pulling rope is designed for low-force cable pullers. It has a low stretch characteristic that makes it suitable for pulls up to 2,000 foot-pounds. Lengths are available from 100' to 1,200'. For heavy cable installations, special-purpose rope is recommended because it does not stretch appreciably and can withstand the friction heat generated by a power tugger.

WARNING!   Any equipment associated with the pull must have a working load rating in excess of the force applied during the pull. All equipment must be used and mounted in strict accordance with the manufacturer's instructions.

Care must be used when selecting the proper line or rope for the pull, and then every precaution must be taken to ensure that the cable pulling force does not exceed the rope capacity. Normally, tensile strength ratings are provided on the pull line or rope box by the manufacturer.

5.3.0 CABLE PULLING EQUIPMENT

Except for short cable pulls, hand-operated or power-operated tuggers or winches are used to furnish the pulling power. In general, cable reels are set in place at one end of the pathway system and the tugger is set up at the opposite end. One end of the previously-installed pulling line or rope is attached to a clevis, basket, or the cable. The other end of the rope (at the cable tugger) is wrapped around the rotating drum on the tugger (*Figure 39*).

When pulling in conduit, pulling lubricant—sometimes referred to as **soap**—is inserted into the empty conduit as well as applied liberally onto the front of the cable. One or more operators must be on hand to help feed the cable, while one worker is usually all that is required on the pulling end.

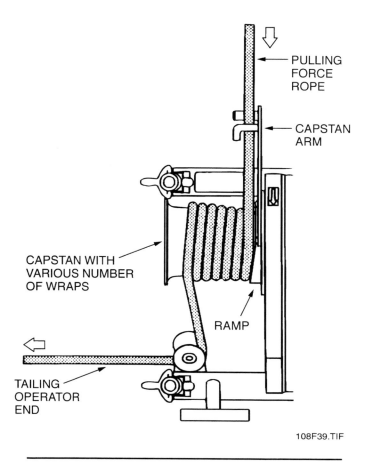

Figure 39. Basic Parts Of A Power Cable Tugger

| **WARNING!** | Always use a pulling lubricant that is compatible with the type of cable being pulled. Check with the cable manufacturer for their recommendations and always contact the lubricant manufacturer about the compatibility of their products with specific cables. |

Table 7 lists some lubricants suitable for most cable pulls.

The number of wraps on the puller drum decides the amount of force applied to the pull. For example, the operator needs to apply only 10 pounds of force to the pulling rope in all cases. With this amount of force applied by the operator, and with one wrap around the rotating drum, 21 pounds of pulling force will be applied to the pulling rope; with 2 wraps, it will be 48 pounds; with 3 wraps, it will be 106 pounds, etc. This principle is known as the **capstan** theory and is the same principle applied to block-and-tackle hoists.

Name of Lubricant	Manufacturer
Flaxoap	Murphy-Phoenix, Cleveland, OH
Gel Lube 7/5	Richards Manufacturing, Irvington, NJ
Ivory Snow	Procter & Gamble, Cincinnati, OH
Polywater A, C, G & J	American Polywater, Stillwater, MN
Quelube	Quelcor, Media, PA
Slip X-300	American Colloid, Skokie, IL
Slipry Loob SWP	Thomas/Jet Line Industries, Matthews, IL
Slipry Loob MWP	Thomas/Jet Line Industries, Matthews, IL
Wire Lube & Aqua Gel	Ideal Industries, Sycamore, IL
Wirepull	MacProducts, Kearney, NJ
Wire-Wax	Minnerallac Electric, Addison, IL
Y-er Eas	Electro Compound, Cleveland, OH
U.S. Gypsum	United States Gypsum, Des Plaines, IL

Table 7. Cable Pulling Lubricants Suitable For Most Applications

Table 8 gives the amount of pulling force with various numbers of pull line wraps when the operator applies only 10 foot-pounds of tailing force. Reducing the tailing force reduces the pulling force proportionately (i.e., for two wraps, one foot-pound of operator force would produce 4.8 foot-pounds of pulling force).

Number of Wraps	Operator Force (Ft.-Lbs.)	Pulling Force (Ft.-Lbs.)
1	10	21
2	10	48
3	10	106
4	10	233
5	10	512
6	10	1,127
7	10	2,478

Table 8. Typical Pulling Forces For Various Wraps

5.3.1 Pulling Safety

Adhere to the following precautions when using power cable pulling equipment:

- Read and understand all instructions and warnings before using any tool.
- Use compatible equipment (i.e., use the properly-rated cable puller for the job, along with the proper rope and accessories).
- Always be prepared for the unexpected. Make sure that your footing and body position are such that you will not lose your balance in any unexpected event. Keep out of the direct line of force in case of rope failure.
- Ensure that all cable pulling systems, accessories, and rope have the proper rating for the pull.
- Inspect tools, rope, and accessories before using; replace damaged, missing, or worn parts.
- Personally inspect the cable pulling setup, rope, and accessories before beginning the cable pull. Ensure that all equipment is properly and securely rigged.
- Ensure that all electrical connections are properly grounded and adequate for the load.
- Use cable pulling equipment only in uncluttered areas.

CAUTION: Make absolutely certain that all communication equipment is in working order prior to the pull. Place personnel at strategic points with operable communication equipment to stop and start the pull as conditions warrant. Anyone involved with the pull has the authority to stop the pull at the first sign of danger to personnel or equipment.

5.4.0 VERTICAL AND HORIZONTAL PATHWAY CABLE PULLS

The following paragraphs outline the general methods and precautions for pulling cable in vertical and horizontal pathways.

WARNING! Make sure goggles, gloves, and a hardhat are used when pulling cable.

CAUTION: When handling, installing, bending, or creating service loops with backbone cable, the minimum bending radius is normally 10 times the diameter of the cable. For high-performance four-pair cable, the minimum bend radius is four times the diameter of the cable. For optical fiber cable, the minimum bend radius can range from 10 to 30 times the diameter, depending on the cable. Manufacturer's recommendations for maximum pulling tensions must also be followed.

Note: It is extremely important that during the installation of backbone or horizontal cable, adequate service loops for each cable be provided in the equipment room and each telecommunications closet to allow for corrections or future changes to the wiring. Service loops for cables entering these rooms should be long enough to reach the farthest corner of the room and, at that corner, also reach from the floor to the ceiling.

5.4.1 Vertical Backbone Cable Pulls From The Top Down

This is the simplest method of placing backbone cables in vertical pathways because no power pulling equipment is required and long, heavy cables can be readily accommodated. After the cable reels or boxes are positioned and supported at the highest floor and the vertical pathways are determined to be clear, complete the following procedure:

Step 1 Double check the length of the pathways to make sure adequate cable will be available.

Note: Communication cable should not be spliced.

Step 2 Ensure that adequate support facilities (brackets, clamps, etc.) are installed or available on each floor to secure the cable(s) after the pulling is completed. In some open pathways, steel cables are secured to the top and bottom floors, and the communication cables are clamped to the steel cable for support. In this type of system, wire ties that pass through the steel cable strands are used to secure the communication cable every 3' or 4' before the steel cable is put under tension.

Step 3 Set up a reel brake (*Figure 40*).

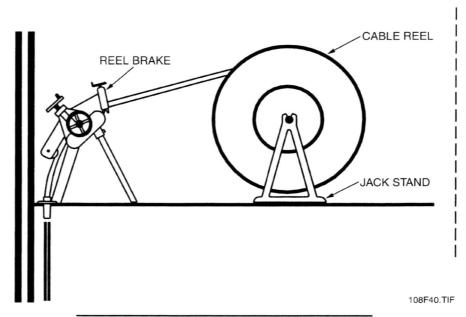

Figure 40. Typical Reel Brake For Cable Reel

Step 4 If the weight and length of the cable or cable bundle is excessive, set up a pulley or bullwheel (*Figure 41*) to guide the cable(s) entering the pathway to prevent damage to the cable(s). Ensure that the pulley or bullwheel is securely fastened to the building structure or to a swing fixture anchored to the floor or building structure.

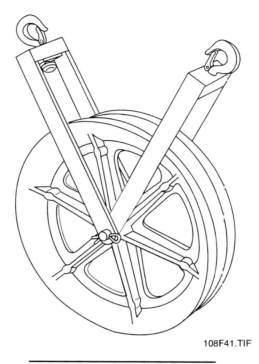

Figure 41. Typical Bullwheel

Step 5 Prepare the end(s) of the cable(s) with an appropriate pulling device if a factory pulling eye has not been installed.

Step 6 As necessary, fish an adequate pull line or rope through the pathway and connect it to the end of the cable(s) to be pulled.

Step 7 Position co-workers with communication devices on each floor to view, aid, and communicate the progress of the descending cable(s). On long pulls or with heavy cable(s), temporary restraining devices for the cable(s) should be installed. One of the most common methods uses **sheaves** to secure the rope to the building structure. These are attached to the descending cable on every floor or every three or four floors using a rolling hitch knot (*Figure 42*) and are used to **snub** (catch) the descending cable.

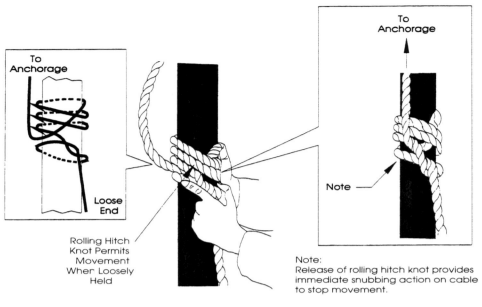

Figure 42. Rolling Hitch Knot Used To Snub Cable In Vertical Pulls

WARNING! If the cable(s) are very heavy and/or if many floors must be crossed and the accumulated weight of the cable(s) will exceed the tension strength of the cable(s), the descending cable(s) may have to be lowered in increments of every 4 to 10 floors or so. This can be accomplished by snubbing the cable with a rolling hitch at a selected floor and feeding a certain length of cable out of the pathway onto the floor above. The cable is then snubbed above that floor and the accumulated length is fed down to the lower floors. This process is repeated every 4 to 10 floors until the cable(s) reach the lowest level desired.

Step 8 Lubricate any sleeves or conduit that the cable(s) will pass through unless inner duct is installed and used.

Step 9 Commence the pull slowly, with co-workers carefully observing the process. Lubricate the cable(s) as needed. When the cable(s) reach the lowest level, make sure that enough is pulled out to allow termination. Secure the cable(s) at that level with the appropriate clamps or other devices (*Figure 43*). Progressing upward to each floor, pull enough cable down at each floor to create a service loop in the telecommunications closet and then secure the cable(s) at that floor.

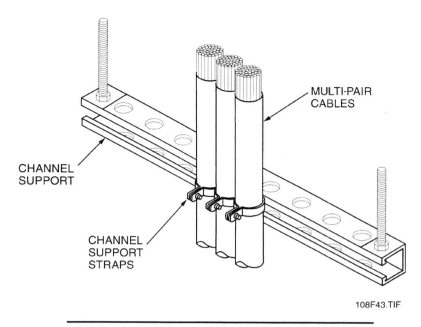

Figure 43. Typical Support Channel With Straps

Note: Another method of securing cable at each floor uses a vertical split mesh grip that is placed around the cable and closed by weaving a pin through the mesh. A ½" (13mm) steel rod is inserted in the top loop of the grip. The cable is then allowed to slide down until the rod rests on the pathway opening. The mesh grip then tightens, holding the cable. Alternatively, the top loop of the grip can be anchored to the building structure. The cable(s) may also be tied to steel cable that is stretched from the top floor to the bottom floor.

Step 10 Label the cable(s) with their identification at all appropriate locations. Document the vertical backbone cable information on the building plans/blueprints and cable(s) as follows:

- Type(s) of cable(s) installed
- Origin and termination of each cable
- Conduit (or other pathway) used for each cable
- Application (use) for each cable

5.4.2 Vertical Backbone Cable Pulls From The Bottom Up

This method of placing backbone cables in vertical pathways is more difficult because power pulling equipment is required and long, heavy cables cannot be readily accommodated. After the cable reels or boxes are positioned and supported at the lowest floor and the vertical pathways are determined to be clear, proceed as follows:

Step 1 Double check the length of the pathways to ensure that adequate cable will be available.

Note: Communication cable should not be spliced.

Step 2 Ensure that adequate support facilities (brackets, clamps, etc.) are installed or available on each floor to secure the cable(s) after the pulling is completed. In some open pathways, steel cables are secured to the top and bottom floors and the communication cables are clamped to the steel cable for support. In this type of system, wire ties that pass through the steel cable strands are used to secure the communication cable every 3' or 4' before the steel cable is put under tension.

WARNING! If any multi-pair cable(s) to be used are very heavy and/or if many floors must be crossed and the accumulated weight of the cable(s) will exceed the tension strength of the cable(s), smaller cable or cable bundles combined with multiple pulls may be required to eliminate the problem. A more time consuming method would be to pull the total amount of the cable(s) required for the vertical pathway (including all required slack) to an intermediate level. The cable would be secured below that level. Then, the power tugger would be moved to the next intermediate level and the process repeated until the top level is reached.

Step 3 Set up pulleys and a bullwheel between the reel(s) and the pathway to guide the cable(s) entering the pathway to prevent damage to the cable(s). Ensure that the bullwheel is securely fastened to the building structure or to a swing fixture anchored to the floor or building structure.

Step 4 Prepare the end(s) of the cable(s) with an appropriate pulling device if a factory pulling eye has not been installed.

WARNING! Make sure that the rope used is rated to handle twice the weight of the amount of cable that will be lifted in any one pull.

Step 5 As necessary, fish the appropriate rope through the pathway and connect it to the end(s) of the cable(s) to be pulled.

Step 6 Mount the power tugger securely to the floor at the desired pulling level. If possible, position the tugger so that it is far enough from the pathway so that a reasonable amount of slack cable can be pulled into the area (*Figure 44*).

WARNING! Make sure the power tugger is securely bolted to the floor. If it breaks loose during a heavy cable pull, it may cause severe injury or death, as well as property damage.

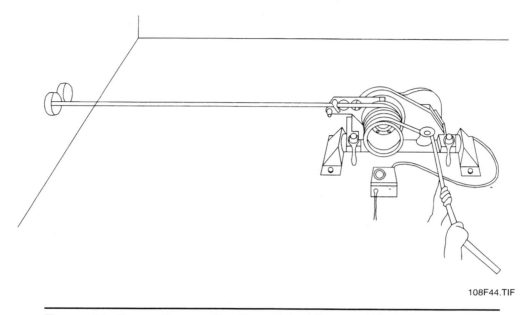

Figure 44. Power Tugger Positioned And Secured To A Concrete Floor

Step 7 Position co-workers with communication devices on each floor to view, aid, and communicate the progress of the ascending cable(s). For safety and other purposes on long upward pulls or with heavy cable(s), temporary restraining devices for the cable(s) must be installed. One of the most common methods uses sheaves to secure the rope to the building structure. These are attached to the ascending cable with a rolling hitch knot at the top floor, on every floor, or on every three or four floors and are used to snub the ascending cable in case of pulling failures or when the cable must be secured at the end of the pull.

Step 8 Lubricate any sleeves or conduit that the cable will pass through unless inner duct is installed and used.

CAUTION: Do not exceed the rated pulling tension of the cable.

Step 9 Using the tugger, and with the correct number of capstan wraps, commence the pull slowly, with co-workers carefully observing the process. Lubricate the cable(s) as needed. When the cable(s) reach the tugger level, ensure that enough cable is pulled out to allow service loops at each floor, plus termination at the top floor. It may be necessary to restrain the cable at the tugger level, disconnect the pulling rope from the cable eye, and reconnect it to the cable with a rolling hitch knot close to the pathway so that additional cable can be drawn into the tugger level. This action must be repeated until an adequate amount of cable is accumulated. When the pulling is complete, restrain the cable at all temporary locations to prevent it from falling.

CAUTION: Do not allow the cables to wrap around the tugger capstan, as damage to the cables will occur.

Step 10 Secure the cable at the lowest level with the appropriate clamps or other devices. Progressing upward to each floor, lower enough cable at each floor to create a service loop in the telecommunications closet and then secure the cable(s) at that floor.

Note: One method of securing cable at each floor uses a vertical split mesh grip that is placed around the cable and closed by weaving a pin through the mesh. A ½" (13mm) steel rod is inserted in the top loop of the grip. The cable is then allowed to slide down until the rod rests on the pathway opening. The mesh grip then tightens, holding the cable. Alternatively, the top loop of the grip can be anchored to the building structure. The cable(s) may also be tied to steel cable that is stretched from the top floor to the bottom floor.

Step 11 Label the cable(s) with their identification at all appropriate locations. Document the vertical backbone cable information on the building plans/blueprints and cable(s) as follows:

- Type(s) of cable(s) installed
- Origin and termination of each cable
- Conduit (or other pathway) used for each cable
- Application (use) for each cable

5.4.3 Horizontal Backbone Cable Pulls

Horizontal backbone cable is normally run between equipment closets. Like bottom-up vertical pulls, horizontal pulls can be difficult because the entire weight of the cable must be dragged over or through the cable supports. Horizontal backbone may be supported by conduit runs with or without inner duct, cable trays, J-hooks and beam clamps, lay-in pathways, or other such devices. ANSI/TIA/EIA-569-A guidelines require that horizontal pulls for communication cables be no longer than 98' (30m) between pull points.

Usually, large cables must be placed individually to manage their weight. In no case should different size cables be pulled together. The pulling force required for the larger cable can damage the smaller cable.

In some cases, a power tugger may be required due to the cable weight. After the cable reels or boxes are positioned and supported and the pathways are determined to be clear, proceed as follows:

Step 1   Double check the length of the pathways to ensure that adequate cable will be available.

Note: Communication cable should not be spliced.

Step 2   If the pathway is open and the cables will be supported by J-hooks or the equivalent, ensure that the spacing of each hook is such that it will support less than 25 foot-pounds of cable weight.

Step 3   Set up pulleys and bullwheels between the reel and the pathway to guide the cable entering the pathway to prevent damage to the cable. Ensure that the bullwheel is securely fastened to the building structure or to a swing fixture anchored to the floor or building structure. On open pathways, it may be desirable or necessary to mount pulleys along the entire path of the cable at locations near the J-hooks or support devices (*Figure 45*) to ease the pulling of the cable and reduce the stress on the cable. Once pulled, the cable can be transferred to the J-hooks or other support devices and the pulleys can be removed. If cable trays are used, use cable rollers, conveyor pulleys, or bullwheels, as shown in *Figures 46* through *49*.

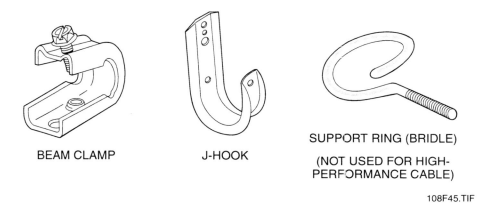

Figure 45. Beam Clamps And J-Hooks

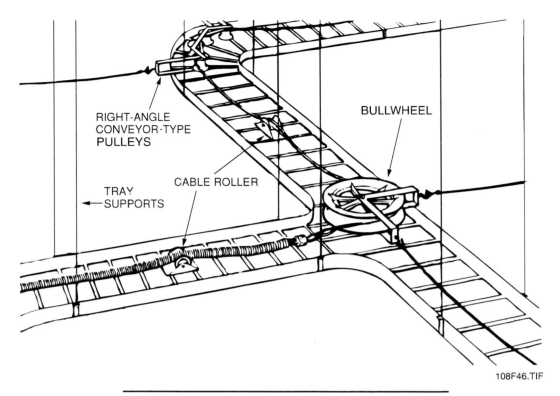

Figure 46. Typical Cable Tray Pulling Arrangement

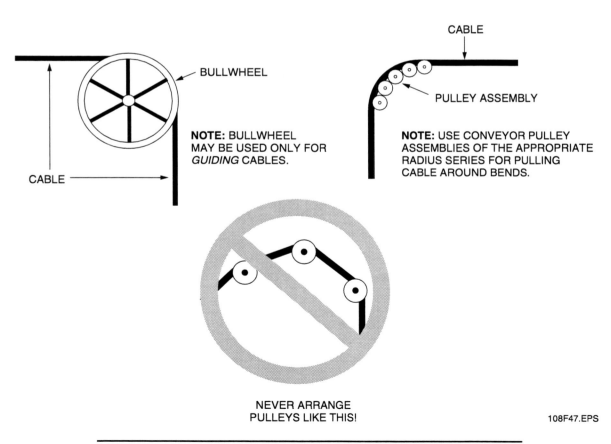

Figure 47. Proper Arrangements Of Cable Bullwheels And Pulleys

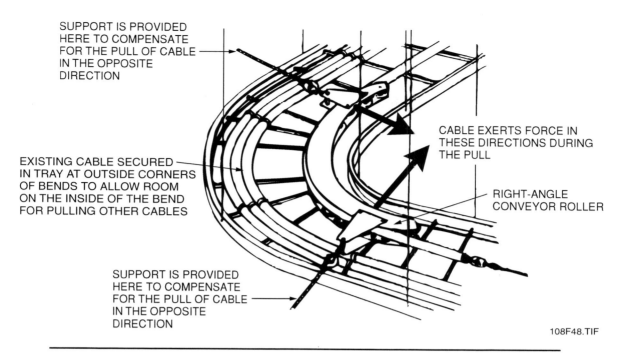

Figure 48. Support Points For Right-Angle Turns And Existing Cable Placement

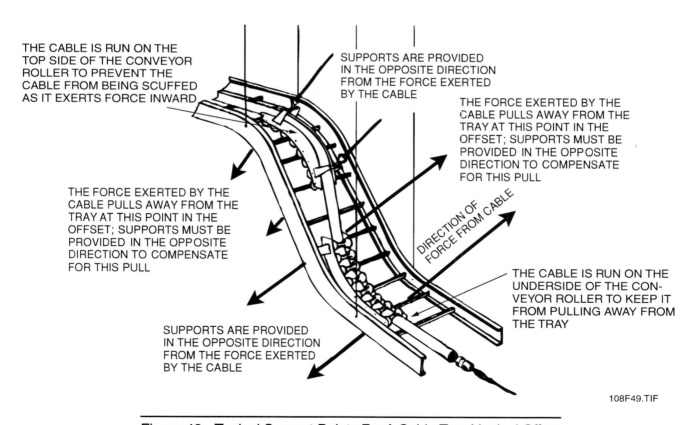

Figure 49. Typical Support Points For A Cable Tray Vertical Offset

Step 4 Prepare the end(s) of the cable(s) with an appropriate pulling device if a factory pulling eye has not been installed.

WARNING! Make sure that an appropriate rope is used and that it is rated to handle the weight of the cable that will be pulled.

Step 5 As necessary, fish the proper type of rope through the pathway and connect it to the end(s) of the cable(s) to be pulled.

Step 6 If it is required for the pull, mount a power tugger securely to the floor at the desired pulling location. If possible, position the tugger so that it is far enough from the pathway so that a reasonable amount of slack cable can be pulled into the area.

WARNING! Ensure that any power tugger used is securely bolted to the floor. If it breaks loose during a heavy cable pull, it may cause severe injury or death, as well as property damage.

Step 7 Position co-workers with communication devices at the reel end of the pull to view, aid, and communicate the progress of the cable.

Step 8 Lubricate any sleeves or conduit that the cable will pass through unless inner duct is installed and used.

Step 9 Commence the pull slowly, with co-workers carefully observing the process. If the cable is being pulled through conduit, lubricate the cable as needed.

CAUTION: Do not exceed the rated pulling tension of the cable.

Step 10 When the cable reaches the tugger (if used), make sure enough cable is pulled out to allow the remaining run to the final destination, plus a service loop and termination slack. It may be necessary to secure the cable with a rolling hitch knot and a rope. After the cable is temporarily secured, disconnect the pulling rope from the cable eye and reconnect it to the cable with a rolling hitch knot close to the pathway so that additional cable can be drawn at the tugger location. This action must be repeated until an adequate amount of cable is accumulated. Make sure that the cable is secured at any open support locations. Create service loops in the telecommunications closets and then secure the cable within the closets using appropriate clamps, brackets, or ties.

CAUTION: Do not allow the cables to wrap around the tugger capstan, as damage to the cables will occur.

Step 11 Label the cable(s) with their identification at all appropriate locations. Document the horizontal backbone cable information on the building plans/blueprints and cable(s) as follows:

- Type(s) of cable(s) installed
- Origin and termination of each cable
- Conduit (or other pathway) used for each cable
- Application (use) for each cable

5.4.4 Optical Fiber Backbone Cable Pulls

Optical fiber cable is generally used as backbone cable. It is usually installed in inner duct, which simplifies the pull. Normally, inner duct is installed in conduit, through sleeves, and on cable trays. However, inner duct can be run in vertical shafts, ducts, plenums, or across open ceilings with appropriate support. If inner duct is installed, make sure that it terminates inside the fiber optic equipment located in the telecommunications closet and that it also has a large service loop inside the closet. After the cable reels or boxes are positioned and supported and the pathways are determined to be clear, proceed as follows:

Step 1 Double check the length of the pathways to make sure adequate cable will be available. Verify the continuity of each fiber while the cable is on the reel (cable must be ordered with access to both ends specified.) A light source must be applied to one end and the other end checked for consistent light output.

WARNING! Never look into the end of a previously installed optical fiber cable. A laser light source that is not visible may be present that can severely damage the retinas of your eyes.

Step 2 If the pathway is open and the cables will be supported by J-hooks or the equivalent, make sure that the spacing of each hook is such that it will support less than 25 foot-pounds of cable weight.

Step 3 Set up pulleys and bullwheels between the reel and the pathway to guide the cable entering the pathway to prevent damage to the cable. Make sure the bullwheel is securely fastened to the building structure or to a swing fixture anchored to the floor or building structure.

Step 4 On open pathways, it may be desirable or necessary to mount pulleys along the entire path of the cable at locations near the J-hooks or support devices to ease the pulling of the cable and reduce the stress on the cable. Once pulled, the cable can be transferred to the J-hooks or other support devices and the pulleys can be removed. If cable trays are used, make sure the inner duct is secured before pulling commences.

Step 5   Prepare the end(s) of the cable(s) with an appropriate pulling device or prepare to attach the strength member of the cable to the pull line.

WARNING!   Ensure that the pull line or rope used is rated to handle the weight of the cable that will be pulled.

Step 6   As necessary, fish a pull line or rope through the pathway and connect it to the end(s) of the cable(s) to be pulled.

Step 7   If it is required for the pull, mount a power tugger securely to the floor at the desired pulling location. If possible, position the tugger so that it is far enough from the pathway so that a reasonable amount of slack cable can be pulled into the area.

WARNING!   Ensure that any power tugger used is securely bolted to the floor. If it breaks loose during a heavy cable pull, it may cause severe injury or death, as well as property damage.

Step 8   Position co-workers with communication devices at the reel end of the pull and at direction transition points to view, aid, and communicate the progress of the cable. Commence the pull slowly, with co-workers carefully observing the process.

Note: On open pulls, station co-workers at direction transitions so that they can help relieve tension on the cable during the pull.

Step 9   When the cable reaches the tugger (if used), make sure enough cable is pulled out to allow the remaining run to the final destination, plus a service loop and termination slack. It may be necessary to secure the cable with a rope and a rolling hitch knot. After temporarily securing the cable, disconnect the pulling rope from the end of the cable and reconnect it to the cable with a rolling hitch knot close to the pathway so that additional cable can be drawn at the tugger location. This action must be repeated until an adequate amount of cable is accumulated.

CAUTION: Do not exceed the rated pulling tension of the cable. Do not allow the cables to wrap around the tugger capstan, as damage to the cables will occur.

Step 10   Ensure that the cable is secured at any open support locations. If inner duct was not used, create service loops in the telecommunications closets and then secure the cable within the closets using appropriate clamps, brackets, or ties. Ensure that the clamps, brackets, or ties are not tight against the cable.

Step 11   Anchor all optical fiber cables to the termination cabinet using the strength member of the cable, as specified in the cabinet manufacturer's instructions.

Step 12 Label the cable(s) with their identification at all appropriate locations. Document the backbone cable information on the building plans/blueprints and cable(s) as follows:

- Type(s) of cable(s) installed
- Origin and termination of each cable
- Conduit (or other pathway) used for each cable
- Application (use) for each cable

5.4.5 Horizontal Work Area Cable Pulls

Horizontal four-pair cable is normally run between a telecommunications closet and each work area outlet on one or more floors. In addition, optical fiber cable may also be run to work area outlets. Sometimes, these types of horizontal pulls can be difficult because they may involve pulling bundles of cables over or through various cable supports. Horizontal work area cable may be supported by conduit runs (with or without inner duct) in walls, floors, and ceilings, or by J-hooks and beam clamps, cable ties, or other such devices, as well as through structural supports in open ceilings. In some cases, a power tugger may be required due to bundled cable weight. After the cable reels or boxes are positioned and supported and the pathways are determined to be clear, proceed as follows:

Note: *Appendices A and B provide various alternate procedures for running cable in unfinished commercial or finished commercial and residential areas.*

Step 1 Double check the length of the pathways to ensure that adequate cable will be available.

Note: Communication cable should not be spliced.

Step 2 If the pathway is open and the cable or cable bundle will be supported by J-hooks or the equivalent, ensure that the spacing of each hook complies with code requirements.

Step 3 If necessary, set up pulleys and bullwheels between the reels or boxes and the pathway to guide the cable(s) entering the pathway to prevent damage to the cable(s). Make sure the bullwheel is securely fastened to the building structure or to a swing fixture anchored to the floor or building structure.

Step 4 On open pathways, it may be desirable or necessary to mount pulleys along the entire path of the cable at locations near the J-hooks or support devices to ease the pulling of the cable and reduce the stress on the cable. Once pulled, the cable can be transferred to the J-hooks or other support devices and the pulleys can be removed.

Step 5 Identify and label the reels/boxes and then label the end(s) of the cable(s) to be pulled with the corresponding reel identification.

Step 6 Prepare the end(s) of the cable(s) or cable bundle for pulling or, if space permits, apply a rolling hitch knot using the pull line or rope. For optical fiber cable, the strength member will be attached to the pull line.

WARNING! Ensure that the pull line or rope used is rated to handle the weight of the cable that will be pulled.

Step 7 As necessary, fish a pull line or rope through the pathway and connect it to the end(s) of the cable(s) to be pulled. Attach a trailing line to the pull line that can be left for future pulls.

Step 8 If it is required for the pull, mount a power tugger securely to the floor at the desired pulling location. If possible, position the tugger so that it is far enough from the pathway so that a reasonable amount of slack cable can be pulled into the area.

WARNING! Ensure that any power tugger used is securely bolted to the floor. If it breaks loose during a heavy cable pull, it may cause severe injury or death, as well as property damage.

Step 9 Position co-workers with communication devices at the reel end of the pull to view, aid, and communicate the progress of the cable.

Step 10 Lubricate any conduit that the cable will pass through unless inner duct is installed and used.

CAUTION: Do not exceed the rated pulling tension of the cable.

Note: The maximum pulling force for one four-pair No. 24 AWG cable is 25 foot-pounds [110 Newtons (N)]. A breakaway swivel or fishline rated for the maximum pulling force is recommended between the pull line and the cable(s).

Step 11 Commence the pull slowly, with co-workers carefully observing the process. If the cable is being pulled through conduit, lubricate the cable as needed.

Step 12 When the cable or cable bundle reaches the tugger (if used), make sure enough is pulled out to allow the remaining run to the final destinations, plus termination slack. Make sure that the cable or cable bundle is secured at any open support locations. Create large service loops in the telecommunications closets and then secure the cable(s) within the closets using appropriate clamps, brackets, or ties.

CAUTION: Do not allow the cables to wrap around the tugger capstan, as damage to the cables will occur.

Step 13 Label the cable(s) with their identification at all appropriate locations before cutting the cable(s) from the reels/boxes. Document the horizontal work area cable information on the building plans/blueprints and cable(s) as follows:

- Type(s) of cable(s) installed
- Origin and termination of each cable
- Conduit (or other pathway) used for each cable
- Application (use) for each cable

Step 14 Apply a temporary label that conforms to the project labeling scheme to each end of all cables. A permanent label must be attached after the cables are terminated.

5.4.6 Conduit Fill For Backbone Cable

ANSI/TIA/EIA-569-A provides fill ratio guidelines for backbone cable installed in conduit. These guidelines (*Table 9*) should be observed when installing cable in conduit, as described in the above procedures. The fill ratios can be increased further by the use of lubricants.

Conduit			Area of Conduit								Minimum Radius of Bends			
					Maximum Occupancy Recommended									
					A		B		C		D		E	
Trade Size (in.)	Internal Diameter*		Area = .79D² Total 100%		1 Cable, 53% Fill		2 Cables, 31% Fill		3 Cables or More, 40% Fill		Layers of Steel Within Sheath		Other Sheath	
	mm	in.	mm²	in.²	mm²	in.²	mm²	in.²	mm²	in.²	mm	in.	mm	in.
¾	20.9	0.82	345	0.53	183	0.28	107	0.16	138	0.21	210	8	130	5
1	26.6	1.05	559	0.87	296	0.46	173	0.27	224	0.35	270	11	160	6
1¼	35.1	1.38	973	1.51	516	0.80	302	0.47	389	0.60	350	14	210	8
1½	40.9	1.61	1,322	2.05	701	1.09	410	0.64	529	0.82	410	16	250	10
2	52.5	2.07	2,177	3.39	1,154	1.80	675	1.05	871	1.36	530	21	320	12
2½	62.7	2.47	3,106	4.82	1,646	2.56	963	1.49	1,242	1.93	630	25	630	25
3	77.9	3.07	4,794	7.45	2,541	3.95	1,486	2.31	1,918	2.98	780	31	780	31
3½	90.1	3.55	6,413	9.96	3,399	5.28	1,988	3.09	2,565	3.98	900	36	900	36
4	102.3	4.03	8,268	12.83	4,382	6.80	2,563	3.98	3,307	5.13	1,020	40	1,020	40
5	128.2	5.05	12,984	20.15	6,882	10.68	4,025	6.25	5,194	8.06	1,280	50	1,280	50
6	154.1	6.07	18,760	29.11	9,943	15.43	5,816	9.02	7,504	11.64	1,540	60	1,540	60

*Internal diameters are taken from the manufacturing standard for EMI and rigid metal conduit.

Table 9. Conduit Fill Requirements For Backbone Cable

These fill requirements do not apply to sleeves, ducts, and straight conduit runs under 50' (15m) with no bends. The percentage of fill applies to straight runs with offsets equivalent to no more than two 90° bends. Column D indicates that conduit bends of 10 times the conduit diameter must exist for cable sheaths consisting partly of steel tape. Column E shows that for trade sizes up to 2", conduit bends should be at least six times the conduit diameter. For trade sizes above 2", the bends should be at least 10 times the conduit diameter.

6.0.0 RESIDENTIAL LOW-VOLTAGE CABLE INSTALLATION

For new construction, ANSI/TIA/EIA Standard 570-A provides guidelines for the installation of cabling for various types and grades of communication services for single and multi-tenant residential structures. Communication services include telephone, satellite, community antenna television (CATV or cable TV), and other data services such as basic, advanced, or multimedia telecommunications services. Other residential low-voltage systems for fire, security, and automation are normally installed in accordance with the manufacturer's instructions. Low-voltage systems must also be installed in accordance with federal, state, and local codes.

6.1.0 RESIDENTIAL UNIT COMMUNICATION/DATA CABLING REQUIREMENTS AND GRADES

For any residential unit in a single tenant or multi-tenant building, the minimum communication outlets recommended by Standard 570-A are one outlet location in each of the following residential unit spaces, as applicable:

- Bedrooms (all)
- Family/great room
- Study/den
- Kitchen

For each outlet location space listed above or any other outlet location spaces, additional outlet locations should be provided within any continuous wall area exceeding 12' (3.7m) in length and at any point in the space that is 25' (7.6m) or more horizontally from all other outlets in the space. The outlet heights are determined by local codes.

The degree of communication services provided to each of the outlet locations in a residential unit is determined by a standard grading system that defines the telecommunications services (*Table 10*) and cabling infrastructures required for the unit.

Grade	Residential Services Supported			
	Telephone	Data	Television	Multimedia
1	Yes	Yes	Yes	No
2	Yes	Yes	Yes	Yes

Table 10. Residential Grades Versus Services Supported

6.1.1 Residential Unit Grade 1 Service Cabling

This grade of service provides a generic cabling system that meets the minimum requirements for telecommunications services. Minimum requirements are for one Category 3, four-pair UTP cable and one 75Ω coaxial cable (types meeting SCTE IPS-SP-001) at each outlet location. Installation of Category 5 cable in place of Category 3 cable is recommended to enable upgrading to Grade 2 service in the future.

6.1.2 Residential Unit Grade 2 Service Cabling

Grade 2 service provides a generic cabling system that meets the requirements for basic, advanced, and multimedia telecommunications services. Minimum requirements are for two Category 5, four-pair UTP cables and two 75Ω coaxial cables (types meeting SCTE IPS-SP-001) at each outlet location. Optionally, installation of one or more dual-optical fiber cable(s) is recommended at each outlet location. Moreover, installation of Category 5e cable in place of Category 5 cable is recommended for performance purposes.

6.1.3 Residential Unit Communication/Data Cable Types

The cable types that can be specified and used for residential unit service are the same as those used for commercial service.

Cable types for outlet cable (horizontal) pathways include:

- Four-pair UTP
- RG-59 (Series 59) for patch and equipment cords only
- RG-6 (Series 6) or RG-11 (Series 11) coaxial
- 50/125µm or 62.5/125µm multi-mode optical fiber
- Single-mode optical fiber

Cable types for backbone pathways include:

- Multi-pair UTP
- RG-59 (Series 59) for patch and equipment cords only
- RG-6 (Series 6) or RG-11 (Series 11) coaxial
- Hard-line trunk coaxial
- 50/125µm or 62.5/125µm multi-mode optical fiber
- Single-mode optical fiber

6.2.0 UNDERSTANDING THE JOB

Because the pathways on a residential job may require onsite planning and layout, the system designer, or the lead electronic systems technician who has been completely briefed on the project, must be present to direct other technicians and see that all the work is done properly. This person must know where each device is going, and what each device is, as well as the appropriate cable from the **head end** (central distribution and/or control equipment) to the devices. This person should also know what equipment comprises the head end and where it is to be located, in addition to any power and ventilation requirements. All dimensions should be known, as well as rough openings for any rack-mounted equipment, etc. The head end, whenever possible, should be located somewhere near the center of the building. This allows the holes that must be drilled for the backbone pathway to be smaller and the cable runs to be shorter.

Once the head end is established, the lead technician and crew must work together to determine the best location(s) for the backbone pathway(s) of the cabling structure. These are the main pathways for all cables leaving the head end. The backbone pathway(s) must be drilled and placed according to all local and national codes, and should be kept well away from power lines/cables; plumbing; and heating, ventilation, air conditioning, and refrigeration (HVACR) equipment.

Special care should be taken to observe the types of construction and materials used on the project and what local and national codes should be applied. It must be remembered that laminated wooden beams cannot be drilled. Also, make sure that other trades do not run their cables through your conduits and holes. Their cables, especially power cables, may cause interference in your system. After the other trades have completed their work and before the finished walls are placed, you should go back to the site and inspect all the cables to make sure none have been damaged. It is also advisable to reinspect at various times throughout the construction process to ensure that the cables and boxes are intact.

6.3.0 RESIDENTIAL CABLE INSTALLATION REQUIREMENTS/CONSIDERATIONS

In new construction, horizontal and vertical pathway cable pulls are accomplished in much the same manner as described earlier for commercial installations. In many cases, cable is merely drawn through holes drilled in wood studs, joists, and plates or through pre-punched and bushed holes in metal studs and joists before the walls and ceilings are covered. In other instances, conduit, raceways, ducts, plenums, or sleeves are used as pathways for cable pulls. In a retrofit to existing construction, drilling and fishing of walls, attics, partitions, etc. will be required. Various methods of fishing cable in existing construction are given in detail in *Appendix A*. When installing residential cable, the following requirements and considerations should be observed:

- When working in enclosed attics, be careful not to work too long in high heat. Try to schedule attic work in the morning when it is cool. Be careful not to move too quickly or you may slip off ceiling joists and fall onto or through the drywall ceiling. Walking planks or plywood sheets (¾" thick) should be used on top of joists. To avoid electrocution, do not touch any bare electrical wires and the box of an electrical fixture at the same time. If the attic has non-covered insulation installed, wear proper respiratory protection and clothing to prevent contact with or inhalation of insulation particles and dust.

- If they are not already installed in new construction, consideration should be given to installing sleeves, metallic or flexible nonmetallic conduits and large boxes (*Figure 50*), cable ducts, or raceways of sufficient size (see ANSI/TIA/EIA-570-A) to accommodate future upgrades or expansion of services. This is especially true for multi-tenant structures or large single-family residential units where altering non-ducted concealed pathways cannot be accomplished without considerable expense once the wall and ceiling coverings are in place.

Figure 50. Typical Flexible Nonmetallic Conduit In Wood Stud Construction

- If conduit and large boxes are not used, the preferred box to use in residential low-voltage new construction is an open-backed box called a *mud ring*, also known as a *dry ring*. These boxes create a frame for drywall mudding, yet are open to make sure that cables are not bent less than their minimum radius or that wires are not crimped or compressed. Make sure that the mud rings are mounted so that they line up with electrical boxes and are not crooked.
- Underground conduit runs must be planned well in advance. If possible, have any low-voltage conduit runs installed by the electrical contractor in parallel with other conduit. In a retrofit, watch out for other buried pipes and cables, etc. Always have other buried utilities located and marked by a locating company prior to digging. Make sure conduit is large enough to pull cables freely and avoid binding. Never pull new cable over old cable in conduit. Always pull the old cable out after attaching a pulling line and then pull all new cable in. Make sure to use the proper type of cable in the conduit and properly space low-voltage cable from other cables to prevent RFI. As a rule, never run power cables and low-voltage cables in the same conduit.
- Do not drill holes in laminated beams.
- Precautions must be taken to eliminate cable stress caused by pulling. Take care not to deform cables as a result of tightly cinching or clamping or exceeding the minimum bend radius of the cables. The bend radius can vary drastically depending on the type of cable used. The minimum bend radius must be observed even when the cable is used inside a junction box or control cabinet.

The minimum bend radii for typical cables are as follows:

- The bend radius for outlet UTP cable shall not be less than four times the cable diameter.
- The bend radius for one- or two-pair optical fiber cable shall not be less than 1" (25.4mm) under no load conditions or 2" (50.8mm) if pulled through pathways under a minimum load of 50 foot-pounds (222N).
- The bend radius for backbone optical fiber cable shall not be less than 15 times the outside diameter if under tensile loading or 10 times the diameter if unloaded, unless the manufacturer recommends otherwise.
- The bend radius for coaxial cable shall not be less than 20 times the outside diameter of the cable under tensile load or 10 times the diameter if unloaded, unless the manufacturer recommends otherwise.

- Cables should not be pulled in excess of their maximum rated tensions. Typical tensions for some types of cable are:
  - Series 6 coaxial cable: 35 foot-pounds (150N)
  - Series 11 coaxial cable: 90 foot-pounds (400N)
  - Hardline trunk coaxial cable: refer to the manufacturer's recommendations
  - Four-pair, No. 24 AWG UTP cable: 25 foot-pounds (110N)

- Precautions must be taken to make sure that no drywall screw can reach the cables. If holes are drilled less than 1½" (38.1mm) from the edge of a stud, steel plates that span the diameter of the hole must be installed on the edge of the stud to prevent drywall screws or nails from being driven into a cable. In addition, cables should be bundled and secured away from the finished wall surface area with the correct fasteners. For instance, metal staples are not acceptable for low-voltage cables. One method is to use cable ties that are fastened to the stud and secured around the cables.

- Cable should be run only on or inside interior walls, partitions, and ceilings to avoid temperature variations that can cause performance problems or damage from siding application.

- When pulling cables, plan ahead. Think about which cables will be pulled before the first pull is started. Pull as many cables as possible at one time. Group the cable reels or boxes together and label the reels/boxes and associated cable ends. Be careful pulling cables over other existing cables. It is very easy to burn the cable insulation and very difficult to detect this once it has been done. Be careful pulling over sharp objects such as air conditioner vent hangers. Do not pull across attics at an angle. Pull along the sides where the cables will eventually reside. Do not place cables where other workers might trip or step on the cables. Always route cables along the proper hangers.

- Long runs of cable require the use of larger wire in order to eliminate excessive voltage drop. The amount of voltage drop can be calculated or various tables can be used to determine the required wire size.

- Service loops and slack lengths similar to those required for commercial cabling should be provided in vertical or horizontal residential pathways. Wires should be cut to sufficient length to ensure enough slack for later connection. This is usually between 12" to 18" (304.8mm to 457.2mm) but may be more if the connection point is away from the wall (typical for audio equipment). Cable slack must be protected until terminated. This can be accomplished by bagging the slack cable or looping it back into the wall box and loosely fastening it to the box. In some cases, cables may be left behind the walls for later retrieval. In this case, the cables must be properly documented and the use of a photograph for later reference is recommended. Typically, cables are looped in a circle at the side of a stud or rafter and secured out of the way using bundling wire.
- Once installed, all cable runs must be properly labeled and documented on the building plans or drawings as follows:
    - Type(s) of cable(s) installed
    - Origin and termination of each cable
    - Cable wire color code assignments
    - Conduit (or other pathway) used for each cable
    - Application (use) for each cable

    Apply a temporary label that conforms to the project labeling scheme to each end of all cables. Permanent labels must be attached after the cables are terminated. Multiple labels applied at both ends are recommended as a precaution against one of the labels coming off and being lost.
- For non-ducted concealed pathways, consideration should be given to conducting any required or desired field tests before the interior wall or ceiling coverings are installed.

6.4.0 DRILLING AND FISHING CABLE IN EXISTING CONSTRUCTION

In retrofits of existing construction (either commercial or residential), drilling into a blind cavity can produce disastrous results if precautions are not taken to prevent penetrating into various hidden obstacles such as pipes, electrical wire, etc. The following are general tips and techniques for drilling to accommodate the cabling methods given in *Appendix B*.

Always observe proper safety procedures when operating power tools:

- Make sure that you have been properly trained for the tools and accessories to be used.
- Wear appropriate personal protective equipment.
- Make sure power cords are not damaged and are approved types.
- Use the right tool and tool bit for the job.

Obstacles are typically located in walls in the following ways:

- Electrical wire is usually located horizontally in walls from receptacle to receptacle at a height of 15" (381mm) to 24" (609.6mm) off the floor. Vertical drops for switches are usually stapled to the center of a vertical stud adjacent to the switch box location.
- Water supply lines are usually vertical through plates in the center of stud cavities or are horizontal and located at a height of between 8" (203.2mm) and 24" (609.6mm) above the floor.
- Waste and vent lines are usually vertical through stud cavities to the roof or are horizontal and located at a height of 16" (406.4mm) to 40" (1.02m) above the floor.

Do whatever is possible to familiarize yourself with the area to be drilled. Examine the area from above, below, and behind, if possible. Look for obstructions. Whenever possible, use specialized tools to locate studs, pipes, electrical wire, etc. before drilling. For walls, one method is to cut a box hole between the studs with a drywall knife and look inside the wall for obstructions. At the base of the wall, remove the baseboard, then cut a hole to look for obstructions at floor level and to use for drilling through the floor. Holes through the bottom plate (or top plate) of a wall can be drilled using a spade bit, bore bit, or flex bit.

7.0.0 INTERIOR LOW-VOLTAGE CABLING INSTALLATION REQUIREMENTS

The NEC specifies requirements for the installation of certain classes of cable in various low-voltage applications in either residential or commercial construction. These requirements are summarized below.

7.1.0 CLASS 1 CIRCUITS

NEC Section 725-21 defines Class 1 circuits as either power-limited or nonpower-limited remote control and signaling circuits. Power-limited circuits (Class 1) shall have a power supply that is not more than 30V and 1,000VA. Nonpower-limited circuits shall not exceed 600V.

Class 1 circuits shall be installed as specified in *NEC Chapter 3*, except as noted in the following paragraphs.

7.1.1 Conductors Of Different Circuits In The Same Cable, Enclosure, Or Raceway

According to *NEC Section 725-26*, two or more Class 1 circuits can occupy the same cable, enclosure, or raceway, either AC or DC, if all conductors are insulated for the highest voltage of any conductor.

With several exceptions, Class 1 circuits and power supply circuits are only permitted to occupy the same cable, enclosure, or raceway when the equipment being powered is functionally associated with the Class 1 circuits.

According to **NEC Section 725-28**, when Class 1 circuit conductors are in a raceway or cable tray, the conductors, including any power supply conductors, must be derated as specified in **NEC Section 310-15** if the Class 1 conductors carry continuous loads in excess of 10% of the conductor **ampacity** and the number of conductors is more than three. If three or more power supply conductors are included with the Class 1 conductors, only the power supply conductors must be derated if the Class 1 conductors do not carry continuous loads in excess of 10% of each conductor. The number of conductors permitted in the raceway or tray is determined by **NEC Section 300-17**.

When Class 1 conductors are installed in cable trays, they must comply with **NEC Sections 318-9 through 318-11**.

7.2.0 CLASS 2 AND 3 CIRCUITS

The installation requirements for Class 2 and 3 circuits (**NEC Section 725-54**) are summarized in the following paragraphs.

7.2.1 Separation Of Class 2 And 3 Circuits From Power Circuits

Class 2 and 3 circuits shall not be placed in cables, compartments, cable trays, enclosures, outlet boxes, device boxes, and raceways with electric power circuits (including electric light, Class 1, nonpower-limited fire protection circuits, and medium power, network-powered broadband communication cables) except as permitted in any one of 1 through 6 below:

1. Installation with power circuits is permitted if the circuits are physically separated by a solid, fixed barrier.
2. Installation with power circuits is permitted if the circuits are connected to the same equipment and as specified in both a and b below:
   a. The conductors and cables of the Class 2 and 3 circuits are separated from the other conductors and cables by at least ¼" (6.35mm), *and*
   b. The circuit conductors operate at 150V or less and comply with (1) or (2) below:
      (1) The Class 2 and 3 circuits are installed using CL3, CL3R, or CL3P or permitted substitutes, provided these Class 3 cable conductors extending beyond the jacket are separated by a minimum of ¼" (6.35mm) or by a nonconductive sleeve or barrier from all other conductors, *or*
      (2) The Class 2 and 3 circuit conductors are installed as Class 1 circuits.
3. The Class 2 and 3 circuits may enter an enclosure through a hole along with power conductors if separated by a continuous and firmly fixed nonconductor such as flexible tubing, provided they enter as permitted above.
4. Installation with power circuits is permitted if the circuits are installed in a manhole and the installation meets specified conditions.
5. Installation with power circuits is permitted if the circuits are contained in a rated hybrid cable as defined by **NEC Section 780-6(a)**.

6. Installation with power circuits is permitted if the circuits are in cable trays and the power circuits are separated by a solid, fixed barrier of a material compatible with the cable tray or the Class 2 and 3 circuits are installed in Type MC cable.

7.2.2 Separation In Hoistways

Class 2 and 3 circuits in hoistways shall be installed in RMC, NMC, IMC, or EMT, except for installations in elevator shafts.

7.2.3 Separation In Other Applications

Class 2 and 3 circuits shall be separated by at least 2" (50.8mm) from all power conductors and medium power, network-powered broadband communication cables except as permitted by 1 or 2 below:

1. Separation of less than 2" (50.8mm) is permitted if either the Class 2 and 3 circuits or the power/broadband communication cables are in raceway, metal sheathed/clad, nonmetallic sheathed, or Type UF cables.
2. Separation of less than 2" (50.8mm) is permitted if the Class 2 and 3 circuits are separated by a continuous, firmly fixed barrier in addition to the circuit insulation.

Note: The separation requirements above are for safety purposes. Additional separation may be required for performance issues such as electromagnetic interference (EMI) suppression.

7.2.4 Support Of Conductors

Class 2 and 3 circuit conductors or cables shall not be strapped, taped, or attached by any means to the exterior of any conduit or other raceway except for raceways dedicated to that purpose.

7.2.5 Installation In Plenums, Risers, Cable Trays, And Hazardous Locations

For Class 2 and 3 circuits, only plenum-rated cables shall be installed in plenums and only riser or plenum-rated cable shall be installed in risers unless installed in compliance with ***NEC Section 300-22***. For one-family and two-family residential installations, CL2, CL3, CL2X, and CL3X shall be permitted for these uses. In hazardous locations, only PLTC-rated cable shall be used and adequately supported, except as noted in ***NEC Sections 501-4, 502-4, and 504-20***. In outdoor cable trays, only PLTC-rated cable may be used.

7.3.0 INSTRUMENTATION TRAY CABLE CIRCUITS

Instrumentation tray cable (***NEC Section 727-4***) shall be permitted for use in industrial establishments only when qualified persons will service the installation. Under this condition, it may be installed in cable trays, raceways, hazardous locations (as defined in the

NEC), and as open wiring when equipped with various versions of a metallic sheath. As open wiring, the cable must be supported and secured at intervals of not more than 6' (1.83m). Nonmetallic-sheathed cable may be used as open wiring to equipment, provided it is less than 50' (15.24m) long and is protected from physical damage. ITC cable that complies with Type MC cable crush and impact requirements may also be used as open wiring to equipment, provided it is less than 50' (15.24m) long. ITC cable may be used as an aerial cable on a messenger or under raised floors in control rooms and rack rooms when protected against damage. ITC cable shall not be used in circuits operating at more than 150V or more than 5A. The cable shall not be installed with electric power circuits of any kind unless its nonmetallic sheath is covered with a metallic sheath or unless it is terminated inside enclosures and separation is maintained by insulating barriers or other means.

7.4.0 NONPOWER-LIMITED FIRE ALARM CIRCUITS

Fire alarm circuits are defined as either power-limited (PLFA) or nonpower-limited fire alarm (NPLFA) circuits. Power-limited fire alarm circuits shall be supplied from a listed, power-limited transformer (Class 3) or power supply. Nonpower-limited fire alarm circuits (***NEC Section 760-25***) shall not exceed 600V and shall be installed as specified in ***NEC Chapters 1 through 4***, except as noted in the following paragraphs.

7.4.1 Conductors Of Different Circuits In The Same Cable, Enclosure, Or Raceway

Some of the NEC requirements for NPLFA circuits are as follows:

- Per ***NEC Section 760-26(a)***, NPLFA and Class 1 circuits can occupy the same cable enclosure or raceway, either AC and/or DC, if all conductors are insulated for the highest voltage of any conductor.

- Per ***NEC Section 760-26(b)***, NPLFA circuits and power supply circuits are only permitted to occupy the same cable, enclosure, or raceway when the equipment being powered is functionally associated with the fire alarm circuits.

- Per ***NEC Section 760-28***, when NPLFA circuit conductors are in a raceway or cable tray, the conductors, including any power supply conductors, must be derated as specified in ***NEC Section 310-15*** if the NPLFA conductors carry continuous loads in excess of 10% of the conductor ampacity and the number of conductors is more than three. If three or more power supply conductors are included with the NPLFA conductors, only the power supply conductors must be derated if the NPLFA conductors do not carry continuous loads in excess of 10% of each conductor. The number of conductors permitted in the raceway or tray is determined by ***NEC Section 300-17***. Also, when Class 1 conductors are installed in cable trays, they must comply with ***NEC Sections 318-9 through 318-11***.

- Per ***NEC Section 760-30***, multi-conductor NPLFA cables may be used on circuits operating at 150V or less and shall be installed:
  - In raceways or exposed on the surface of ceilings and sidewalls or fished in concealed spaces

Note: When installed exposed, maximum protection against damage must be afforded by the building construction. The cable must be securely fastened at intervals of 18" (457.2mm) or less to a height of 7' (2.13m) from the floor. All splices and connections must be made inside listed devices.

- In metal raceway or rigid nonmetallic conduit where passing through a floor or wall to a height of 7' (2.13m) unless other protection is provided
- In rigid metal conduit, rigid nonmetallic conduit, intermediate metal conduit, or electrical metal tubing when installed in hoistways

7.5.0 POWER-LIMITED FIRE ALARM CIRCUITS

Power-limited fire alarm circuits [***NEC Sections 760-52(b) and 760-54***] may be installed in raceway or exposed on the surfaces of walls and ceilings or fished in concealed spaces. Splices and terminations must be made in listed enclosures, devices, or fittings. When exposed, the circuits must be supported and protected against damage by using baseboards, door frames, ledges, etc. When located within 7' (2.13m) of the floor, cables must be fastened at intervals of 18" (457.2mm) or less. The circuits must be installed in metal raceways or NMC when passing through a floor or wall to a height of 7' (2.13m) above the floor unless protection is provided by baseboards, door frames, ledges, etc. or unless equivalent solid guards are provided. When installed in hoistways, the circuits must be installed in RMC, NMC, IMC, or EMT with certain exceptions for elevator shafts or unless other wiring methods and materials are being used to extend or replace conductors/cables in plenums, risers, or other building areas.

7.5.1 Separation Of Power-Limited Fire Alarm Circuits From Power Circuits

Power-limited fire alarm circuits shall not be placed in cables, compartments, cable trays, enclosures, outlet boxes, device boxes, and raceways with electric power circuits (including electric light, Class 1, nonpower-limited fire protection circuits, and medium power, network-powered broadband communication cables) except as permitted in any one of 1 through 3 below:

1. Installation with power circuits is permitted if the circuits are physically separated by a solid, fixed barrier.
2. Installation with power circuits is permitted if the circuits are connected to the same equipment and as specified in both a and b below:
    a. The conductors and cables of power-limited fire alarm circuits are separated from the other conductors and cables by at least ¼" (6.35mm), *and*

b. The circuit conductors operate at 150V or less and comply with (1) or (2) below:
   (1) The power-limited fire alarm circuits are installed using FPL, FPLR, or FPLP or permitted substitutes, provided that power-limited conductors extending beyond the jacket are separated by a minimum of ¼" (6.35mm) or by a nonconductive sleeve or barrier from all other conductors, *or*
   (2) The power-limited circuit conductors are installed as nonpower-limited fire alarm circuits.
3. The power-limited fire alarm circuits may enter an enclosure through a hole along with power conductors if separated by a continuous and firmly fixed nonconductor such as flexible tubing, provided they enter as permitted above.

7.5.2 Separation In Hoistways

Power-limited fire alarm circuits in hoistways shall be installed in RMC, NMC, IMC, or EMT except for certain installations in elevator shafts.

7.5.3 Separation In Other Applications

Power-limited fire alarm circuits shall be separated by at least 2" (50.8mm) from all power conductors and medium power, network-powered broadband communication cables except as permitted by 1 or 2 below:

1. Separation of less than 2" (50.8mm) is permitted if either the power-limited fire alarm circuits or the power/broadband communication cables are in raceway, or in metal sheathed/clad, nonmetallic sheathed, or Type UF cables.
2. Separation of less than 2" (50.8mm) is permitted if the power-limited fire alarm circuits are separated by a continuous, firmly fixed barrier in addition to the circuit insulation.

Note: The separation requirements above are for safety purposes. Additional separation may be required for performance issues such as electromagnetic interference (EMI) suppression.

7.5.4 Support Of Conductors

Power-limited fire alarm circuit conductors or cables shall not be strapped, taped, or attached by any means to the exterior of any conduit or other raceway.

7.5.5 Current-Carrying Continuous Line-Type Fire Detectors

These types of fire detectors, including the insulated copper tubing of pneumatically-operated detectors, are permitted to be used in power-limited circuits. They are installed as specified above for power-limited fire alarm circuits.

7.5.6 Plenums And Risers

For power-limited fire alarm circuits, only plenum-rated cables shall be installed in plenums and only riser or plenum-rated cable shall be installed in risers unless installed in compliance with ***NEC Section 300-22***. For one-family and two-family residential installations, FPL cable shall be permitted for these uses.

7.6.0 OPTICAL FIBER CABLE

In accordance with ***NEC Sections 770-52 and 770-53***, optical fibers are permitted in composite cables with electric power circuits under 600V (including electric light, Class 1, nonpower-limited fire protection circuits, and medium power, network-powered broadband communication cables) when the fibers and conductors are associated with the same function. Nonconductive optical fiber cables are permitted in the same cable tray or raceway with electric power/broadband communication circuits operating at 600V or less. However, conductive optical fiber cables are not permitted. Both conductive and nonconductive optical fiber cables shall be permitted in the same cable tray, enclosure, or raceway with conductors of Class 2 and 3, power-limited fire alarm, communication circuits, CATV, or low power, network-powered broadband communication circuits.

Composite optical fiber cables containing only power circuits operating at 600V or less are permitted to be placed in enclosures or pathways with other power circuits operating at 600V or less except as specified below. Per ***NEC Section 770-52***, nonconductive optical fiber cables are not permitted to be placed in enclosures containing terminations for power/broadband communication circuits except as specified in any one of 1 through 4 below:

1. Occupancy is permitted when associated with the function of the power/broadband communication circuits.
2. Occupancy is permitted if fibers are installed in a factory or field-assembled control center.
3. Occupancy is permitted in industrial settings where maintenance is accomplished by qualified personnel.
4. Composite optical fiber cables are permitted to contain power circuits operating in excess of 600V only in industrial settings where maintenance is accomplished by qualified personnel.

For optical fiber circuits, only plenum-rated cables shall be installed in plenums and only riser or plenum-rated cable shall be installed in risers unless Types OFNR, OFCR, OFNG, OFN, OFCG, and OFC are installed in plenums in compliance with ***NEC Section 300-22*** or Types OFNG, OFN, OFCG, and OFC installed as risers are encased in metal raceway or are located in a fireproof shaft with firestops on each floor. For one-family and two-family residential installations, Types OFNG, OFN, OFCG, and OFC shall be permitted for these uses.

Any listed optical fiber cable may be installed in hazardous locations and in cable trays. Raceway installations of optical fiber cable must comply with *NEC Section 300-17*.

Plenum, riser, and general-purpose raceways listed for optical fiber cable applications may be used [*NEC Sections 770-51(e) through (g)*].

7.7.0 HYBRID CABLE

Conductors of listed hybrid cable [*NEC Sections 780-6(a) and (b)*] can occupy the same enclosure housing terminations of power circuits only if connectors listed for hybrid cable are used.

7.8.0 COMMUNICATION CIRCUITS WITHIN BUILDINGS

In accordance with *NEC Sections 800-48 and 800-52*, communication circuits are permitted in raceways or enclosures with any other power-limited Class 2 and 3, fire alarm, optical fiber, CATV, and low power, network-powered broadband communication circuits.

7.8.1 Separation of Communication Circuits From Power Circuits

Communication circuits shall not be placed in cables, compartments, cable trays, enclosures, outlet boxes, device boxes, and raceways with electric power circuits (including electric light, Class 1, nonpower-limited fire protection circuits and medium power, network-powered broadband communication cables) except as permitted in 1 or 2 below:

1. Installation with power circuits is permitted if the circuits are physically separated by a solid, fixed barrier.
2. Installation with power circuits is permitted if the circuits are connected to the same equipment and the conductors and cables of the communication circuits are separated from the other conductors and cables by at least ¼" (6.35mm).

7.8.2 Separation In Other Applications

Communication circuits shall be separated by at least 2" (50.8mm) from all power conductors and medium power, network-powered broadband communication cables except as permitted by 1 or 2 below:

1. Separation of less than 2" (50.8mm) is permitted if either the communication circuits are enclosed in a raceway or the power/broadband communication cables are in a raceway or in metal sheathed/clad, inner duct sheathed, Type AC, or Type UF cables.
2. Separation of less than 2" (50.8mm) is permitted if the communication circuits are separated by a continuous, firmly fixed barrier in addition to the circuit insulation.

LOW-VOLTAGE CABLING — TRAINEE TASK MODULE 33108

Note: The separation requirements above are for safety purposes. Additional separation may be required for performance issues such as electromagnetic interference (EMI) suppression.

7.8.3 Support Of Conductors

Communication circuit conductors or cables shall not be strapped, taped, or attached by any means to the exterior of any conduit or other raceway.

7.8.4 Cable Trays

All communication cable types except CMX may be used in cable trays.

7.8.5 Plenums And Risers

For communication circuits, only plenum-rated cables shall be installed in plenums or in listed plenum raceways installed in plenums per **NEC Sections 300-22(b) and (c)**. Only riser or plenum-rated cable shall be installed in risers or in listed riser raceways. Exceptions to the preceding are as specified in 1 or 2 below:

1. The installation is permitted if listed cables are encased in metal raceway or are located in a fireproof shaft with fire stops on each floor.
2. The installation of types CM and CMX cable is allowed in one-family and two-family residences.

7.8.6 Plenums, Risers, And General-Purpose Raceways

When communication circuits are installed in raceways, the raceways shall be one of the types covered in **NEC Chapter 3** except that listed inner duct plenum, riser or general-purpose raceway conforming to **NEC Sections 800-51(j) through (l)** and installed in compliance with **NEC Sections 331-7 through 331-14** is permitted.

7.9.0 COAXIAL CATV CABLE INSTALLATION WITHIN BUILDINGS

Coaxial CATV installation within buildings (**NEC Section 820-52**) is summarized in the following paragraphs.

7.9.1 Separation In Raceways And Boxes

Coaxial cables shall be permitted in raceways or enclosures along with any Class 2 and 3, power-limited fire alarm, communication, nonconductive and conductive optical fiber, and/or low power, network-powered broadband communication circuits. Coaxial cable shall not be placed in raceways and boxes along with electric light, power, Class 1, nonpower-limited fire alarm, and/or medium power, broadband communication circuits unless separated by a barrier or unless the power circuits are supplying coaxial distribution equipment. In this case, the coaxial cables shall be routed to maintain a separation from the power circuits of more than ¼" (6.35mm).

7.9.2 Separation In Other Applications

Per *NEC Section 820-52*, coaxial cable shall be separated by more than 2" (50.8mm) from conductors of any electric light, power, Class 1, nonpower-limited fire alarm, and/or medium power, broadband communication circuits except as specified in 1 or 2 below:

1. The circuits are in a separate raceway or in metal-sheathed, metal-clad, nonmetallic-sheathed, Type AC, or Type UF cables or all of the coaxial cables are in a separate raceway.
2. The coaxial cables are permanently separated by a continuous and firmly fixed nonconductor such as porcelain tubes or flexible tubing in addition to the insulation on the wire.

Note: The separation requirements above are for safety purposes. Additional separation may be required for performance issues such as electromagnetic interference (EMI) suppression.

7.9.3 Support Of Conductors

Raceways shall be used for their intended purpose. Coaxial cables shall not be strapped, taped, or attached by any means to the exterior of any conduit or raceway as a means of support.

7.10.0 NETWORK-POWERED BROADBAND COMMUNICATION SYSTEM INSTALLATION WITHIN BUILDINGS

Network-powered broadband communication system installation within buildings *(NEC Section 830-58)* is summarized below.

7.10.1 Separation In Raceways And Boxes

Some of the NEC requirements for separation in raceways and boxes are as follows:

- Low and medium power, network-powered broadband communication cables shall be permitted in the same raceway or enclosure.
- Low power, network-powered broadband communication cables shall be permitted in raceways or enclosures along with any Class 2 and 3, power-limited fire alarm, communication, nonconductive and conductive optical fiber, and/or community antenna television and radio distribution system circuits.

- Medium power, network-powered broadband communication cables shall not be permitted in raceways or enclosures along with any Class 2 and 3, power-limited fire alarm, communication, nonconductive and conductive optical fiber, and/or community antenna television and radio distribution system circuits.
- Low and/or medium power, network-powered broadband communication cable shall not be placed in raceways and boxes along with electric light, power, Class 1, nonpower-limited fire alarm, and/or medium power, broadband communication circuits unless separated by a barrier or unless the power circuits are supplying network distribution equipment. In this case, the network-powered broadband communication cables shall be routed to maintain a separation from the power circuits of more than ¼" (6.35mm).

7.10.2 Separation In Other Applications

Low and/or medium power, network-powered broadband communication cable shall be separated by more than 2" (50.8mm) from conductors of any electric light, power, Class 1, and nonpower-limited fire alarm circuits except as specified in 1 or 2 below:

1. The circuits are in a separate raceway or in metal-sheathed, metal-clad, nonmetallic-sheathed, Type AC, or Type UF cables, or all of the network-powered broadband communication cables are in a separate raceway.
2. The network-powered broadband communication cables are permanently separated by a continuous and firmly fixed nonconductor (such as porcelain tubes or flexible tubing) in addition to the insulation on the wire.

Note: The separation requirements above are for safety purposes. Additional separation may be required for performance issues such as electromagnetic interference (EMI) suppression.

7.10.3 Support Of Conductors

Raceways shall be used for their intended purpose. Low and/or medium power, network-powered broadband communication cables shall not be strapped, taped, or attached by any means to the exterior of any conduit or raceway as a means of support.

8.0.0 ELECTROMAGNETIC INTERFERENCE CONSIDERATIONS

When installing low-voltage cabling, electromagnetic interference (EMI) is one of the primary considerations during the selection and routing of cable. Electromagnetic interference may cause erratic operation of low-voltage circuits. Electromagnetic interference can be the result of coupled or induced currents created from nearby power lines, transformers, fluorescent ballasts, motors, office/factory equipment, or other electrical equipment. Harmonics of 60Hz power can also produce circuit noise. Low-voltage circuits can also be affected by radiated EMI from various noise sources such as variable-speed AC motor drives, generators, electronic equipment, fluorescent lamps, etc. In some cases, it may be man-made interference such as that caused by radio and television transmission signals or signals used for deliberate jamming purposes.

Adequate physical separation between the low-voltage circuits and any potential EMI sources is the primary solution in most installations. Along with separation, the use of high-performance UTP cable should eliminate most EMI problems. In especially harsh environments (e.g., industrial facilities) various types of shielded cable or optical fiber cable can be used to reduce or eliminate EMI. Check with cable manufacturers to determine the level of noise immunity offered by various grades of cable.

The following major precautions should be considered to reduce interference from sources of EMI:

- Use grounded metallic pathways to limit inductive or radiated EMI.
- Minimize emission from power conductors by using sheathed power cables or other construction such as taping, twisting, or bundling the conductors to prevent separation of the line, neutral, and grounding conductors.
- Maintain a minimum of 24" (609.6mm) between electrical power cables and electrical or electronic equipment and unshielded low-voltage cables.
- If at all possible, cross power cables at 90° angles.
- For unintentional or deliberate high-noise environments, shielding and/or EMI filters may be required.

8.1.0 EMI GUIDELINES

The following general guidelines detail some of the precautions required for increased protection against EMI:

- Twisted-pair cable results in less loop area for inductive pickup and lower inductance per foot of running cable.
- Multiple-conductor cable should consist of twisted conductors.
- Multiple-conductor cable used for transmission of several individual signals should consist of twisted-pair signal leads that are isolated from each other.
- Multiple-conductor cables should have an overall shield to further improve EMI.
- Terminate unused conductors at both ends or remove them altogether.
- Similar signals should be run together and not intermixed (e.g., run power with power and data with data).
- Closely space power leads to maximize cancellation of magnetic fields. Where possible, twist conductors that carry alternating currents.
- Low-level data transmission lines should not be run parallel to high-level power lines.
- Use localized magnetic barriers when signal lines are found close to switchgear.

- Conduit should not be buried below high-voltage transmission lines or in areas with high ground currents.
- Where unshielded cables of different signal conditions must cross, they should do so at right angles, not gradually over long distances.
- Pull boxes and junction boxes may require special attention, such as the need for ground barriers between circuits operating at different signal levels.
- Conduit separation in cable tray should be considered as a potential problem area.
- Tray networks in the floor provide better noise protection than other installations.
- System modifications and additions should be viewed as possible areas of concern.
- Signal return wire should not be shared by more than one signal path. Individual signal return wires must be used.
- EMI source suppression should be considered. Dealing with noise at the source helps to eliminate major corrective action on cabling systems.
- Avoid installing low-voltage circuits in pathways with power circuits, even if separated.
- Do not use isolated grounding circuits unless the equipment manufacturer mandates using isolated circuits.
- Whenever possible, use grounded conduits and enclosures for low-voltage circuits.
- Always follow the equipment manufacturer's instructions for the use and grounding of any cable shields.

SUMMARY

Knowing the different types of low-voltage cables, their construction, how they are rated, and their uses is important to the electronic systems technician when cable must be selected, routed, and pulled for a particular job. The technician must be familiar with the equipment, safety procedures, and techniques of pulling various types of cable in residential and commercial buildings. The procedures for installing low-voltage cable and the various requirements of the NEC and other standards or considerations covering the installation of low-voltage cable must also be understood.

References

For advanced study of topics covered in this Task Module, the following books are suggested:

National Electrical Code Handbook, Latest Edition, National Fire Protection Association, Quincy, MA.

Telecommunications Cabling Installation Manual, Latest Edition, BICSI, Tampa, FL. www.bicsi.org

Telecommunications Distribution Methods Manual, Latest Edition, BICSI, Tampa, FL. www.bicsi.org

TIA/EIA Telecommunication Building Wiring Standards, Latest Edition, Global Engineering Documents, Englewood, CO.

ACKNOWLEDGMENTS
Figures 10, 11, 13–17, 23, 24, 25, 32, 35, 37, 38, and 40–45 courtesy of BICSI, Tampa, FL
Tables 4, 5, 6, and 9 courtesy of BICSI, Tampa, FL
Figure 50 courtesy of Carlon Electrical Products, Cleveland, OH
Appendices A, B, and C courtesy of Labor $aving Devices®, Inc., Commerce City, CO

REVIEW/PRACTICE QUESTIONS

1. Low-voltage cable typically contains solid wire sizes ranging from _____ AWG.
   a. No. 28 to No. 14
   b. No. 26 to No. 12
   c. No. 22 to No. 10
   d. No. 20 to No. 8

2. Which of the following is most commonly used for conductor insulation and cable jackets?
   a. Polyvinyl chloride
   b. Polyethylene
   c. Rubber
   d. Crosslinked polyethylene

3. In an optical fiber cable, the _____ creates a reflective boundary.
   a. buffer layer
   b. acrylate coating
   c. cladding
   d. jacket

4. Single-mode optical fiber core is normally _____ in diameter.
   a. 8 to 9 microns
   b. 50 microns
   c. 62.5 microns
   d. 100 microns

5. Communications circuits are covered in _____.
   a. *NEC Article 760*
   b. *NEC Article 770*
   c. *NEC Article 780*
   d. *NEC Article 800*

6. A power-limited fire alarm cable that can be used in plenums is marked as _____.
   a. FPL
   b. FPLP
   c. FPLR
   d. CM

7. A limited-use communication cable that can only be used in residences and raceways is marked as _____.
   a. CMX
   b. CM
   c. NP
   d. CMG

8. On a shielded cable, floating the ground means _____.
   a. removing the shield
   b. not grounding the shield
   c. grounding both ends of the shield
   d. grounding one end of the shield

9. Category 4 UTP cable has a bandwidth up to _____ MHz.
   a. 10
   b. 16
   c. 20
   d. 100

10. The minimum grade of UTP cable acceptable for new installations is Category _____.
    a. 2
    b. 3
    c. 4
    d. 5

11. The UTP cable tip color for the first five pairs is normally _____.
    a. red
    b. blue
    c. yellow
    d. white

12. A power fishing system is used primarily to blow or vacuum a _____ through a raceway system.
    a. cable
    b. wire
    c. pull line
    d. heavy rope

13. A vertical cable pull from the top down does not require _____.
    a. a pull line
    b. reel brakes
    c. a rope tugger
    d. temporary take-up devices

14. Horizontal pulls for backbone communication cable installations must not exceed _____ between pull points.
    a. 98'
    b. 295'
    c. 328'
    d. 984'

15. Which of the following residential spaces would *not* usually receive a telecommunications outlet?
    a. Bedrooms
    b. Kitchen
    c. Garage
    d. Family room

16. Grade 1 residential service does not provide _____ service.
    a. telephone
    b. data
    c. television
    d. multimedia

17. In new residential construction where conduit is not used, cables can be brought through the finished wall by the use of a _____.
    a. mud ring
    b. closed box
    c. sleeve
    d. ceramic tube

18. A _____ is used to prevent cables from being damaged by drywall screws or nails.
    a. wire tie
    b. steel plate
    c. mud ring
    d. sleeve

19. In accordance with the NEC, power-limited fire alarm circuits are normally separated from non-related electrical power and medium power, network-powered broadband conductors/cables by a minimum space of _____.
    a. ¼"
    b. 1"
    c. 2"
    d. 3"

20. To reduce EMI, it is recommended that consideration be given to separating low-voltage circuits from power circuits by a minimum distance of _____.
    a. 2"
    b. 12"
    c. 24"
    d. 36"

ANSWERS TO PRACTICE/REVIEW QUESTIONS

Answer	Section Reference
1. b	2.1.0
2. a	2.3.0
3. c	3.0.0
4. a	3.0.0
5. d	4.1.0
6. b	4.1.0/Table 3
7. a	4.1.0/Table 3
8. d	4.2.0
9. c	4.3.1
10. b	4.3.1
11. d	4.3.1/Table 4
12. c	5.0.0
13. c	5.1.2
14. a	5.4.3
15. c	6.1.0
16. d	6.1.0/Table 10
17. a	6.3.0
18. b	6.3.0
19. c	7.5.3
20. c	8.0.0

APPENDIX A

INSTALLING CABLE IN UNFINISHED AREAS OF COMMERCIAL CONSTRUCTION

The following procedures and illustrations are provided courtesy of Labor $aving Devices®, Inc. The tools referred to in the procedures are illustrated and described in *Appendix C*.

Note: In the section to follow, we have addressed some of the most common situations facing wire/cable installers. Listed are the basic tools required for the best results, along with a brief explanation of how to proceed. It is purely informative, and in no event can Labor $aving Devices®, Inc. be liable for any special, incidental, or consequential damages.

IT REMAINS THE SOLE RESPONSIBILITY OF THE INSTALLER TO OPERATE IN ACCORDANCE WITH ANY APPLICABLE FEDERAL, STATE, COUNTY, CITY, AND LOCAL BUILDING OR OTHER CODE(S), AND TO FOLLOW OFFICIAL REQUIREMENTS ON THE USE OF SAFETY EQUIPMENT.

Over False (Grid) Ceilings

1. LONG RUN (UP TO 125 FEET AT A TIME), SMALL ACCESS AREA, OPEN SPACE

How to proceed:
1. Remove tile from false ceiling; aim the yoke of the **Sling-A-Line** holding it HORIZONTALLY. Push the button on the reel to release the line and put the weight in the leather sling. Shoot.
2. At the other end, attach your wire to a length of **PullCord**, tie the PullCord to the lead weight and reel it back.
3. In areas where cable cannot be resting on the ceiling tiles, attach one **2nd Man Pulley** above the drop ceiling to receive your cable.

Note: Do not remove the ceiling tile in the area you are shooting towards: the weight could drop through the opening and cause injury or property damage. If the line gets tangled over any object, pull the line very slowly with your hand to allow the weight to unravel itself.

2. MEDIUM RUN (UP TO 40 FEET), OPEN SPACE

How to proceed:
1. Remove a tile, rest the wire on the outer "V" of the **Grabbit 18's Ztip**.
2. Push the wire to its destination by extending the Grabbit section by section; if working a two person crew with two Grabbits, the second installer, 40 ft. away from you, will grab the wire with the knifed edges side of the Ztip and pull the wire in collapsing the sections of the Grabbit, one by one.

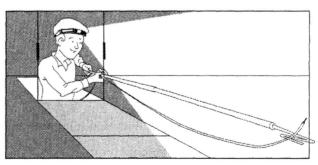

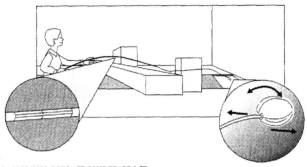

3. MEDIUM RUN, CROWDED SPACE

How to proceed:
1. Attach as many of the **CZ30 sections** as you will need, using pliers to snug the connectors together.
2. With a simple flick of your wrist, the rods will jump up over air ducts, lighting fixtures, cable raceways, etc.
3. The wisp head also raises the end of the Creep-Zit so it can slide smoothly over the grid dividers; turn the end of the Creep-Zit in your hand and the wisp head at the other end will walk from side to side.
4. Attach the wire to the wisp head and bring it back to you, or to the bullnose at the back end.

Note: If pulling multiple cables at once, we recommend you tie them into a **PullSleeve** that you attach to the end of the PullCord; also, run with your cables at least one extra length of nylon cord for subsequent cable runs. Also, in areas where cables cannot be resting on ceiling tiles, attach one **2nd Man Pulley** above the drop ceiling to receive your cable.

4. SHORT RUN (UP TO 25 FEET), SEMI-CROWDED SPACE

How to proceed:
1. When working above ceiling in occupied office, the **YFT 30-E**, contained in its round carrying case, can be maneuvered without risk of property or bodily damage.
2. To direct wisp head from side to side. Remove the YFT entirely from its case.
3. Because of its extreme flexibility, the YFT will require some practice to "jump" over obstacles.

5. LONG RUN (100 FEET AND OVER), SEMI-CROWDED SPACE

How to proceed:
1. The CF50 **Fishtix** is a heavy duty version of the Creep-Zit kit, giving you the added rigidity required to pull run 100 feet and longer, yet operating the same way the Creep-Zit does (see **medium run, crowded space**).
2. When you have extended the Fishtix to its final destination, either attach the male bullnose to the last section, then your cable or PullCord to it and go to the other end to pull it to you, or replace the wisp head up front with the female bullnose, attach cable or pull cord, and pull towards you from the back end. Hint: When pulling long cable runs, it is suggested to pull at least one **PullCord** with the cable to facilitate your work, should additional cable can be pulled at a later date.
3. In areas where cable cannot be resting on the ceiling tiles, attach one **2nd Man Pulley** above the drop ceiling to receive your cable, pass the cable through the pulley first, then attach it to the Fishtix's bullnose. If using cables on reel, place the reels on the **Decoil-Zit**, just below the ceiling opening.

Over Pipes or Beams, from the Ground

How to proceed:
1. If working alone, attach at the end of the wire or **PullCord** a weight (i.e., a golf ball), then "rest" the wire in the outer "V" of the Ztip.
2. Extending the **Grabbit** section by section, pass the wire (PullCord) and the weight over the pipe or beam and release it from the Ztip (the weight will keep the wire from falling back).
3. Now, using the knifed edge side of the Ztip, pull the wire over the beam. Repeat the steps for the next beams.

Note: If working a two person crew with two Grabbits, ignore the weight and just pass the wire to one another over the beam. If pulling multiple cables, tie them in the **PullSleeve**, and attach the Pull-Sleeve to the end of the PullCord.

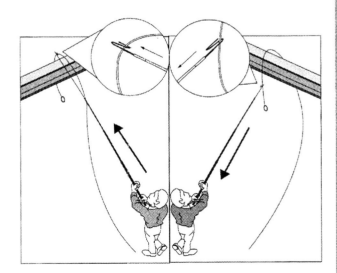

In Race Ways or Over Pipes

**IN RACE WAYS OR OVER BEAMS AND PIPES
NOT MORE THAN 6 FEET APART**

How to proceed:
1. Attach the wisp head to the first section, then as you start pushing the **Fishtix** in the raceway or over the beams, snap as many sections together as needed for your run.
2. Attach the male bullnose at the end of the last section (or the female bullnose at the front of the first section after having removed the wisp head), then attach the Pull-cord (or cable) to the bullnose and pull back.

Note: If pulling multiple cables at once, we recommend you tie them into a **PullSleeve** that you attach to the end of the **PullCord**; also, run with your cables at least one extra length of nylon cord for subsequent cable runs.

Horizontal Turns and Elevation Changes

How to proceed:
1. Snap the hanging strap of the **2nd Man Pulley** off the axle pin and the retaining latch. Attach the pulley to the proper support device, snap the strap back on the axle pin and in the retaining latch, in the same manner, position a 2nd Man Pulley at each horizontal turn, elevation change, each up and down location.
2. Pull the cable through each 2nd Man Pulley.

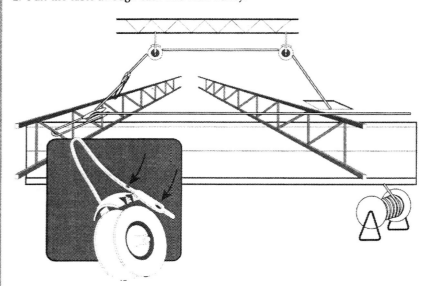

APPENDIX B

INSTALLING CABLE IN RESIDENTIAL AND FINISHED AREAS OF COMMERCIAL CONSTRUCTION

What?...Where?...When?...How?...Wow!

In the section to follow, we have addressed some of the most common situations facing wire/cable installers. Listed are The basic tools required for the best and fastest results, briefly explaining how to proceed. It is purely informative, and in no event can Labor $aving Devices®, Inc. be liable for any special, incidental or consequential damages.

IT REMAINS THE INSTALLERS SOLE RESPONSIBILITY TO OPERATE IN ACCORDANCE WITH ANY APPLICABLE FEDERAL, STATE, COUNTY, CITY AND LOCAL BUILDING OR OTHER CODE(S), AND TO FOLLOW OFFICIAL REQUIREMENTS ON THE USE OF SAFETY EQUIPMENT.

Fishing Wires Inside Walls

In Empty (Inside) Walls

1. DOWN, FROM THE TOP PLATE TO THE BASE OF THE WALL, ON CONCRETE FLOOR

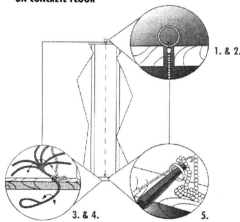

How to proceed:
1. In the attic, drill a hole in the top plate with the selected drill bit (**Spear-Zit, Bell Hanger,** or **Drill-Eze**).
2. Attach the "stop ring" to one end of the ball chain of the WR24; drop the chain in the drilled hole.
3. In the room, remove the baseboard and drill a hole at the base of the wall.
4. Bend the flexible retriever of the WR24 and insert in the hole. Move it left to right until the magnet catches the ball chain.
5. Bring the ball chain to the hole; with the **LBS** (or the **Hook-Zit 9"**) pull the chain out. Attach the wire to the chain. (If pulling coax cable or large wires, attach a nylon **PullCord** to the chain, then the cable to the PullCord.)
6. From the top plate, pull the ball chain and the wire. (Hint: with concrete floors, run the wire behind the baseboard or between baseboard and tack strip if there is carpet.)

2. DOWN, FROM THE TOP PLATE TO THE CRAWL SPACE OR UNFINISHED BASEMENT

How to proceed:
1. In the attic, drill a hole in the top plate with the selected drill bit.
2. Attach the "stop ring" to one end of the ball chain of the WR24; drop the chain in the drilled hole.
3. Instead of pulling down the molding, this time drill up through the bottom plate from the basement or crawl space.
4. Keep the flexible handle of the WR24 almost straight, with just a light bend on the magnet side. Insert the drilled hole and fish the ball chain. Proceed as above.

3. MULTIPLE CABLE RUNS, DOWN FROM THE TOP PLATE TO LOWER FLOOR(S)

How to proceed:
1. First select the smallest Drill-Eze bit size that will accommodate your cable run. Start drilling through the top plate from above, having selected the 2-ft. shaft.
2. Pull the bit back to you and replace the 2-ft. shaft by the smallest next size up (4- or 6-ft.) that will let you reach the next "plate" inside the wall.
3. Once you have drilled through the second plate, pull the bit back to you or add to the shaft the smallest extension that will let you drill through the third "plate".
4. Use extreme caution when drilling the second or following plates in order to avoid drilling through the wall; it is suggested to "rest" the shaft (or extension) against the wall of the first hole.
5. Once you have reached the lower destination point, attach to the end of the last extension a nylon **PullCord**. If pulling multiple cables or pre-connected ones, we recommend the use of the **PullSleeve**.
6. From downstairs, pull the bit, the nylon PullCord and cable will follow.

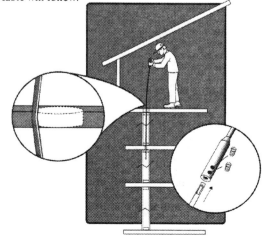

LOW-VOLTAGE CABLING — TRAINEE TASK MODULE 33108

In Wall with Stapled Bat Insulation

1. DOWN, FROM THE TOP PLATE TO THE BASE OF THE WALL ON CONCRETE FLOOR

How to proceed:
1. Use the **StudSensor** to locate the inside edge of the stud (or use the **VideoScanner** to locate the stud and check for hot electrical lines along the stud).
2. Cut a 4" **Wire Bit** and at the inside edge of the stud, drill through the ceiling, where it meets with the wall. Remove the drill motor, leaving the wire bit poking through the ceiling.
3. In the attic, using the wire bit sticking up as a reference, measure 3/4" and drill a hole (use **Spear-Zit** or **Bell Hanger**). You will be right in the void between the dry wall and the stapled insulation.
4. Drop the ball chain of the **WR24** in the drilled hole.
5. In the room, first, remove the wire bit (the hole is so small that it will remained unnoticed), then remove the molding and drill a hole at the base of the wall. Insert the flexible handle of the **WR24** and bring the ball chain to the hole. Pull the chain with the **LBS** (or **Hook-Zit**).
6. Attach the wire to the chain and pull from the attic.

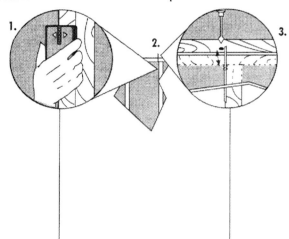

2. DOWN, FROM THE TOP PLATE TO THE CRAWL SPACE OR UNFINISHED BASEMENT

How to proceed:
1. Use the **StudSensor** to locate the inside edge of the stud (or use the **VideoScanner** to locate the stud and check for hot electrical lines along the stud).
2. Cut a 4" wire bit and at the inside edge of the stud, drill through the ceiling, where it meets with the wall. Remove the drill motor, leaving the wire bit poking through the ceiling.
3. In the attic, using the wire bit sticking up as a reference, measure 3/4" and drill a hole. You will be right in the void between the dry wall and the stapled insulation.
4. Drop the ball chain in the drilled hole.
5. Instead of pulling the molding in the room, this time drill up through the bottom plate from the basement.
6. Keep the flexible handle of the **WR24** almost straight, with a light bent on the magnet side. Insert in the hole and fish the ball chain. Attach the wire to the chain and pull from the attic.

In Wall with Insulation with Moisture Barrier

1. DOWN, FROM TOP PLATE TO BASE OF WALL ON CONCRETE FLOOR... OR... TO CRAWL SPACE OR UNFINISHED BASEMENT

How to proceed:
Same procedure as for Stapled Bat Insulation.

2. DOWN, FROM THE MIDDLE TO THE BASE OF THE WALL, ON CONCRETE FLOOR

How to proceed:
1. Use the **SSZ9 StudSensor** to locate the middle between two studs and mark the wall.
2. Using a small pointed object, poke a hole in the dry wall at a "down angle", making sure not to pierce the moisture barrier.
3. Attach the **WNH-1** to the bullnose of the **LZsec** and enter it through the hole, in the wall, between the drywall and the moisture barrier, until the bullnose touches the bottom plate.
4. Having dropped the base molding, run the **LZO** compass alongside the base of the wall to locate the **LZsec** magnet.
5. When the red arrow points at the wall, move the compass upward; if the white arrow points this time at the wall, you have located the magnet, not a nail or a pipe.
6. Drill a hole (with **Spear-Zit** or **Bell Hanger**) right where the compass located the magnet, and, with the **LBS** retriever (or **HZ9 Hook-Zit**), pull the ball chain out.
7. Attach the wire (or **PullCord** if running cable) to the ball chain. Pull the rod up.

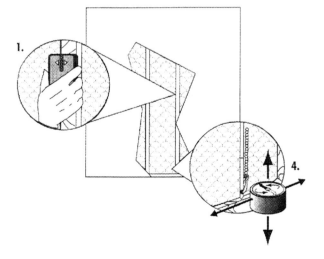

In Empty or Insulated Walls, Down from the Middle of Wall, to Carpeted Floor

How to proceed:
1. Pull the carpet and padding from the wall; with a chisel and a hammer, cut a 2" piece of the tack strip.
2. Attach the **BBZ Base-Boar-Zit** to the high speed drill motor, place it against the wall where you cut the tack strip and start drilling, holding the unit firmly against the floor... The drill will go through the molding, the drywall, up the bottom plate (note: always start the bit in a wood molding for it to go up).
3. Push and pull the bit several times to clean the hole, always leaving the drill motor in forward. Remove the bit.
4. Attach the **WNH-1 ball chain** to one of the eyelet of the .047 fish wire provided in the **Base-Boar-Zit Kit** and insert in the hole. (mark the fish wire with tape at the length you need)
5. When the fish wire is in the wall up to the marking tape, insert the **HZ9** or **LBS** in the access hole and fish the ball chain; bring it to the hole and attach the wire or a PullCord. Pull the fish wire from the bottom.
6. Run wire between molding and tack strip (see **fishing wire under carpet, between molding and tack strip**) or down to the basement (see **fishing wire in crawl space, from upstairs with carpeted floor**); replace the padding and the carpet; they will hide the hole in the molding and the wire.

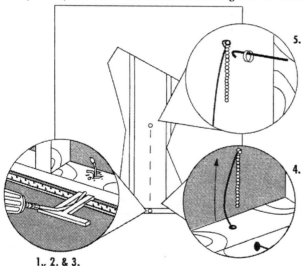

1., 2. & 3.

In Wall with Blown-in Cellulose

1. FROM UNFINISHED BASEMENT, UP TO THE MIDDLE OF THE WALL

How to proceed:
1. Cut a 4" Wire Bit and against the wall, drill down through the floor; remove the drill motor, leaving the wire bit in the floor.
2. In the unfinished basement, locate the wire bit sticking through the floor and, with the **Spear-Zit flex bit** or **Bell hanger** and the **directional tool**, drill up the bottom plate immediately behind the wire bit at an angle towards you. This will later direct the **LZsec rod** towards the front drywall. Drill with extreme care so the bit does not go through the wall! (Note: you could also use a right angle drill motor.)

3. Attach the **WNH-1** 1-ft. ball chain to the bullnose of the LZsec; push it up the drilled hole using a jabbing motion; the rod will "ride" against the drywall. (Remember to mark the rod at the length you want to stop.)
4. Back in the room, use the **LZO compass** to locate the magnet of the LZsec; poke a hole right in front of the magnet and with the **Hook-Zit 9"** or **LBS** pull the ball chain. Attach the wire or a PullCord and pull the rod down from the basement.

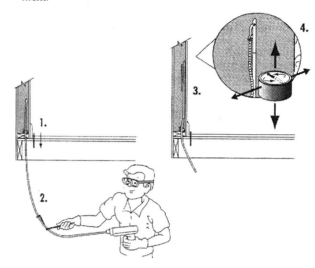

2. DOWN, FROM THE MIDDLE OF THE WALL, TO BASE OF WALL ON CONCRETE FLOOR

How to proceed:
1. With the **StudSensor**, locate the inside edge of the stud. Using a **Spear-Zit** or **Bell Hanger** drill a hole in the wall at a down angle (it will direct the fiberglass fish tape on the back side of the wall cavity, against the stud).
2. Attach the **WNH-1** to the bullnose of the **fiberglass fish tape** (rod) and enter it in the drilled hole, and push it down using a jabbing motion.
3. Remove the molding, use the StudSensor to locate the inside edge of the same stud and drill a hole right next to the stud edge. Use the **Hook-Zit** to grab the ball chain and bring it to you. Attach the wire or a PullCord to the chain. Bring the fiberglass rod up.

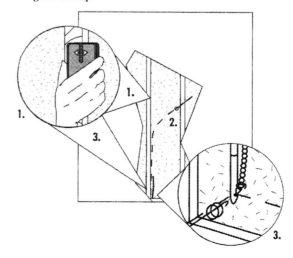

LOW-VOLTAGE CABLING — TRAINEE TASK MODULE 33108

Inside Wall, Down from In-Wall Speaker Opening or up from System's Outlet Opening

How to proceed:
1. With the **StudSensor** locate the two studs in between which you will locate the speaker and mark the wall.
2. If installing a rectangular speaker, use the manufacturer's provided template and mark the contour (use the **pocket level** to guarantee perfect alignment).
3. Adjust the depth gauge of the **Spira-Cut** bit 1/8" longer than the thickness of the material to cut and start the bit into the material at a 45 degree angle. Slowly bring it to 90 degrees and begin to cut clockwise (start and end at the top of the hole). Note: if cutting a round opening, adjust the circular guide of the Spiracut to the diameter to cut, then proceed as above.
4. To run your cable, follow the procedures described above, based on the type of wall (insulation), you are dealing with. If you need to cut an opening for an outlet box proceed as for rectangular speakers: if the box is already installed and you need to cut the drywall covering it, plunge the Spirabit inside the box, guide it to the right until it touches the inside edge of the box; pull the bit out far enough so it is now against the outside of the box and move the Spiracut counter clockwise.

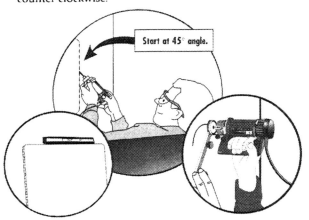

Fishing Wires In Finished Basement

Basement Window Contact or In-Wall Speaker, up to Upper Floor

How to proceed:
1. From the window header, drill up into the cavity between floor and ceiling using the **Spear-Zit** or **Bell Hanger**; before going upstairs, measure the distance from the drilled hole to the outside wall perpendicular to the one you are working on. If installing in-wall speaker in the basement, use the **Spiracut** to cut the speaker opening (refer to application *inside wall, down from in-wall speaker or up from system's outlet*) then drill up into wall.
2. Upstairs, measure the same distance, from the same outside wall in order to be located between the same floor joists. Pull the carpet and drill down, between molding and tack strip.
3. Attach the stop-ring to the ball chain of the **WR24** and drop chain down the drilled hole; it will pile up on top of the ceiling.
4. Downstairs, insert the flexible handle of the WR24 in the hole drilled through the window header and fish the ball chain in a "twist-pull-back" motion. Bring the chain down to you and attach your wire.
5. Upstairs, pull the ball chain and the wire; from one corner of the room, send the **YFT-10 fiberglass fish tape** between the tack strip and the molding. Attach the wire to the bullnose of the YFT-10; pull it to its destination. If the upstairs floor is not carpeted, run you wire behind the molding or up the wall.

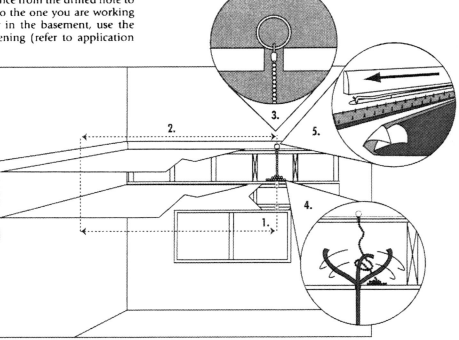

Crawl Space or Unfinished Basement

From Upstairs, with Carpeted Floors

How to proceed:
1. Upstairs, pull the carpet and with **Spear-Zit** or **Bell Hanger** drill a hole down, between molding and tack strip.
2. Attach the stripped end of the wire through the hole of the **GR3 GloRod** and flip it back on the flat of the rod (or pass it through the cross hole of the bullnose of the **GR3BB** or **GR3L**).
3. Holding the wire and the GloRod together, pushed them in the drilled hole; still holding onto the wire, jab the push rod to "quick release" the wire. Pull the rod back, the wire will stay.
4. In the crawl space, wearing the **TSP-1** headlight if necessary, extend the **GR12 Grabbit** telescoping pole towards the dangling wire; grab it with the knifed edges side of the Ztip and bring the wire to you.

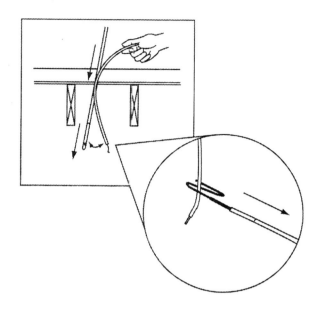

From Upstairs, with Uncarpeted Floors

How to proceed:
Same procedure as above, except that you will have to remove the molding to drill down and run your wire behind it or up the wall.

Fishing Wire Under Carpet

Between Base Molding and Tack Strip

How to proceed:
Pull the carpet in the corner of the room with a pair of needle nose pliers; feed the **YFT fiberglass** rod between the base molding and the tack strip to where your wire is. Attach the stripped wire end to the bullnose of the rod. Pull the rod and the wire back from the corner of the room. Repeat the steps as necessary. The space between molding and tack strip is a "natural" channel to run wires, without having to move furniture.

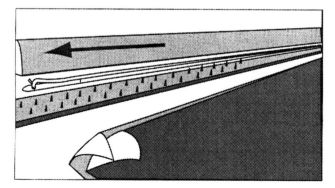

Between Carpet and Padding

How to proceed:
1. If you must run wire under carpet, always run it between carpet and padding to create a double cushion to protect the wire. Never staple or tape the wire as traffic could create damage to the wire jacket resulting in a short.
2. Insert the rolled-up end of the **UCT25 Under Carpet Tape** between carpet and padding from the destination place of the wire; when reaching the other side, strip the end of the wire and attach it to the rolled end of the UCT. Roll back the UCT25 in its case as it brings the wire to you. (The rounded end of the UCT25 is for running over waffle type padding.)
Note: Do not staple or tape the wire; avoid traffic areas.

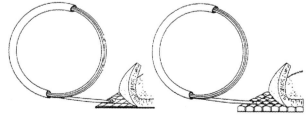

APPLICATIONS—Residential and Finished Areas of Commercial

LOW-VOLTAGE CABLING — TRAINEE TASK MODULE 33108

Fishing Wires Between Floors

Smooth (Sheet Rock) Ceiling Material

How to proceed:
1. With the **SSZ9 StudSensor**, locate two joists in between which the device will be located and cut a hole a least 2" in diameter with **Spiracut**. Then measure the distance between the hole and the outside wall parallel to the joists.
2. Upstairs, measure the same distance from the same outside wall in order to be between the same joists, against an inside wall perpendicular to the outside wall.
3. Pull the carpet and drill a hole between tack strip and molding (use **Spear-Zit**, **Bell Hanger**, or **Drill-Eze**). Attach the "stop-ring" to the **ball chain of the WR24** and drop it in the hole. It will pile up on top of the ceiling.
4. Back downstairs, insert the **Walleye** in the hole cut in the ceiling and locate the "pile" of ball chain. Direct the **CZD Drag-Zit** magnet (screwed on a **CZX section of the CZ-30**) to the pile. Pull the rod slowly back to you so it will feed the ball chain to you.
5. Once you have the chain to you attach to it a minimum of 10 feet of **PullCord**, the wire or cable to the cord.
6. Upstairs, pull the ball chain, the PullCord... The wire (or cable).

Rough (Lath & Plaster) Ceiling Material

How to proceed:
1. With the **SSZ9 StudSensor**, locate two joists in between which the device will be located and cut a hole a least 2" in diameter with **Spiracut**. Then measure the distance between the hole and the outside wall parallel to the joists.
2. Upstairs, measure the same distance from the same outside wall in order to be between the same joists, against an inside wall perpendicular to the outside wall.
3. This time feed a length of thin gauge wire (or **PullCord**) down the drilled hole instead of the ball chain.
4. Back downstairs, screw the **Ztip** on the **CZX rod** of the **Creep-Zit**. With the **Walleye**, locate the wire (or PullCord) hanging down and grab it with the knifed edges side of the Ztip. Bring it to you, tie the wire and proceed as above.

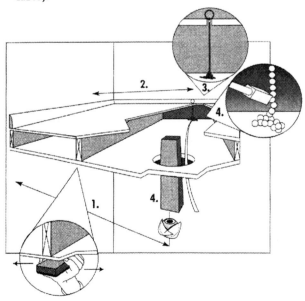

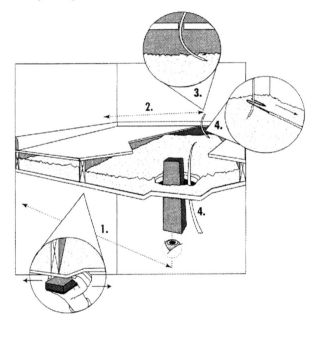

Fishing Wires To and In an Attic

1. LIGHT OR MEDIUM INSULATION, UP FROM WALL DOWNSTAIRS

How to proceed:
1. Using the **Spear-Zit** or **Bell Hanger** drill bit, drill a hole up the top plate.
2. Put the stripped end of the wire through the hole of the **GloRod** and flip it on the flat of the rod. Holding both the rod and the wire, feed it up the hole until it touches the roof line. Holding the wire, pull the rod back a little in order to create a loop in the attic. This can be repeated as often as you have wire runs to bring up the attic.
3. In the attic (wearing the TSP-1 headlight if necessary) and from a comfortable and convenient location, extend the GR-12 (12-ft.) Grabbit towards the first looped wire. With the knifed edges side of the Ztip, grab the wire and bring it to you. The rod will fall back downstairs. Repeat for the following runs. Remember to collect the rods downstairs.

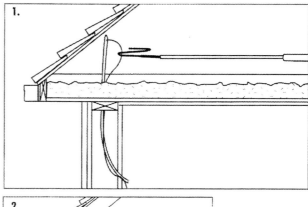

2. MEDIUM INSULATION, FROM A WINDOW HEADER

How to proceed:
1. Using the **Spear-Zit** or **Bell Hanger** drill bit, drill a hole up the window header and the top plate.
2. Insert the 24" or 36" (depending on ceiling height) stainless tube of the **FEK-1 Fish-Eze Kit** into the drilled hole until it its the roof sheeting, then pull back a little.
3. Remove the coiled fish wire of the Fish-Eze kit from its holder and feed it into the tube; the fish wire will coil in the attic, on top of the insulation.
4. Remove the tube and attach both the wire and a nylon line to the eyelet of the fish wire (operating as a two person crew, you will use the nylon line to bring the fish tape back to you for the next run). If operating alone, disregard the nylon line and repeat the above steps using several coiled fish wires rather than the same one over and over.
5. In the attic, extend the **GR-8** or **GR-12 Grabbit** with Jtip towards the first coiled fish wire. Hit it so the coils lay flat and won't kink. Grab all the loops the wire with the Jtip so it uncoils as you bring it to you. Pull the fish wire, it will bring the wire to you.

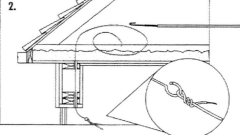

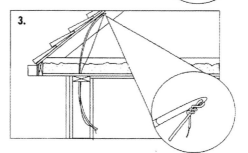

3. THICK INSULATION (18" AND MORE), FROM WALL BELOW

How to proceed:
1. Using the **Spear-Zit**, drill a hole up through the top plate.
2. Tie a 30-lb. nylon fishing line to one of the bullnose of the **YFT fiberglass fish tape** and feed it in the drilled hole. When it hits the roof sheeting, pull on the nylon line. The YFT will bend and follow the roof line.
3. You can now go in the attic and pull the YFT and the wire to you or tie the wire to the YFT and pull it back down.

4. IN ATTIC, FROM ONE END TO THE OTHER

How to proceed:
1. Attach as many of the **CZ30 sections** as you will need, using pliers to snug the connectors together.
2. With a simple flick of your wrist, the rods will jump up over air ducts, lighting fixtures, cable raceways, etc.
3. The wisp head also raises the end of the **Creep-Zit** so it can slide smoothly over the grid dividers; turn the end of the

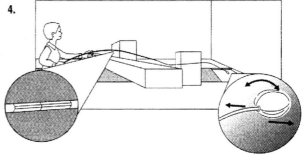

Creep-Zit in your hand and the wisp head at the other end will walk from side to side.
4. Attach the wire to the wisp head and bring it back to you, or to the bullnose at the back end.
5. If running coax, cat 5 cables or large speaker wires, we recommend you attach a **PullCord** to the wisp head or bullnose of the Creep-Zit sections, then attach your cable (or the **PullSleeve**) to the PullCord.

APPENDIX C

TOOL ILLUSTRATIONS AND DESCRIPTIONS

"Jobber" Bellhanger Drill Bits

Unlike conventional bellhanger bits, the "jobber" bellhangers are aircraft drills Type A specifications for a wider variety of drilling applications:
- Specially designed for drilling through *steel plates*, as well as wood;
- 118 degree point provides more gradual breakthrough; limits the grabbing occurring when using 135 degree point;
- Split point substantially reduces bit "walking";
- Slow design spiral improves bit self-cleaning, eases "pull back".

PART #	MODEL #	DESCRIPTION
		1/4" BIT
65-118	FF1418F	FREEFORM 1/4 X 18
65-124	FF1424F	FREEFORM 1/4 X 24
65-136	FF1436F	FREEFORM 1/4 X 36
65-148	FF1448F	FREEFORM 1/4 X 48
65-158	FF1458F	FREEFORM 1/4 X 58
65-172	FF1472F	FREEFORM 1/4 X 72
		3/8" BIT
65-218	FF3818F	FREEFORM 3/8 X 18
65-224	FF3824F	FREEFORM 3/8 X 24
65-236	FF3836F	FREEFORM 3/8 X 36
65-248	FF3848F	FREEFORM 3/8 X 48
65-258	FF3858F	FREEFORM 3/8 X 58
65-272	FF3872F	FREEFORM 3-8 X 72
		1/2" BIT
65-318	FF1218F	FREEFORM 1/2 X 18
65-324	FF1224F	FREEFORM 1/2 X 24
65-336	FF1236F	FREEFORM 1/2 X 36
65-348	FF1248F	FREEFORM 1/2 X 48
65-358	FF1258F	FREEFORM 1/2 X 58
65-372	FF1272F	FREEFORM 1/2 X 72

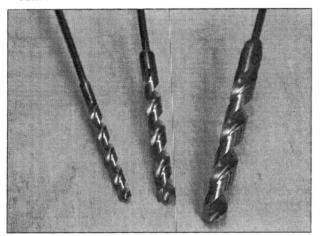

Applications: Any wood drilling application, at any angle; will not catch on and unravel carpet, will go through knots and nails. See pages 37, 38, 39, 40, 41, 42, 43.

Drill-Eze™ Drill Bits

Designed with the systems contractor in mind the Drill-Eze bit concept is a family of 3 components, each offered in different sizes. It is the ONLY solution for pulling pre-connected cables or large bundles of wires in pre-wire AND retrofit situations.

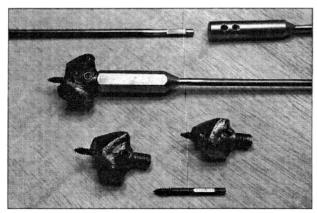

Wood boring self-feeding bits: precision ground tool steel specially heat treated for sharp, long-lasting cutting edges; 3 different sizes: 1", 1 1/4" and 1 3/8" diameter with replaceable self-feeding screw.

PART #	MODEL #	DESCRIPTION
64-100	DE-100	1" DRILL-EZE DRILL BIT
64-125	DE-125	1-1/4" DRILL-EZE DRILL BIT
64-138	DE-138	1-3/8" DRILL-EZE DRILL BIT
64-150	DEFS6M	REPLACEMENT FEED SCREW FOR DRILL-EZE BIT
64-202	DEFS-2	2FT FLEX SHAFT FOR DRILL-EZE BIT
64-204	DEFS-4	4FT FLEX SHAFT FOR DRILL-EZE BIT
64-206	DEFS-6	6FT FLEX SHAFT FOR DRILL-EZE BIT
64-302	DEFSE-2	2FT FLEX EXTENSION FOR FLEX SHAFT
64-304	DEFSE-4	4FT FLEX EXTENSION FOR FLEX SHAFT
64-306	DEFSE-6	6FT FLEX EXTENSION FOR FLEX SHAFT
64-450	DEFkit	DRILL-EZE FLEX KIT: DE-100, -125, -138; DEFS-2, -4, -6; DEFSE-2, -4, -6

Screw-on shafts: high torque, extra flexible shafts are available in 2-ft., 4-ft. and 6-ft. lengths; just screw on the selected diameter bit and drill!

Extensions: to cover all possibilities, 2-ft., 4-ft. and 6-ft. high torque, extra flexible extensions are available to increase the reach of the screw-on shafts.

Note: Do not set in reverse. It would "unscrew" the bit from the shaft. Not recommended in insulated walls.

Applications: See pages 37, 42.

Spear-Zit™ Drill Bits

The Spear-Zit wood drill bit is a perfect diamond shaped head drill bit, allowing the user to drill at any angle. They are fully guaranteed against breakage, and easily field-resharpenable with a file.

Hint: Use any 3/8" Spear-Zit bit as a pilot bit for hole saws to cut holes from 9/16" to 2 5/8" diameter through a top or bottom single plate, with an "up-to-6-ft.-long" shaft hole saw!

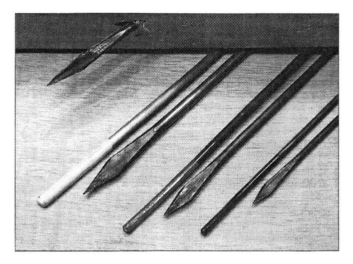

Applications: Any wood drilling application, at any angle; will not catch on and unravel carpet, will go through knots and nails.

Because the spear shape of the head will prevent it from walking, use the directional tool to align the head of a 60" or 72" flex bit in the desired direction.

PART #	MODEL #	DESCRIPTION
61-206	SZ1/4X6	1/4" X 6" SPEAR-ZIT
61-212	SZ1/4X12	1/4" X 12" SPEAR-ZIT
61-218	SZ1/4X18	1/4" X 18" SPEAR-ZIT
61-224	SZ1/4X24	1/4" X 24" SPEAR-ZIT
61-236	SZ1/4X36	1/4" X 36" SPEAR-ZIT
61-248	SZ1/4X48	1/4" X 48" SPEAR-ZIT
61-260	SZ1/4X60F	1/4' X 60" SPEAR-ZIT, FLEXIBLE SHANK
61-272	SZ1/4X72F	1/4" X 72" SPEAR-ZIT, FLEXIBLE SHANK
61-306	SZ3/8X6	3/8" X 6" SPEAR-ZIT
61-312	SZ3/8X12	3/8" X 12" SPEAR-ZIT
61-318	SZ3/8X18	3/8" X 18" SPEAR-ZIT
61-324	SZ3/8X24	3/8" X 24" SPEAR-ZIT
61-336	SZ3/8X36	3/8" X 36" SPEAR-ZIT
61-348	SZ3/8X48	3/8" X 48" SPEAR-ZIT
61-360	SZ3/8X60F	3/8" X 60" SPEAR-ZIT, FLEXIBLE SHANK
61-372	SZ3/8X72F	3/8" X 72" SPEAR-ZIT, FLEXIBLE SHANK
61-406	SZ1/2X6	1/2" X 6" SPEAR-ZIT
61-412	SZ1/2X12	1/2" X 12" SPEAR-ZIT
61-418	SZ1/2X18	1/2" X 18" SPEAR-ZIT
61-424	SZ1/2X24	1/2" X 24" SPEAR-ZIT
61-436	SZ1/2X36	1/2" X 36" SPEAR-ZIT
61-448	SZ1/2X48	1/2" X 48" SPEAR-ZIT
61-460	SZ1/2X60	1/2" X 60" SPEAR-ZIT
61-472	SZ1/2X72	1/2" X 72" SPEAR-ZIT

The 3/8" (1/4" shank) Spear-Zit can be used as a pilot for deep hole saws, to drill larger holes in single plate.

The flexible extensions (below) will give you added flexibility in allowing you to make one drill bit any length you want. Also, see pages 37, 38, 39, 40, 41, 42, 43.

Flexible Extension

These extensions (model SZE6-...) are made of special *high torque*, extra flexible steel. They are available in a 6-ft. length, and receive 3/16", 1/4" or 5/16" bit shanks. Ideal for extending the length of your favorite bit.

PART #	MODEL #	DESCRIPTION
61-516	SZE3/16-6	6FT EXTENSION FOR 3/16" SHANK BITS (1/4" SPEAR-ZIT)
61-526	SZE1/4-6	6FT EXTENSION FOR 1/4" SHANK BITS (3/8" SPEAR-ZIT)
61-536	SZE5/16-6	6FT EXTENSION FOR 5/16" SHANK BITS (1/2" SPEAR-ZIT)

Directional Tool

This directional tool (model AT7) is a must for drilling with any type of flexible shaft drill or extension.

PART #	MODEL #	DESCRIPTION
61-000	AT7	DIRECTIONAL TOOL FOR FLEX BITS

Door Drill Guide (Guide-Zit)

Safe, fast and easy-to-use door drilling guide, essential when connecting electrified locks to power transfer hinges. Designed to guide a 3/8" (1/4" shank) Spear-Zit - the *only* bit guaranteed not to "walk" - through any commercial solid wood doors, from hinge to lock.

Application: Remove old hinge; loosen bolt of the guide holding block to insert the 3/8" Spear-Zit (48" or 60") in the guiding tube (end first); Slide the fixture over the door and tighten the bolt (for extra security, screw the fixture onto the door); align guiding tube/Spear-Zit with center of the lock and tighten the holding block's bolt; start drilling. Run wire through the door with push/pull rod (see page 15).

PART #	MODEL #	DESCRIPTION
61-050	DDG-Z	DOOR DRILL GUIDE

Wire Bits

The wire bit is a .047" spring wire 36" long that can be cut (at an angle) any length to make it the smallest "reference" bit.

Applications: Its primary application is to be used as a reference point to measure from when having to drill in a specific wall cavity (i.e. drilling down through top plate in wall with stapled bat insulation, up from basement in wall with blown-in cellulose or Styrofoam insulation). The hole created is so small that it will remain unnoticeable, even on a ceiling. See page 38.

PART #	MODEL #	DESCRIPTION
84-547	PW36047	PACK OF 10, 36" LONG .047" DIAM. SPRING WIRE

Base-Boar-Zit®

The Base-Boar-Zit is a specialty drill bit designed to drill through a baseboard, up the bottom plate, without having to remove the baseboard. If there is carpet, it will hide the hole.

The kit consists of a steel "T" shaped foot with a curved tubing housing a 5/16" Spear-Zit "paddle" type bit assembled on a flexible shaft. It also includes a 10-ft., .047" fish wire (coiled in a carrying case) to pull the wire.

Applications: Use it to drill through a baseboard and up the wall. See page 39.

PART #	MODEL #	DESCRIPTION
63-516	BBZ5/16	BASE-BOAR-ZIT KIT COMPLETE WITH 10FT FISH WIRE
63-510	BBZ-A	BASE-BOAR-ZIT ASSEMBLY WITHOUT FISH WIRE
63-016	BBZ-B	BASE-BOAR-ZIT REPLACEMENT BIT ASSEMBLY
85-510	ICFW10047	REPLACEMENT 10FT .047" FISH WIRE COILED IN A CASE

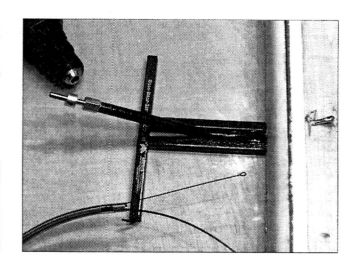

Rebore-Zit™ Drill Bit

Rebore-Zit is the only bit that can re-drill and enlarge a hole with a wire already present. It is a counter-sink equipped bit with a pilot and swivel-eyelet to tie the existing wire on.

Applications: Use in pre-wire situations, when the hole needs to be enlarged.

PART #	MODEL #	DESCRIPTION
62-338	REB3/8	3/8" REBORE-ZIT WITH 1/4" PILOT
62-348	REB1/2	1/2" REBORE-ZIT WITH 3/8" PILOT
62-368	REB3/4	3/4" REBORE-ZIT WITH 3/8" PILOT
62-388	REB1	1" REBORE-ZIT WITH 3/8" PILOT
62-000	REB-P1/4	REPLACEMENT 1/4" PILOT FOR REBORE-ZIT
62-005	REB-P3/8	REPLACEMENT 3/8" PILOT FOR REBORE-ZIT

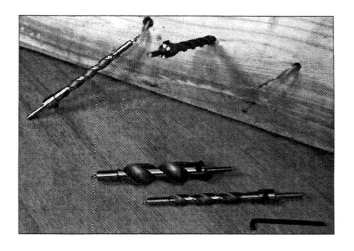

LOW-VOLTAGE CABLING — TRAINEE TASK MODULE 33108

Hole Saws

Bi-metal shatterproof construction, 4/6 variable tooth provides faster, smoother cuts. 3/16" thick backing plate. The deep hole models (1 3/4" deep) will drill through a regular 2×4 in a single pass.

Note: To drill hole 9/16" or larger inside wall, use 3/8" Spear-Zit as pilot drill for your hole saw.

PART #	MODEL #	DESCRIPTION
60-002	BM9/16DH	9/16" DEEP HOLE SAW
60-004	BM5/8DH	5/8" DEEP HOLE SAW
60-005	BM11/16DH	11/16" DEEP HOLE SAW
60-006	BM3/4DH	3/4" DEEP HOLE SAW
60-007	BM7/8DH	7/8" DEEP HOLE SAW
60-010	BM1DH	1" DEEP HOLE SAW
60-011	BM11/8DH	1 1/8" DEEP HOLE SAW
60-012	BM11/4DH	1 1/4" DEEP HOLE SAW
60-014	BM11/2DH	1 1/2' DEEP HOLE SAW
60-020	BM2DH	2" DEEP HOLE SAW
60-021	BM21/8SD	2 1/8" STANDARD DEPTH HOLE SAW
60-025	BM25/8SD	2 5/8" STANDARD DEPTH HOLE SAW
60-102	BM1/4M	REPLACEMENT 1/4" MANDREL FOR HOLE SAW
60-121	BM7/16M	REPLACEMENT 7/16" MANDREL FOR 2 1/8" HOLE SAW

Applications: Use to cut holes from 9/16" to 2 5/8" through everything from wood and plastic to aluminum, brass, copper, cast iron, stainless steel.

De-Plug-Zit™

A *must* for anybody using hole saws on a regular basis, the De-Plug-Zit is a hole saw arbor, complete with pilot bit and adaptor to fit any size threaded hole saw that eliminates the problem of removing wood "plugs" or steel "washers" from the saw. The saw threads up and down the mandrel, exposing the "plug" on the pilot bit. Just remove the plug, reposition the saw on the mandrel, and be ready to cut the next hole.

PART #	MODEL #	DESCRIPTION
60-700	DPZ-7	DE-PLUG-ZIT HOLE SAW MANDREL

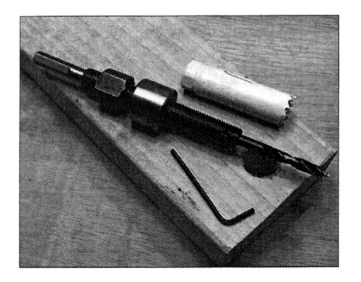

Spiracut® Spiral Saw

The Spiracut makes fast, smooth, clean cuts in everything from plaster to wood to ceramic tiles. It lets you see where the tool is slicing, when you are doing it, not when you are done. Powered by a 4.0-amp, 30,000 RPM electric motor, it is lightweight (2.7 lbs.) and has a removable handle to easily cut in tight places.

The SPIRACUT spiral saw kit is comprised of:
1. Spiral saw 4.0-amp, 30,000 RPM, complete with its removable handle, collet wrench, bayonet style removable base and depth guide.
2. Circle cutting guide for fast and clean circles from 3 1/2" to 14".
3. Ten each full fluted sabrecut bits 1/8", the most versatile bit for all types of drywall, wood, wood composite to 1", fiberglass, vinyl siding, plastics and laminates.
4. Two each "tilecut" carbide bits 1/8" to cut ceramic wall tile, cement board and plaster (will not cut floor tile).
5. "Store & Carry" heavy duty molded case.

Applications: See pages 40, 42.

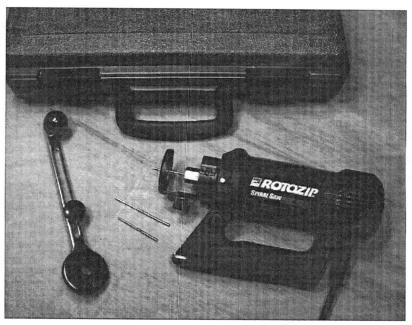

PART #	MODEL #	DESCRIPTION
56-010	SCS01K	SPIRACUT SPIRAL SAW KIT
56-011	SCSCE5	5-PACK SABRE CUT BIT 1/8"
56-012	SCTC1	CARBIDE BIT 1/8"
56-013	SCCCG	CIRCLE GUIDE 3 1/2" TO 14"
56-014	SCCASE	SPIRACUT CARRYING CASE

Drill Doctor® Bit Sharpener

The Drill Doctor® precision drill bit sharpener will give you industrial quality performance to renew dull bits to a precision, like-new sharpness with pencil sharpener ease. Sharpen high-speed, cobalt or carbide drill bits with 118 degree or 135 degree angles, conventional points, split-points or masonry bits... or turn a conventional 118 degree points into 135 degree split-points! Will sharpen complete range of drills from 3/32" to 1/2". Super abrasive, long lasting 1/2" diamond grinding wheel does not require dressing and is good for 300 to 500 resharpenings.

PART #	MODEL #	DESCRIPTION
56-050	500SP	DRILL DOCTOR BIT SHARPENER
56-052	500SP-GW	1/2" GRINDING WHEEL
56-053	750SP-C	1/2" TO 3/4" CHUCK

StudSensor

Possibly one of the most "must have" tools for the installer: being non-magnetic, the StudSensor is looking for changes in density, not metal, finding hidden wood and metal studs through drywall, plaster, wood, ceilings.

Applications: Use to locate studs and joists prior to fishing inside walls insulated with stapled bat insulation, moisture barrier, blow-in cellulose or Styrofoam, when fishing wires between floors (in ceilings). See Pages 38, 39, 40, 42.

PART #	MODEL #	DESCRIPTION
57-100	SSZ9	STUDSENSOR

VideoScanner™

The VideoScanner is an upscale version of the StudSensor: it is an accurate and sturdy electronic tool that features three different scanning modes, plus a continuous "hot" electrical wire detection.

Applications: See pages 38, 39, 40, 42.

PART #	MODEL #	DESCRIPTION
57-105	VSZ5	VIDEOSCANNER

Wet Noodle & Retriever

The Retriever component of the WR24 is a flexible, insulated 24" handle with a neodenium magnet attached to one end. This rare earth magnet, unlike an alnico magnet, will not loose its magnetism if it is hit. The "Wet Noodle" consists of 10 feet of .090" (half the size of a regular ball chain) smooth nickel plated steel ball chain, a "stop ring" that will prevent the chain from falling through the drilled hole, a #6 in-ring connector to pull the wire and a #3 "b" coupling to attach a second chain if needed.

Applications: See pages 37, 38, 40, 42.

PART #	MODEL #	DESCRIPTION
85-124	WR24	WET NOODLE AND RETRIEVER
85-024	WN10	REPLACEMENT WET NOODLE AND CONNECTORS
85-120	R24	REPLACEMENT FLEXIBLE INSULATED RETRIEVER W/ MAGNET
85-030	WNC	REPLACEMENT CONNECTORS (10 EA. RING, 3B COUPLING, #6 IN-RING)

Mini Scoping Retriever

The Mini Scoping Retriever (model LBS) is a pocket size penclip telescoping retriever with a hook at the end, small enough to pull the chain through a 1/4" hole – only 4" long when collapsed, it extends to 18". See pages 37, 38, 39.

PART #	MODEL #	DESCRIPTION
53-310	LBS	MINI SCOPING POCKET RETRIEVER

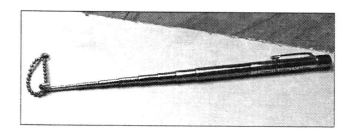

Hook-Zit™

Unlike the LBS pocket retriever, the Hook-Zit is flexible, being made of spring steel, .072" in diameter. The small hook at one end and the larger "L" shape at the other end, are faced the same way, so you always know which way the "fishing" hook is facing, even in blind fishing conditions.

Applications: The mini hook end is designed to fish ball chain or wire inside a wall, while the larger "L-shape" hook end is more for fishing a ball of wires in a restricted space, i.e., inside a panel. In pre-wiring situations, the 18" or 36" Hook-Zit is ideal to pull wire from stud to stud, even 7 feet high, without the use of a step ladder. See pages 37, 38, 39.

PART #	MODEL #	DESCRIPTION
84-109	HZ9	9" LONG HOOK-ZIT
84-118	HZ18	18" LONG HOOK-ZIT
84-136	HZ36	36" LONG HOOK-ZIT

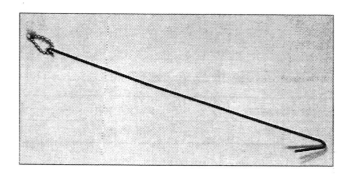

Steel Fish Wires

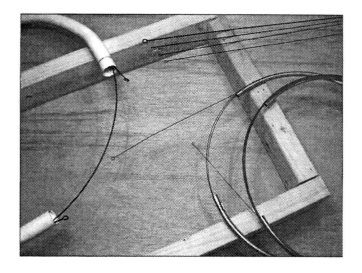

An economical, yet limited, version of the fiberglass push/pull rods, the spring steel fish wires are offered in different diameters, depending on the "flex" required. Each fish wire has a centered eyelet at each end to attach wire or pull cord.

PART #	MODEL #	DESCRIPTION
84-047	FW36047	.047" X 36" SLIM FISH WIRE
84-062	FW36062	.062" X 36" MEDIUM FISH WIRE
84-072	FW36072	.072" X 36" HEAVY FISH WIRE
84-092	FW72092	.092" X 72" HEAVY FISH WIRE
85-410	ICFW10092	.092" X 10FT HEAVY FISH WIRE IN CASE
85-510	ICFW10047	.047" X 10FT SLIM FISH WIRE IN CASE
85-525	ICFW25047	.047" X 25FT SLIM FISH WIRE IN CASE
85-550	ICFW50047	.047" X 50FT SLIM FISH WIRE IN CASE
85-600	ICFW100047	.047 X 100FT SLIM FISH WIRE IN CASE

LOW-VOLTAGE CABLING — TRAINEE TASK MODULE 33108

Creep-Zit™

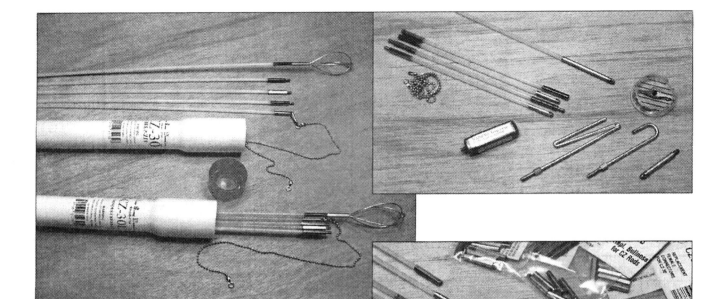

The tool that makes conventional fish tape obsolete., the Creep-Zit kit consists of five 6-ft. green fiberglass rods (.159" in diameter):
- Four with male/female screw-on connectors
- One with female on one end and a bullet shaped head with cross hole (bullnose) at the other end to attach the wire
- A 1-ft. ballchain attached to the bullnose to grab it easily when inside a wall
- An "eggbeater" shaped head (wisp) used as a guide to keep the rods from getting stuck on obstacles and to provide a "wheel" to "creep" from side to side.

The model CZ-30L is made of luminescent epoxy fiberglass rods (stronger and slightly stiffer) that truly glow in the dark after being "charged" in full light. The luminescent rods are interchangeable with the regular green rods.

Multiple accessories are designed for specific applications:
- *Drag-Zit™ magnet:* fish a ball chain between ceiling and upper floor. See page 42.
- *Ztip:* fish wire way back in a cavity where rough surface prevents the use of a Drag-Zit magnet and ball chain. See page 42.
- *Jtip:* uncoil coax or other cable.
- *Locate-Zit™ rod (or Locate-Zit screw-on tip):* special bullnose with magnet incorporated to be used in conjunction with the LZO compass to fish wires in insulated walls with moisture barrier or blown-in cellulose; locates the tip of the rod before you drill a hole. See pages 38, 39.
- *Adapt-Zit™ kit:* combination of short fiberglass sections with multiple connector combinations allowing fast conversion of the Creep-Zit section ends.

Applications: See pages 39, 40, 41, 42, 43, 44.

PART #	MODEL #	DESCRIPTION
81-130	CZ30	CREEP-ZIT GREEN FIBERGLASS WIRE RUNNING KIT
81-230	CZ30-L	CREEP-ZIT LUMINOUS FIBERGLASS WIRE RUNNING KIT
81-106	CZX	6FT GREEN ROD W/ MALE/FEMALE CONNECTORS
81-116	CZB	6FT GREEN ROD W/ FEMALE/BULLNOSE CONNECTORS
81-206	CZX-L	6FT LUMINOUS ROD W/ MALE/FEMALE CONNECTORS
81-216	CZB-L	6FT LUMINOUS ROD W/ FEMALE/BULLNOSE CONNECTORS
81-310	CZH	REPLACEMENT WISP (EGGBEATER) HEAD FOR CREEP-ZIT & YFT30-E
81-320	CZCOB	PACK OF 5 REPLACEMENT BULLNOSE END ONLY
81-330	CZCOF	PACK OF 5 REPLACEMENT FEMALE CONNECTOR ONLY
81-340	CZCOM	PACK OF 5 REPLACEMENT MALE CONNECTOR ONLY
81-350	CZCONN	PACK OF 5 FEMALE, 5 MALE AND 2 BULLNOSE
85-035	WNH-1L	PACK OF 5 1FT BALL CHAIN, LARGE SNAPHOOKS, #6AA COUPLING
82-350	Ztip	ACCESSORY Ztip FOR CREEP-ZIT AND GRABBIT
82-370	Jtip	ACCESSORY Jtip FOR CREEP-ZIT AND GRABBIT
81-360	CZD	ACCESSORY DRAG-ZIT CERAMIC MAGNET
81-121	CZA	ACCESSORY ADAPT-ZIT KIT
81-125	LZBN	ACCESSORY LOCATE-ZIT SCREW-ON BULLNOSE
81-126	LZSEC	ACCESSORY LOCATE-ZIT 6FT SECTION
81-127	LZO	ACCESSORY COMPASS FOR LOCATE-ZIT

Fishtix Wire & Cable Kit

The Fishtix (model CF-50) is a highly diverse cable installation system, designed to meet the many needs of today's cable installation world. It is ideally suited for such environments as drop ceilings, cable trays, raceways, wire mold and raised floors.

The Fishtix kit consists of:
1. Twelve 4-ft. tubular sections (7/8" diameter).
2. One "Cable Glide" or wisp head, used to provide horizontal and vertical steering control and easy tangle-free passage through busy cable ways.
3. One Bullnose section with stainless steel female end.
4. One Bullnose section with quick release button snap male end.
5. One 3/16 Quick Link.
6. One heavy duty Corex (corrugated polypropylene) Carrying Case.

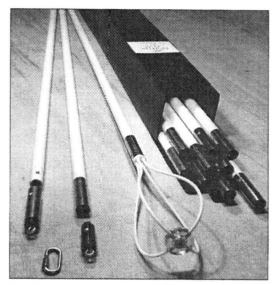

PART #	MODEL #	DESCRIPTION
81-500	CF-50	FISHTIX WIRE & CABLE RUNNING KIT
81-502	CFS4	4FT SECTION FOR FISHTIX
81-504	CFW	WISP HEAD FOR FISHTIX
81-506	CFBF	FEMALE BULLNOSE FOR FISHTIX
81-508	CFBM	MALE BULLNOSE FOR FISHTIX
81-510	CFQL	QUICK LINK FOR FISHTIX
81-512	CFCC	CARRYING CASE FOR FISHTIX

The Fishtix has extreme tensile strength while retaining its pliancy. It is capable of handling over 200 pounds of pulling force, and when used in cable trays it can virtually be extended to infinity; it will easily ride over joists, beams, etc., even with a 5-ft. span, jump obstacles and pull around corners up to 60 degrees.

Applications: See pages 45, 46.

Fiberglass Fish Tape

Probably the first fiberglass fish tapes that retain their memory to straight. Made of yellow epoxy fiberglass, the YFTs have a diameter of .110", making them capable of handling a much sharper bend that the green fiberglass rods or even the luminous ones. (Initially designed to handle a 90 degree bend–not elbow–in a 3/4" conduit). Offered in three different lengths (10-ft., 15-ft., 30-ft.), these fiberglass yellow fish tapes have a bullnose at each end so wires can be pulled from either direction. The 30-footer is also available with the Creep-Zit wisp head. Each model is coiled in a PVC storage case, easy to carry over the shoulder, making it easier to handle in crowded places.

PART #	MODEL #	DESCRIPTION
85-310	YFT10	10FT YELLOW EPOXY FIBERGLASS FISH TAPE IN CASE
85-315	YFT15	15FT YELLOW EPOXY FIBERGLASS FISH TAPE IN CASE
85-330	YFT30	30FT YELLOW EPOXY FIBERGLASS FISH TAPE IN CASE
85-335	YFT30-E	30FT YELLOW FIBERGLASS FISH TAPE, WITH WISP HEAD
85-351	R-YFT10	10FT REPLACEMENT YFT FISH TAPE W/ BULLNOSES
85-352	R-YFT15	15FT REPLACEMENT YFT FISH TAPE W/ BULLNOSES
85-353	R-YFT30	30FT REPLACEMENT YFT FISH TAPE W/ BULLNOSES
85-354	R-YFT30-E	30FT REPLACEMENT YFT-E FISH TAPE W/ BN & FEM-BN
85-355	YFTBN	PACK OF 5 REPLACEMENT BULLNOSE FOR YFT
85-356	YFTFBN	PACK OF 3 REPLACEMENT FEMALE BULLNOSE CONNECTORS FOR YFT30-E
85-036	WNH-1S	PACK OF 5 1FT BALL CHAIN, SMALL SNAPHOOKS, # 6AA COUPLING

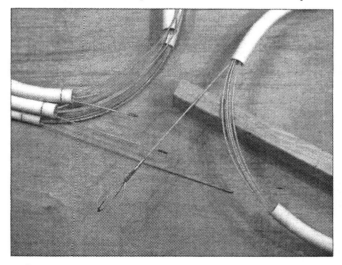

Applications: See pages 39, 40, 41, 43, 44.

GloRod™ Wire Pusher

The GloRod wire pushers are fluorescent (reflects light shined on them) green fiberglass rods, .159" in diameter, flexible, yet retain their memory to straight after being flexed. Available in 2-ft., 3-ft., 4-ft., 5-ft. and 6-ft. lengths, the 3-ft. rod is the most popular. Designed to push *(not pull)* wire through top or bottom plate

Applications: Use to push wire through top plate to attic or bottom plate to unfinished basement and pull the wire with a Grabbit; to push wire through a wall. See pages 41, 43.

PART #	MODEL #	DESCRIPTION
84-202	GR2	2FT GLOROD WIRE PUSHER
84-203	GR3	3FT GLOROD WIRE PUSHER
84-204	GR4	4FT GLOROD WIRE PUSHER
84-205	GR5	5FT GLOROD WIRE PUSHER
84-206	GR6	6FT GLOROD WIRE PUSHER

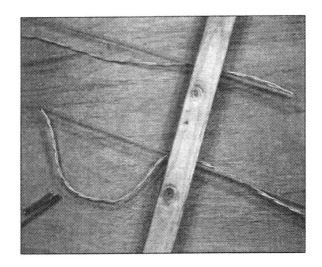

Fiberglass Push/Pull Rods

The fiberglass green and luminous push/pull rods have a bullet shape tip with cross hole (bullnose) on both ends, and, unlike the GloRod wire pushers, these can be used to pull wire as well. The luminous rods are made of phosphorous epoxy fiberglass truly glowing in the dark after being "charged" in full light. The epoxy fiberglass makes them stronger than the regular "green" rods, therefore a little stiffer.

PART #	MODEL #	DESCRIPTION
84-232	GR2BB	2FT GREEN PUSH/PULL ROD
84-233	GR3BB	3FT GREEN PUSH/PULL ROD
84-234	GR4BB	4FT GREEN PUSH/PULL ROD
84-235	GR5BB	5FT GREEN PUSH/PULL ROD
84-236	GR6BB	6FT GREEN PUSH/PULL ROD
84-212	GR2-L	2FT LUMINOUS PUSH/PULL ROD
84-214	GR4-L	4FT LUMINOUS PUSH/PULL ROD
84-216	GR6-L	6FT LUMINOUS PUSH/PULL ROD

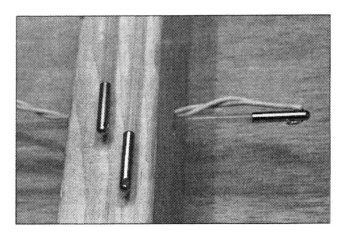

Applications: Same as GloRods, plus pull wires through walls. See pages 41, 43.

Grabbit™

The Grabbits are lightweight, shockproof, sturdy friction self-locking telescoping fiberglass poles, indispensable for running (pushing or pulling) wires or cables indoors as well as outdoors, in residential or commercial installations. The inner "V" of the patented "Ztip" has a double knife edge designed to literally "grab" the wire, while the outer "V"–without knife edge–is designed to push wires. The GR12 is made of three sections, and the GR18 has five sections.

PART #	MODEL #	DESCRIPTION
82-108	GR8	8FT GRABBIT
82-112	GR12	12FT GRABBIT
82-118	GR18	18FT GRABBIT
82-045	GR12-1	REPLACEMENT SECTION 1 FOR GRABBIT 12, 18
82-046	GR12-2	REPLACEMENT SECTION 2 FOR GRABBIT 12, 18
82-047	GR12-3	REPLACEMENT SECTION 3 FOR GRABBIT 12, 18
82-074	GR18-4	REPLACEMENT SECTION 4 FOR GRABBIT 18
82-075	GR18-5	REPLACEMENT SECTION 5 FOR GRABBIT 18
82-350	Ztip	REPLACEMENT PUSH/PULL Ztip FOR GRABBIT 8, 12, 18
82-360	Ptip	ACCESSORY PULL TIP (TIGHT SPACE) FOR GRABBIT 8, 12, 18
82-370	Jtip	ACCESSORY "UNCOIL" & PULL TIP (FISH-EZE WIRE, CABLES) GRABBIT 8, 12, 18
82-380	GRC	ACCESSORY RUBBER TIPS FOR SECTION 2, 3 AND 4

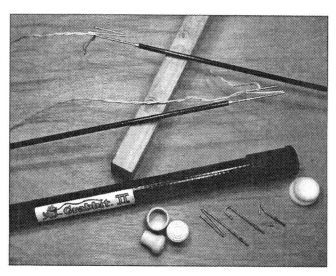

Applications: See pages 41, 43, 44, 45.

Sling-A-Line Slingshot

The Sling-A-Line slingshot yoke is made of heavy duty Baltic birch. A fishing reel (with a 10-lb. test line, a #7 swivel and a 3/8 oz. lead weight) is mounted on the yoke. Neither a toy nor a gimmick, it is the fastest tool for pulling long wire runs in open space with limited access area. It is also the very first tool ever offered by Labor $aving Devices.

Hint: For better accuracy, always shoot holding the yoke horizontally.

Applications: See pages 44.

PART #	MODEL #	DESCRIPTION
85-202	SAL202	SLING-A-LINE WIRE PULLER
85-210	SALR	REPLACEMENT RUBBER ASSEMBLY

2nd Man Cable Pulley™

The 2nd Man Cable Pulley device $aves time and money by eliminating the need for a second technician to feed the cable up through the ceiling or around turns for a cable pull. Just *snap* the hanging strap *off* the axle pin and the retainer piece, attach the pulley to a proper support device, *snap* the strap back *on* the axle pin and in the retainer latch. Position a 2nd Man Pulley above the drop ceiling to receive the cable from the supply side of your pull, at each horizontal turn, each elevation change, each up and down location. The 2nd Man Pulley will accommodate up to eight (8) four pair, Category 5 or like-size cables, protect and preserve the integrity of the cable and virtually eliminate friction on the cable.

Applications: See pages 44, 45, 46.

PART #	MODEL #	DESCRIPTION
85-142	DC2M	PACK OF 2-2ND MAN CABLE PULLEYS

PullSleeve

The PullSleeve is a very flexible, heavy duty, "double wall" weave monofilament sock designed to pull pre-connected coax and twisted pair, single or bundled cables. Its unique double wall construction reduces the fraying of the sleeve itself, while adding extra protection for the connectors. The sleeve is fused into a bullet shape, high resistance bullnose head with cross hole to make pulling these connected cables as easy as unfinished wires.

PART #	MODEL #	DESCRIPTION
85-950	PS1/2	1/2" PULLSLEEVE FOR PRE-CONNECTED CABLES
85-975	PS3/4	3/4" PULLSLEEVE FOR PRE-CONNECTED CABLES
85-990	PSNTW	PACK OF 25, 1/4"X.10 REPLACEMENT NYLON TIE WIRES

Applications: See pages 37, 43, 44, 45, 46.

PullCord

The PullCord is a nylon twine used primarily when running heavy wires or cables. It makes it easier to pull wire (cable) through holes and corners, and will protect your pulling device. Ideal when using the PullSleeve.

Applications: See pages 37, 38, 42, 43, 44, 45, 46.

PART #	MODEL #	DESCRIPTION
70-501	PC135	PULLCORD 135LBS BREAKING STRENGTH
70-504	PC400	PULLCORD 400LBS BREAKING STRENGTH

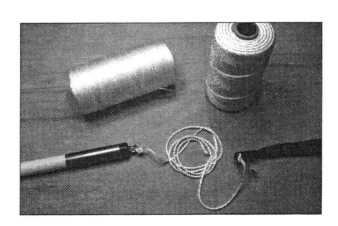

Undercarpet Tape

The Undercarpet Tape is 25 feet long annealed stainless steel, 3/4" wide with one end curled up and the other rounded. Run wire between padding and carpet.

Applications: See page 41.

PART #	MODEL #	DESCRIPTION
85-125	UCT25	UNDER CARPET TAPE, 25FT

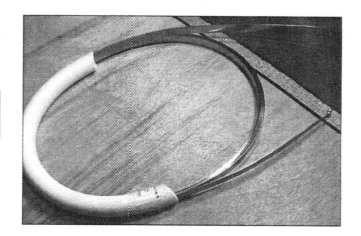

Fish-Eze™ Kit

Makes running wire from a window header up to an attic easy! The kit includes two stainless steel tubes, less than 1/4" diameter, 24" (for 8-ft. ceiling) and 36" long (for 9-ft. ceiling), stored in a fiberglass housing, a 12-ft. spring wire, heat treated in a coil .047" diameter with eyelet on both sides, in a circular stainless holder.

Applications: The kit is primarily designed to run wire from a window header up to an attic. See page 43.

PART #	MODEL #	DESCRIPTION
83-110	FEK-1	FISH-EZE ATTIC FISHING KIT
83-112	FEKW12	FISH-EZE WIRE IN A STAINLESS STEEL HOLDER
83-120	FEKWOH	REPLACEMENT FISH-EZE WIRE WITHOUT HOLDER
83-124	FEKT24	REPLACEMENT 24" STAINLESS HOLLOW TUBE FOR FISH-EZE
83-136	FEKT36	REPLACEMENT 36" STAINLESS HOLLOW TUBE FOR FISH-EZE
83-140	FEKH	REPLACEMENT STAINLESS STEEL WIRE HOLDER

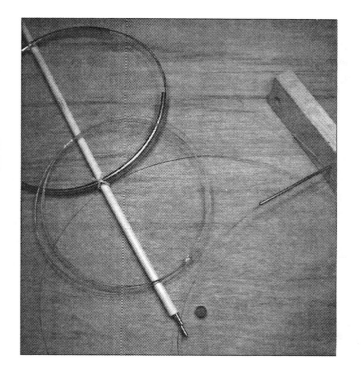

LOW-VOLTAGE CABLING — TRAINEE TASK MODULE 33108

NCCER CRAFT TRAINING USER UPDATES

The NCCER makes every effort to keep these manuals up-to-date and free of technical errors. We appreciate your help in this process. If you have an idea for improving this manual, or if you find an error, a typographical mistake, or an inaccuracy in the NCCER's Craft Training Manuals, please write us, using this form or a photocopy. Be sure to include the exact module number, page number, a description of the problem, and the correction, if possible. Your input will be brought to the attention of the Technical Review Committee. Thank you for your assistance.

Instructors – If you found that additional materials were necessary in order to teach this module effectively, please let us know so that we may include them in the Equipment/Materials list in the Instructor's Guide.

Write: Curriculum Revision and Development Department
National Center for Construction Education and Research
P.O. Box 141104
Gainesville, FL 32614-1104
Fax: 352-334-0932

Craft _____ Module Name _____

Copyright Date _____ Module Number _____ Page Number(s) _____

Description of Problem
_____
_____
_____
_____

(Optional) Correction of Problem
_____
_____
_____

(Optional) Your Name and Address
_____
_____
_____